LIFE ASSESSMENT OF HOT SECTION GAS TURBINE COMPONENTS

Also from IOM Communications

B726 Cyclic Oxidation of High Temperature Materials

B708 Advanced Heat Resistant Steel for Power Generation

B689 Advances in Turbine Materials, Design and Manufacturing

B728 Materials for Power Generation and High Temperature
Process Plant Applications

B667 Microstructural Development and Stability in
High Chromium Ferritic Power Plant Steels

B693 High Temperature Surface Engineering

B736 Parsons 2000

B723 Microstructural Stability of Creep Resistant Alloys for
High Temperature Plant Applications

Life Assessment of Hot Section Gas Turbine Components

Proceedings of a Conference
Held at Heriot Watt University
Edinburgh, UK
5–7 October 1999

Edited by
R. Townsend, M. Winstone,
M. Henderson, J.R. Nicholls,
Alan Partridge, B. Nath,
M. Wood and R. Viswanathan

Book B731
First published in 2000 by
IOM Communications Ltd
1 Carlton House Terrace
London SW1Y 5DB

IOM Communications Ltd
is a wholly-owned subsidiary of
The Institute of Materials

ISBN 1–86125–108–4

Typeset in the UK by
Dorwyn Ltd, Rowlands Castle, Hants

Printed and bound in the UK at
Cambridge University Press

Contents

Introduction

I am writing these few lines of introduction shortly after the tragic crash of Concorde in Paris on 25 July 2000. Although engine failure is now looking less likely to have been the cause, there can surely be no more poignant event to focus our minds on the importance of maintaining the safety and integrity of gas turbines. This was the main theme of our conference, covering the important issues for both aero and industrial engines at the conference held at Heriot Watt University, Edinburgh. Over three days in October 1999, British and International engineers and scientists debated the critical issues associated with design, creep and fatigue crack growth, coating technologies, new materials, repair and maintenance of hot section components. Throughout these discussions it was clear that plant manufacturers, plant operators, repair companies and research institutes all have different attitudes and perspectives on the need for component life assessment and the capability or otherwise for life extension. That the conference was unable to resolve these differences is perhaps understandable, but at least it enhanced understanding on all sides of the important issues. I commend this conference proceedings as an important step in the resolution of this debate.

R.D. Townsend
Conference Chairman

Design Issues for Aeroengines

F. KIRKLAND and R. CAVE

Rolls–Royce plc, PO Box 31, Derby DE24 8BJ, UK

ABSTRACT

The selection of materials in gas turbine engines is governed by the need to meet customer demands for higher efficiencies, higher thrusts, lower weights, low costs and lower levels of noise and emissions.

This paper considers the effect on material operating conditions that result from the requirement to meet each of the above design goals. The requirements are then individually considered and current and future methods of achieving these new demands are covered. These include both the use of materials already widely utilised in the gas turbine engine in new and existing applications and the use of materials not presently ordinarily used. The use of existing materials to meet the more strenuous demands through the use of better modelling and lifing to allow the material to be utilised closer to it's limit and the use of protective coatings is discussed. The effect on materials selection of new applications such as blisks and the use of different manufacturing technologies is considered. Finally the way in which the material requirements can be met through further use of materials not currently widely used in gas turbines is covered. This can consist of the development of new specialist materials or the increasing use of other materials as their properties become more fully understood.

1   INTRODUCTION

Many gas turbine performance improvements can be attributed primarily to advances in materials and their manufacturing technology. This paper looks briefly at the historical improvements in aero gas turbine performance and discusses the contribution that materials have had. The future requirements of the aeroengine markets are then discussed, including legislative and customer demands on the performance of future engines. The design issues that arise as a result of these performance demands are then covered, along with possible ways in which materials technology may meet these demands. Examples of some of these approaches to material improvement and the design problems arising from their use is then discussed.

2   HISTORICAL TRENDS IN GAS TURBINE PROPULSION

Since the introduction of gas turbine engines there have been enormous improvements in all aspects of engine performance. The application of advanced materials has played a vital role in this development with an estimated 50% of the increase in efficiency and performance being directly attributed to material devlopment.

Fuel burn has been reduced by around 60% relative to the first commercial jet aircraft. Approximately 45% of this is due to improvements in engine specific fuel consumption (sfc), with the remaining 15% due to improvements in aircraft design. Of this 45% improvement in sfc, 15% can be attributed to gains in propulsive efficiency, which is largely due to the bypass ratios used with modern civil turbofan engines.

The high bypass ratio fan blade only became a design possibility with the availability of strong lightweight materials such as composites and titanium. A low weight fan blade is necessary because the front structure of the engine must be able to withstand the large out of balance forces that would result from a fan blade failure.

The remaining 30% improvement in sfc is a result of increases in thermal efficiency, which results from better component efficiencies, increases in turbine entry temperature and increases in overall pressure ratio. Over the last 30 years the temperature of the gas entering the high pressure turbine has risen by approximately 500°C. Early jet engines had overall pressure ratios of as low as 4 : 1, whereas modern civil engines have an overall pressure ratio approximately 10 times higher.

The materials used in the gas turbine have played a major part in achieving improved et engine performance. Of the 500°C increase in turbine entry temperature, 250°C is as a result of the development of directionally solidified and single crystal nickel superalloys, which the remainder has come from blade cooling technology.

The trend to increase overall pressure ratios has resulted in increases in compressor exit temperatures, with the consequence that the final stages of the HP compressor are now typically made of nickel alloys.

Significant improvements to engine life, reliability and maintainability have provided significant reductions in cycle cost. The introduction of large twin engined aircraft for long-haul service has driven large improvements in reliability, notably the Boeing 777 which required a step improvement in engine reliability from entry into service. Increasing the reliability has required much more accurate prediction of component lives. Much of this has been provided by better understanding of material properties and the development of modelling techniques such as finite element analysis, which allow the stresses in the component to be determined more accurately.

Increased component performance is often dependent on the strength of the material, with specific strength being especially important to rotating components. The use of lighter stronger materials has allowed an increase in the maximum speed at which components may rotate, whilst still reducing the weight of the engine.

Steel was a major component in early jet engines, making up over half of the engine by weight. Steel components have been replaced for a lot of applications, usually with materials to allow either higher temperatures or weight savings. Aluminium, which was used in early engines, has been phased out almost entirely, being replaced with titanium and organic matrix composite materials.

These advances to the modern gas turbine engine have helped to make air travel one of the safest, and certainly the most convenient, form of long distance travel. Looking to the future however, presents new challenges for the producers of aeroengines.

3   FUTURE REQUIREMENTS

Whilst the expansion in world air transportation fluctuates according to global economic and political developments these perturbations have not historically affected the underlying growth. The increase in world air transportation, measured in available passenger kilometres, is likely to continue at about 5% per annum, leading to a doubling of air traffic over the next 15 years. The civil aerospace market over the next 20 years will be worth around $1250 billion, with around $420 billion of this in propulsion systems alone.

Whilst this increase in air transport offers great opportunities to aeroengine manufacturers it will also result in tougher design conditions. Since 1960 there has been a 75% reduction in aircraft noise (20 Effective Perceived Noise dB (EPNdB)). The increasing rate of take-off and landings and the trend towards larger aircraft as well as tightening controls at many airports mean that noise reduction will continue to be a major driver.

Gaseous emissions legislation, first introduced in 1974, addressed the elimination of smoke and reduction of unburned hydrocarbon and carbon monoxide. Once these problems had been addressed the focus changed to the reduction of oxides of nitrogen (NO_X) with the first legislation appearing in 1986. By the year 2002 NO_X levels for new aircraft will be set 36% below the 1986 level, or 16% below the current level. More recently the focus has been on global warming due to carbon dioxide (CO_2) release. Although aircraft account for only 3% of global CO_2 emission, this relatively small proportion will increase with air traffic growth. Presently there is no legislation to control CO_2 emissions but the industry must anticipate future requirements and be responsive to the potential impact that CO_2 may be having on the upper atmosphere.

Safety is a major ongoing concern that is highlighted by the increase in air travel. Although the rate of accidents per flying hour is reducing, the increase in travel may cause a rise in the overall number of accidents in the future. This is unlikely to be acceptable to the public and may deter potential travellers, despite the high statistical safety of flying.

The major economic considerations for the purchase of an aeroengine are fuel burn, first cost and maintenance costs. As world-wide supplies of conventional fuels decrease prices are likely to rise faster than inflation due to increasing demand and financial disincentives imposed to reduce fuel use.

The prospect of moving towards 'power by the hour' type arrangements whereby airlines effectively lease the propulsion system, with the engine manufacturer being responsible for the maintenence and servicing, will require a step change in engine design and lifing. Engine spare part sales currently account for a large part of engine manufacturers profits. This means that it is not always in the manufacturers best interests to extend the life of components beyond that required by the original contract. Engines designed for power by the hour will have much more emphasis placed on long service intervals and ease of service, as well as extremely high levels of reliability. This will require not only changes in design but also improvements in modelling and lifing techniques to obtain the maximum life from components which are approaching their material limits. Accurate lifing will be required to allow predictive support, whereby all maintenance is routinely planned and excess spares are avoided.

Although defence budgets have been severely reduced since the end of the Cold War era there is still a large market for new military aeroengines, predicted to be worth $106 billion over the next 20 years. With the reduced budgets airforces are looking for more flexible aircraft capable of performing dual or multi-role missions. The engines on such aircraft are expected to require much less frequent servicing and have a higher level of reliability.

Many of the requirements for civil engines are also applicable for military engines although not necessarily for the same reasons. For example, emissions from fighter engines must be kept low, not because of legislation but instead to meet the requirement for low observability. Weight is especially important, with one of the most important performance parameters for a military jet engine being thrust to weight ratio.

4 TECHNOLOGY REQUIREMENTS TO MEET FUTURE NEEDS

The above objectives lead to many requirements for aeroengine gas turbine design, many of which conflict. One effective method of reducing noise is to increase the bypass ratio of the engine. The search for improved specific fuel consumption to date has resulted in the increase in bypass ratios, benefiting noise reduction. The penalties of added weight and drag associated with larger fan diameters mean that the bypass ratio of current large civil engines is approaching the optimum in terms of fuel burn for high subsonic aircraft speeds. Further significant increases in bypass ratio with the aim of reducing noise will have the effect of increasing fuel burn, something which is clearly undesirable. Increasing the fan diameter also decreases the optimum speed of the Low Pressure (LP) spool, reducing the efficiency of the LP turbine. The efficiency disadvantages of a slow shaft speed for the LP turbine can be compensated for by using more LP turbine stages, however these stages are both heavy and expensive.

Decreasing the LP spool speed has the effect of increasing the torque, and hence diameter and weight, of the LP shaft. This has further consequences for the HP turbine disc, which has a design that is constrained by the discs bore diameter, and hence the diameter of the LP shaft.

One possible method of overcoming the problems of the conflicting requirements for a slow fan speed and fast LP turbine speed may be the introduction of geared fans. If the push for low noise becomes strong enough then the disadvantages of increased cost, weight, drag and unreliability may become justified.

Fuel burn is a major influence for a variety of reasons. Whilst the operating costs of fuel used is relatively important, fuel burn is of further significance for other reasons. Environmental concerns about the emission of CO_2 into the upper atmosphere will have a direct effect on future fuel consumption aims as CO_2 output is almost solely a function of fuel burn. The fuel carried by an aircraft also represents a major portion of the take off weight, up to approximately half for some long range aircraft. Reducing the fuel consumption obviously offers the operator greater flexibility, either in carrying a larger load or in increasing the aircraft's range.

It is estimated that a further 200°C increase in turbine entry temperature will be required over the next 20 years to meet airlines' demands for improved performance. The overall pressure ratio of engines will also rise to meet these demands, which will contribute towards the rise in turbine entry temperature. While the rise in pressure ratio and turbine entry temperature will help increase engine efficiency, the higher temperatures in the combustor will tend to increase the production of NO_X. Thus striving for less fuel burn, which will result in lower emissions of CO_2, may result in the increase of the production of another pollutant.

It is also beneficial to reduce the quantity of cooling air to the HP turbine as the use of this air reduces the efficiency of the gas turbine. This means that as well as using cooling air more effectively, solutions must be provided to allow the components to operate in a hotter environment. This may take the form of improved materials and/or protective coatings of some description.

The new requirements for materials that last longer at higher temperatures and stresses and contribute to a lower engine weight can be met in a number of ways. These can be broken down essentially into two forms; the easiest being the steady development of a known material system through iterations of material and processing developments. This is the route taken with nickel based superalloys which have had their temperature capability increased by 350°C since the 1940s. The alternative is the introduction of new types of advanced material tailored to meet the specific requirement of an application within the gas turbine engine. This approach is much more difficult and also involves a much greater element of risk due to the lack of experience in the use of the material involved. There are several approaches towards the use of new and existing materials and the type of application to which they may be put and examples of some of the different approaches are given below.

5   METAL MATRIX COMPOSITES

Titanium Metal Matrix Composites (TMCs) will play a significant role in the future of gas turbine aeroengine development. TMCs offer significant improvements over monolithic metals with very high specific strength and stiffness. Whilst the incorporation of TMCs into production aeroengines is largely dependent on lifing and production issues, they are also a good example of the design issues arising from the introduction of a new material into an aeroengine.

In order to reduce weight in the compressors of aeroengines, some military engines now have an integral bladed disc (blisk) in place of the traditional disc with individual blade attachments. With the removal of the heavy blade fixings the weight of the disc and blades can be reduced by approximately 50%.

The use of TMCs has potential in bladed rings (blings), in which the cob of a blisk can be removed due to the greater specific strength and stiffness of TMCs over traditional monolithic metals. This can result in weight savings of up to 40% of a blisk, or a total of 70% over a traditional disc and blades with individual fixings. Careful design of the bling can insure that load is almost entirely in the hoop direction with very little load in the transverse direction. This is ideally suited to the properties of uni-directionally reinforced TMCs as the maximum use can be made of the superior axial fibre properties.

TMC has the potential to replace steel in engine shafts. It's high stiffness and strength in the fibre direction coupled with the reduced density can give weight reductions of around 20–30%, with improved whirling performance and torque capability. The design of the shaft is a major problem however, especially the ends of the shaft where the torque is transmitted. Fibre orientation along the shaft also requires careful consideration as the stresses resulting from torque and axial loads are not coincident.

Casings are another part that will require significant design work to realise the benefits of TMCs. The numerous bosses, holes and flanges in current casings are all difficult to accommodate in TMCs. The use of TMCs in casings offers not only weight benefits but also the possibility of improved aerodynamic performance due to reduced tip leakage. The reduction in leakage could come about not just from the increased stiffness of TMCs, but also from the ability to tune the coefficient of thermal expansion by varying the volume fraction of fibres to match the designers requirements.

6 γ-TiAl ALLOYS

Gamma Titanium aluminides offer many potential benefits to gas turbine aeroengines with the combination of low density, high stiffness, good high temperature strength, burn resistance and good oxidation and corrosion resistance. The ductility and toughness of γ-TiAl alloys are much lower than those of nickel based alloys however, and this presents challenges for their use in aeroengines.

The most attractive feature of γ-TiAl alloys is their low density (3600–4000 kg m^{-3}), with substitution for nickel and steel parts offering large weight savings. This is especially true for rotating components where a large proportion of the stress on the component is due to centrifugal loading. The weight savings that come directly from the substitution of γ-TiAl for heavier materials can also have a large knock-on effect. The use of γ-TiAl in rotating blades for example, greatly reduces the stresses on the disc, which gives the designer a number of options including reducing the size of the disc, increasing the life of the disc or increasing the speed that the disc is operating at. Replacing three nickel alloy low pressure turbine stages with γ-TiAl alloy on a large civil engine could potentially save approximately 350kg.

Another advantage of γ-TiAl for aerofoil blade design is the high specific stiffness. This has the effect of raising the natural vibration frequencies of the blade, making it easier for the designer to ensure that they are above the lower engine order excitation frequencies. The high specific stiffness and low coefficient of thermal expansion of γ-TiAl alloys means that they are also suited to applications in which tolerances are critical. γ-TiAl alloys are susceptible to Foreign Object Damage (FOD) which is likely to preclude their use in rotor stages especially susceptible to this, such as the first stage of the HP compressor.

A different design approach is required to γ-TiAl alloys as they represent a step change from conventional nickel, steel and titanium materials. Using existing design methods tends to base fatigue life on crack growth rates. The extremely fast crack growth means that this method gives unrealistic lives. Most design criteria involve a fixed strain/time boundary for a given component. The common high primary creep rates for gamma aluminides when operating near their temperature limit necessitate either a change in the standard criteria or a pessimistic component life prediction. It is likely that initial utilisation of γ-TiAl alloys will be with components operating well within the materials properties to provide the lowest risk introduction route.

As an initial application the use of γ-TiAl alloys for compressor stators offers many advantages. Design rules with the aim of preventing titanium fires currently restrict the use of titanium in stators above defined temperature limits. The use of γ-TiAl for such an application would mean that steel or nickel parts could be replaced operating at a relatively modest temperature (300–400°C), allowing the accumulation of service experience in a comparatively mild environment. While the modest weight benefits are unlikely to justify the large cost increase inherent in the initial use of a new material, the experience gained would pave the way for γ-TiAl's use in more critical parts such as compressor and turbine blades, where they offer large potential benefits.

7   ALTERNATIVE NICKEL ALLOYS FOR COMBUSTOR USE

In order to meet the damands for reductions in NO_X emissions from aeroengines there is a desire to reduce the cooling air to the combustor. A 20% reduction in cooling air in the liner of the combustor typically gives approximately a 10% reduction in NO_X emissions.

One method of reducing the cooling air requirement of the combustor is to increase the temperature capability of the combustor material. A possible option is the production of combustor components using spray casting of nickel superalloys developed for use in turbine blades. While these alloys have a potentially higher temperature capability, from a design viewpoint one possible difficulty is the weldability of such parts. It is desirable that materials used for combustion and structural applications are weldable using the TIG process in order to enable joining of parts and the repair of manufacturing defects and service failures. If welding is not possible then parts have to be joined by complex mechanical joints, adding weight and cost to the engine.

8   CERAMICS

Ceramics as a class of material have huge potential for use in the gas turbine engine. The temperature capability of ceramics is much above the temperature at which super alloys are molten. Ceramics are also generally resistant to corrosion and oxidation and have a low density, as well as being hard with very good wear resistance. The largest obstacle to the use of ceramics in gas turbine engine design is their lack of toughness.

The incorporation of continuous ceramic fibres to form Ceramic Matrix Composite (CMC) materials offers a large improvement in toughness whilst still retaining most of the benefits of monolithic ceramics. Most of the recent work on ceramics for gas turbine engines has focused on CMCs but there are still many obstacles to widespread service in production gas turbine engines. Although CMCs are much tougher than monolithic ceramics the critical flaw size is still much lower than that of metals, making it difficult to detect critical flaws. Other difficulties in the use of CMCs in gas turbine engines include the lack of repair techniques, the long production cycle times and the extremely high cost of some fibres.

The use of ceramics is targeted at those areas where there is a large benefit to engine performance possible due to their high temperature capabilities. Some applications that fit this criterion are stator and rotor blades for the turbine, liners and the combustion chamber. The use of ceramics in the turbine to allow either a higher Turbine Entry

Temperature (TET) or to reduce the amount of cooling required to the HP turbine could offer significant improvements in engine efficiencies. Incorporating ceramics into the combustion chamber with the aim of reducing cooling air would allow a significant reduction in NO_X production, as well as helping to provide a more uniform temperature gradient that is of benefit to the turbine.

A major design problem in the use of CMC materials is the joining of CMCs to metals. The large mismatch in coefficients of thermal expansion and the high stiffness of CMCs can result in high stresses in a joint, especially for the extremely high temperature applications for which ceramics are being investigated. Another obstacle to the widespread use of CMCs in aeroengine parts is the lack of long term material data. CMCs have been employed for high temperature components at low structural loads such as reheat systems in military engines. The gradual introduction of CMCs in this way to gain service experience will allow their implementation into combustor and turbine components.

9   FURTHER DEVELOPMENT OF NICKEL SUPERALLOYS

The development of Nickel superalloys has resulted in a temperature capability increase of 350°C over the last 35 years. This has involved the development of both materials composition and manufacturing methods to allow current advanced single crystal alloys. As current alloys are already operating at up to 85% of their melting point there is only limited scope for further increases in temperature capability. A lot of future work is likely to concentrate on both improving other material properties for specific applications and reducing cost. Cost is an especially important factor as currently raw material accounts for approximately 20% of the cost of a turbine blade, a proportion that has increased with increasing amounts of expensive alloying elements.

Despite the above, small increases in temperature capability can justify the cost for certain applications within the engine. For example the development of a 3rd generation single alloy allowed the cooled 2nd generation alloy IP turbine blade on the Trent 800 to be replaced with an un-cooled 3rd generation alloy blade. The elimination of blade cooling has the result of increasing turbine efficiency. As temperatures continue to rise, so will the need for increased material capability if this material is to remain un-cooled. The use of Thermal Barrier Coatings (TBCs) which insulate the blade alloy from the hot gas stream will provide some of the future increase in temperature capability.

10   CONCLUSIONS

The market for aeroengines, both civil and military, offers large potential for manufacturers. In order to capture this market the demands for higher thrusts, lower weights, lower purchase and running costs and lower noise and emissions must be met. Materials technology will play an important part in meeting these requirements and will do so through a combination of evolutionary development of existing materials and the revolutionary incorporation of others into the jet engine. Whilst future engines are likely to contain large amounts of existing and improved nickel and titanium alloys the introduction of new materials into non-vital parts will occur. The service experience gained from such usage will allow the widespread incorporation of such materials eventual where their properties have the potential to give significant gains in engine performance.

11 REFERENCES

1. C.P. Beesley, 'The Application of CMCs in High Integrity Gas Turbine Engines,' *CMMC 96*, September 1996.
2. A.J. Bradley and R.J. Flatman, 'Future Trends in Aeroengine Propulsion Design and Technology', *Advances in Turbine Materials, Design and Manufacturing, Proc. 4th Int. Charles Parsons Turbine Conf.*, A. Strang, W.M. Banks, R.D. Conroy and M.J. Gouletteeds, IOM communications, 1997, 173–185.
3. P.J. Doorbar, 'The Introduction of Reinforced TMC Material into Rotating Machinery – The Safe Approach', *Design Principles and Methods for Aircraft Gas Turbine Engine*, May 1998.
4. M.G.J.W. Howse, 'Technology Trends for Large Gas Turbine Engines,' Royal Aeronautical Society Inaugural Sir Roy Fedden Lecture, March 1999.
5. P.C. Ruffles, 'The Future of Aircraft Propulsion', *Proc.IMechE C:J.Mech.Eng.Sci.*, 2000, **214**(1), 289–305.
6. D. Rugg, *Titanium Aluminides, Great Potential – But Not Yet Flying*, 1998.

Lifing Strategies for High Temperature Fracture Critical Components

G.F. HARRISON and M.B. HENDERSON

Defence Evaluation and Research Agency, Farnborough, UK

ABSTRACT

The development of gas turbine aeroengines continues to make demands for increased levels of thrust with reduced weight and higher efficiencies, though more emphasis is being placed on reducing the cost of ownership for both the military and civil customers of engine manufacturers. Targeted improvements in life cycle costs are achieved by reducing the number and frequency of inspection and maintenance intervals and by extending the in-service life times of components. However, the costs of engine removal and the consequences of grounding aircraft due to engine failure are many times greater than the replacement component costs. It is imperative, therefore, that component service lives are maximised, whilst ensuring that airworthiness is not compromised. The trends in engine performance requirements and operating practices will mean components having to operate at increased levels of stress and higher temperatures for longer periods. Often, this will push the materials used beyond the regimes for which they were originally developed and in to regimes where the effects of creep and environmental damage increasingly influence the conventional fatigue damage mechanisms leading to premature failure. Improved component performance and service life estimation will only be achieved by improvements in the understanding of the materials used and the component's response when exposed to high temperature and highly aggressive loading conditions. This paper briefly describes the strategies currently available for lifing fracture critical aeroengine components and reviews a programme of research that studied the high temperature low cycle fatigue and crack growth behaviour of the nickel-based disc alloy IN718. To assess the lifing strategies, crack initiation and propagation models were developed and validated against high temperature spin pit testing of model discs designed to have a rim feature as the life limiting critical location.

1   INTRODUCTION

Demands for increases in the flight capabilities of modern military and civil aircraft are set to continue. The main drivers for engine designers are to increase the thrust to weight ratio levels, improve engine efficiency and reduce the total life cycle costs to the customer, as well as reduce manufacturing costs. The prime requirement for engine fleet operators is to reduce life-cycle-costs, whilst maintaining flight safety and engine performance. The introduction of advanced turbomachinery and combustion technologies and advanced high strength-high temperature materials aims to satisfy these objectives. To improve thrust for the same weight, turbine discs will have to bear higher stresses and temperatures as engines run faster and hotter. There is a danger, however, that life cycle costs could escalate as a consequence of reduced engine reliability. As a consequence, a disc lifing strategy needs to encapsulate the engineering and materials technologies used to produce these key components. Consideration must be given to component design, projected and actual mission cycles, the materials and manufacturing processes used and the characteristic microstructures, defect distributions and residual stresses that effect the requisite mechanical properties. Equally, a thorough understanding of the analysis methods for specimen and component test results is essential if safe and economical life times are to be declared without the need for excessive numbers of high cost component tests. Future design strategies will aim to minimise the dependence on component test results as a means of reducing costs. This places an increased emphasis on improved finite element stress analysis techniques, materials modelling, statistical analysis and risk assessment.

Gas turbine engine discs possess enormous destructive potential by virtue of their rotational energies. It is not feasible to design aeroengine casings capable of containing the disc fragments should failure occur, so that service failure of a disc endangers the integrity of the whole aircraft. The statutory regulations governing the procedures used to determine the service lives of these 'fracture critical' components stipulate that the possibility of failure shall be extremely remote. The major failure mode in aeroengine discs is by the initiation and growth of cracks under the cyclic loading conditions experienced during service operation. Lifing procedures therefore centre on the creation and validation of appropriate methods for describing the fatigue process and estimating operational lives. It is necessary to quantify appropriate 'exchange rates' for the flight profiles experienced. They are evaluated in terms of the fatigue life consumed through the accumulation of damage from both major and minor load excursions in the ground-air-ground cycle and are expressed in terms of equivalent numbers of major stress cycles. Appropriate statistical weighting is needed to ensure that aircraft safety is not compromised.

The turbine disc's main functions are to locate the blades within the hot gas stream that emerges from the nozzle guide vane section at temperatures above 2000 K, and to transmit the power generated to the compressor driveshaft. Operational speeds can be in excess of 20 000 rpm, such that highly accurate component design is required to avoid excessive wear, vibration and poor efficiency. The disc must withstand high centrifugal stresses and the axial loads arising from the blade set and be able to accommodate the thermal transient and vibrational loadings imposed during operation. Under steady-state conditions, turbine disc temperatures can vary from approximately 300°C in the cob, to in excess of 650°C close to the rim. The rim region contains high stress concentration features for mechanical fixing of the blades and is susceptible to the combination of high temperature low-cycle fatigue, creep-fatigue and multi-axial loadings under an oxidising environment. The cob or bore region is subjected to lower temperature-high stress low cycle fatigue that places a requirement on the tensile strength and burst resistance characteristics of the material. Account must be taken of the effects of the basic material properties on the general thermal and mechanical stress levels within the component and of the influence of these stress levels on the failure modes. Additionally, there is a need for analytical assessment of the effects of any possible defects on local stress concentrations and, where this may lead to particle cracking, on associated local stress intensities. Low cycle fatigue resistance is the principal property controlling turbine disc life and to meet the operational parameters requires high integrity materials having a balance of key properties, as follows:

- high stiffness and tensile strength to ensure accurate blade location and resistance to overspeed burst;
- high fatigue strength and crack propagation resistance;
- high creep strength to avoid distortion and growth;
- oxidation and corrosion resistance and resistance to fretting damage at mechanical fixings.

The high cost of manufacture and the fracture critical nature of their operation requires them to have long life-times, typically in excess of 15 000 engine flight cycles.

A series of progressively higher strength nickel-base superalloys has been developed to meet these demands. Figure 1 shows a typical microstructure for a high strength powder U720Li disc material.

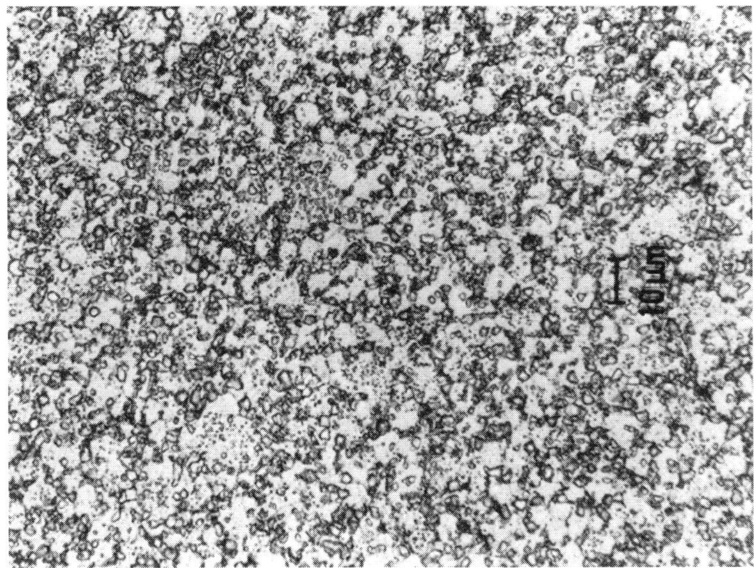

Fig. 1   Micrograph of HIP and extruded superalloy disc material U720Li.

2   DISC LIFING METHODOLOGIES

Lifing methodologies aim to declare safe component life times whilst optimising materials utilisation without unnecessary inspection and maintenance procedures. The life limiting features within a component are dependent on the specific mission type performed. For example, during take-off transient thermal stresses within the disc cob reach a maximum that can result in this being the most highly stressed region of the disc. Hence when flights are short and large numbers of take-off/shut-down sequences are experienced, the cob could be the life limiting region. Conversely, rapid manoeuvres during formation flying and interdiction will induce high rim loads and operating temperatures. These combined with the accumulation of many minor cycles associated with in-flight throttle movements can cause significant damage at stress concentration regions.

13

2.1 LIFE-TO-FIRST-CRACK (LTFC)

The declared service lives of a large proportion of current civil and military aeroengine discs have been calculated using the LTFC methodology. Under current UK Military Defence Standards[1] and European Civil Joint Airworthiness Requirements (JAR-E),[2] service lives can be declared on the basis of testing full size engine discs in a spin-rig facility, using stresses and temperatures similar to those found in service. It is assumed components enter service defect free and are rejected before the appearance of a fatigue initiated 0.38 mm deep 'engineering crack'. For all conventional aeroengine materials, fatigue lives (for identical load conditions) are assumed to be distributed according to a log-normal density function and that the ratio of the lives at the ± 3σ points is less than 6.[3] As illustrated in Fig. 2, certification requires that the calculated safe cyclic life is declared at a suitably remote point on the LTFC distribution, specified as the 1/750 quantile (-3σ). A predicted safe cyclic life (PSCL) is declared by applying statistically derived safety factors to the test results such that at this life, to a 95% confidence level, not more than 1 in 750 discs would be expected to contain a crack of surface length greater than 0.75 mm (0.38 mm deep). Given the assumed fatigue life scatter factor of 6, the life corresponding to the lower 1/750 quantile is a factor of √6 = 2.449 below the geometric mean (GM) reference LTFC obtained from the sample. With tests costing up to £250 000, the number of available test results is generally small and rarely exceeds 5 or 6. The Lifing Regulations define the procedure to be used in interpretation of these results and in calculation of the declared safe service lives. Fuller descriptions of the lifing regulations, statistical analysis procedures and necessary safety factors are given elsewhere.[3, 4]

The declared LTFC service life takes account of material variability and ensures that the weakest component is withdrawn from service whilst having adequate remnant life. In

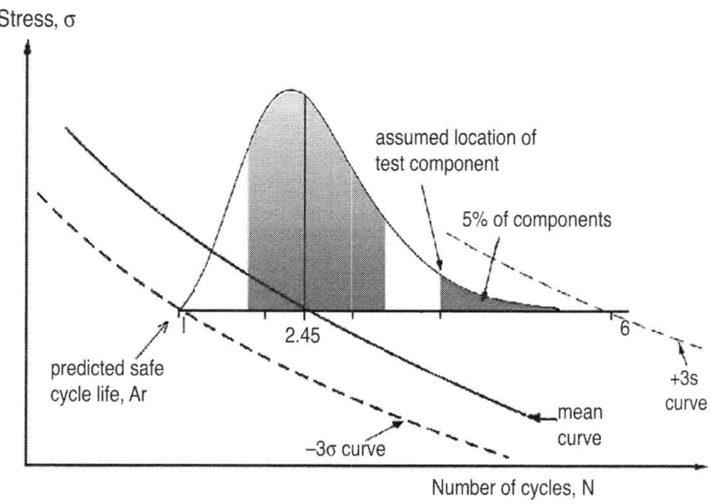

Fig. 2 Schematic illustration of the regulatory statistical analysis procedures used to define life-to-first-crack. All fatigue results are assumed to exhibit a log normal distribution and have a 6 times scatter in life between the ±3σ points on the failure distribution. The declared safe service life is calculated from the minus 3σ (1/750) quantile.

14

addition to the statistical safety factors used in the analysis, LTFC provides a further safety margin arising from the propagation life necessary to grow the crack to a dysfunction (failure) size and has been shown to be conservative; disc failures are 'extremely remote' and occur only once or twice every 100 million flying hours. A corollary being that significant potential life is being wasted, leading to higher than necessary life cycle costs.

Under FAA regulations,[5] service lives can be declared solely on the basis of laboratory specimen data. Since such tests are relatively cheap, it is feasible to generate a large test database. Although it may be possible to reproduce the standard of component surface finish for the test specimens, the residual stresses due to manufacture and heat treatment cannot be so easily represented. Also, for subsurface failure locations, as induced under complex thermomechanical loading, the volume of material subjected to the peak stresses is important and may not be represented fully by specimen testing methods.

2.2 '2/3 Dysfunction' Philosophy

In the LTFC approach, an acceptable but variable margin of safety arises from defining a 0.38 mm deep 'engineering crack'. However, the crack size that provides a constant safety factor is both geometry and material dependent. In the UK, this has led to the introduction of the '2/3 dysfunction' criterion and in practice the crack size at '2/3 failure' is often of similar value to the 'first crack' value of 0.38 mm. The constant ratio between the declared life and the burst life ensures a more consistent safety margin. In high strength disc alloys, operating at high temperatures and stress levels, the critical crack size for the onset of rapid fatigue crack growth can be smaller than the 0.38 mm engineering crack. Application of '2/3 dysfunction' enables these alloys to be exploited safely.

Relative to an identical major loading sequence, minor cycles are more damaging during crack propagation than during the LTFC period. For any particular aircraft mission or sortie profile, the ratio of the damage imposed by all the cycles incurred to the damage induced by the major (reference) cycle is called the 'cyclic exchange rate'. For an identical mission profile, the exchange rate which operates during the period that the crack is propagating (β_P) is typically a factor of 2.5 to 3 times greater than the ratio during crack initiation (β_i). For a component in which the dysfunction life equals 1.5 times the LTFC, application of either '2/3 dysfunction' or LTFC will release the same service life. However, if β_P equals $2.5\beta_i$, although in reference cycles the release life has an additional crack propagation safety factor with that equals 50% of the LTFC, in terms of the actual missions prior to dysfunction, this results in only a 20% margin. This additional safety value is always a known quantity in the 2/3 dysfunction approach, but is an unknown variable quantity in the LTFC approach.

2.3 Fracture Mechanics Life Prediction Procedures

Due to the large number of different disc designs, modifications and operating conditions it is impossible to combine LTFC results into a common database. Fracture mechanics based lifing methods, however, are able to rationalise disc lifing data by assuming that failure occurs from either initiation and growth of cracks induced by the imposed loading sequences, or as a result of crack propagation from inherent defects. Currently there are three approved fracture mechanics based methodologies as follows:

2.3.1   Databank Lifing

A fracture mechanics crack growth model is used to bring the results from different disc designs and features into a common databank. This can account for variations in component geometry, stress field and crack shape and offers a means of combining the results from different disc designs and large specimens. All discs in a given material produced using identical procedures are treated as part of a common data set and are assumed to contain small defects or pre-existing cracks that grow in a predictable manner under fatigue loading from the first cycle, as indicated in Fig. 3. Provided the geometry specific stress intensity factors and final crack lengths are available for the features for which failure lives are known, a crack growth equation and statistical analysis can be used to back calculate a maximum effective 'pseudo' crack size present in the whole component/test piece population at cycle 1. This initial crack size can be used to predict the maximum allowable life to dysfunction. Since the number of component and specimen test results within a databank is considerably greater than the 5 or 6 results available for any single disc design much smaller confidence intervals are associated with any particular confidence level and hence in general, it enables longer service lives to be declared.

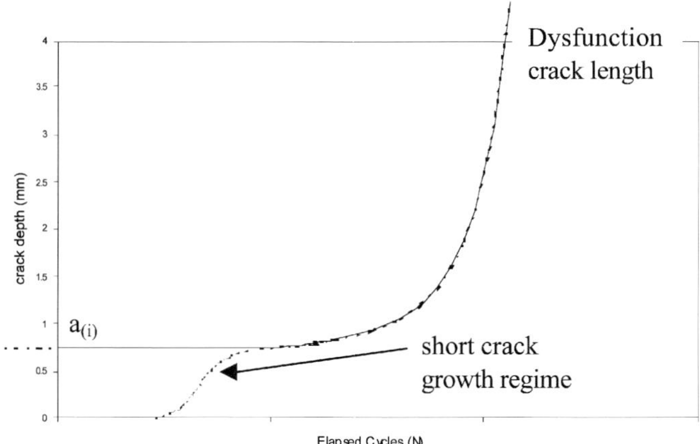

Fig. 3   Database lifing assumes all discs contain inherent 'pseudo' defects whose size (a_i) is determined by back calculation from a known crack size end point using fracture mechanics crack growth analysis. Defects grow in a predictable manner from cycle 1.

Additional results can be added to the database and, since a declared life does not depend on a small number of randomly selected samples, an accurate estimate of component life can be determined at the design stage. This allows greater design optimisation and more effective use of materials. Both CAA and FAA Airworthiness Authorities have approved fracture mechanics database methods for declaring civil engine component lives.

2.3.2 Damage Tolerance Lifing

It is assumed that some form of initial damage cannot be discounted in newly manufactured components. This assumed inherent damage is based on a proven NDI capability limit that is used in a fracture mechanics lifing calculation to determine the residual life of the component by calculating the number of cycles required to grow the starter crack to a critical size. The Engine Structural Integrity Programme (ENSIP)[6] is the standard lifing methodology applied to all US Airforce aeroengines. The declared service life to disfunction is calculated, using mean crack growth data, from an NDI crack detection size set to achieve a detection level of 90% to 95% confidence limit. The declared service life is often set at half this value. It is assumed that even if a crack of this level is missed, the component will be re-inspected or retired from service before the crack has achieved a critical level. As illustrated in Fig. 4, Damage Tolerance can be used to provide additional guarantees in component integrity. In the UK this approach has been applied in military applications using a full probabilistic damage tolerance analysis, backed by component testing, to determine the available safe life. This reflects the maximum crack size that could be missed on inspection and is taken as the starter crack size.

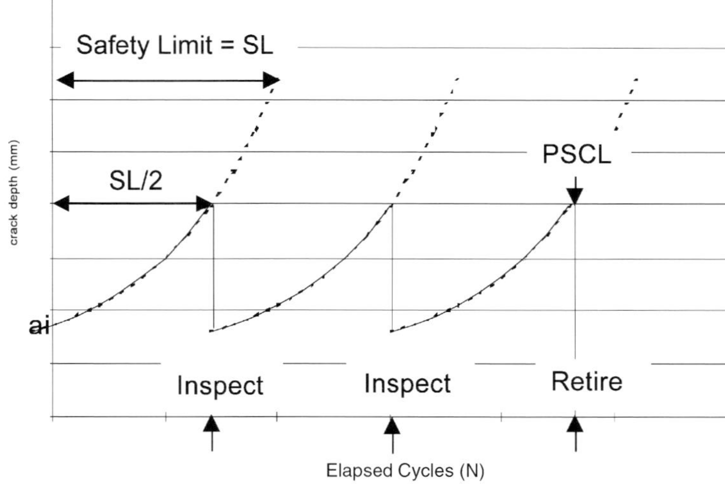

Fig. 4 Schematic representation of combined Damage Tolerant-Predicted Safe Cyclic Life approach. Damage is assumed to exist in newly manufactured discs and an inherent crack size, based on a proven NDI capability, is used in lifing calculations.

Problems associated with the statistical distribution of the crack initiation process are avoided, however, these are replaced by NDI uncertainties. Achievement of an acceptable level of safety depends both on the ability to detect cracks and on how often cracks or defects of a given size occur in practice. Probability of detection (POD) curves are constructed for a representative set of cracked components and test specimens and appropriate statistical procedures applied.[7]

2.3.3   Retirement For Cause

For both LTFC and Damage Tolerant lifing, the vast majority of engine discs are retired from service having a large proportion of their potential lives unused. It can be shown that at least 80% of discs have over one full PSCL still available when retired from service. The US Airforce has been instrumental in the introduction of a Retirement For Cause methodology[8] which aims to maximise disc life using component inspections to declare progressive life extensions, as illustrated in Fig. 5. Although the procedure seems straight forward, there are complex issues associated with calibration of the safe inspection intervals. To maintain safety margins, the NDI reliability needs to increase as the number of potentially cracked discs increases. This can only be achieved by increasing the detection crack size. However, based on this larger crack size, the inspection interval has to be reduced. This situation rapidly becomes non-viable on safety and cost grounds. The inspection intervals are determined using damage tolerance and risk analysis techniques. The risk of failure between two inspections is calculated by Monte Carlo analysis that takes account of the deviations in crack growth rate properties, different mission types and the probability that a growing crack will be missed during inspection. This is determined by the probability of a single crack of a given length being missed multiplied by the number of components that contain such a crack. In practice, as the fleet ages, the number of components that contain a growing crack increases drastically and the risk of a catastrophic failure reaches an unacceptable level long before each component contains a crack. Hence there is a

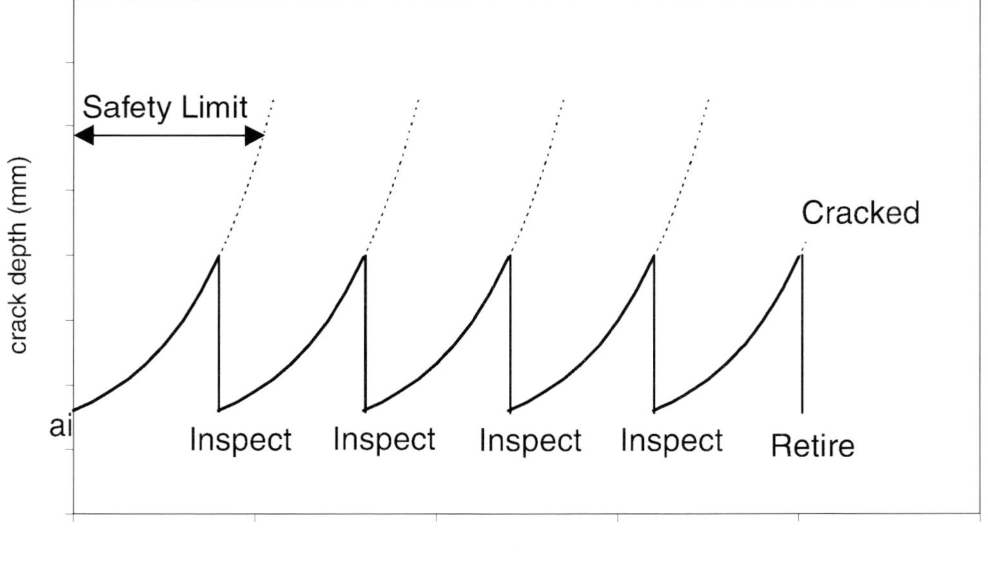

Fig. 5   Schematic representation of the retirement for cause approach taken to extremes, wherein successive inspections are used to determine the presence of a crack and extend damage tolerance by declaring progressive life extensions.

determined life limit, namely the life where the risk is just acceptable according to the Monte Carlo analysis.

3   COMPONENT BEHAVIOUR RESEARCH

To improve the understanding of the failure mechanisms in gas turbine engine discs and assess the validity of the lifing methods available, DERA has participated in an international collaborative research programme studying the advanced lifing strategies for engine discs operating under high temperature creep-fatigue conditions. A co-ordinated laboratory testing and materials behaviour modelling programme was used to generate, develop and validate high temperature crack initiation and growth models for the prediction of service lives of IN718 discs. The second part of this paper describes briefly the full scale component modelling and validation testing carried out by DERA.

3.1   FINITE ELEMENT ANALYSIS

Finite Element (FE) elastic, static, cyclic plastic and creep stress analyses of the programme's test specimen and model disc geometries were conducted to provide realistic predictions of the initial and shakedown stress distributions. Figure 6 shows a 7° segment of the 400 mm diameter, rim featured model disc, which contained 24 equally spaced V-shaped radially aligned notches. Preliminary elastic analysis of the disc was conducted to establish the test conditions at 600°C necessary to achieve a crack initiation life of between

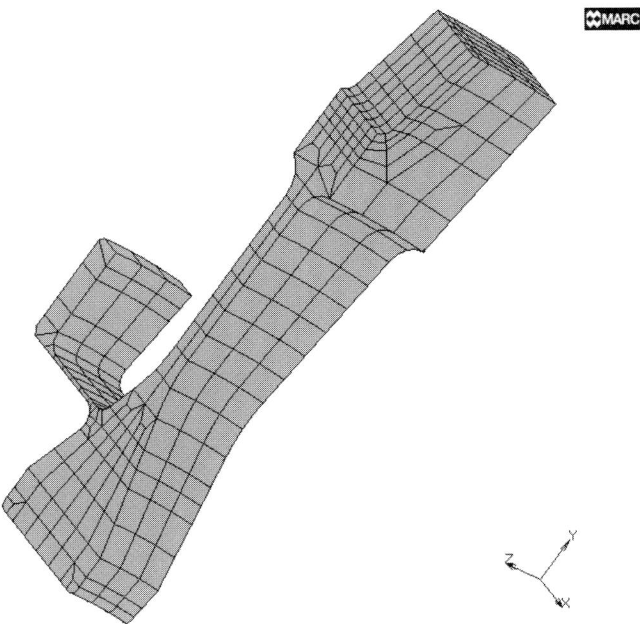

Fig. 6   Refined finite element analysis mesh of 7.5° segment of model rim disc.

10 000 to 20 000 cycles and a crack propagation life of between 5 000 to 10 000 cycles. The analysis indicated that a maximum speed of 15 000 rpm held for 30 seconds before braking to 100 rpm for 1 second was sufficient. Ramp rates of 90 rad s^{-1} gave a total cycle time of 67 seconds (17.5 seconds ramps).

A second 3D model of the 7° disc segment was generated to enable prediction of the linear elastic fracture mechanics stress intensity parameter K for a range of crack types (aspect ratio) and sizes. The K-calibration curves derived for the disc were used in the analysis of the crack propagation phase.

3.2   Lifing Models

Cack initiation and propagation phases are generally modelled separately and fall, there-fore, into two distinct classes identified as either low cycle fatigue (LCF) crack initiation or fatigue crack growth life prediction methods.

3.2.1   LCF Crack Initiation Life Prediction

LCF crack initiation life prediction is based on elastic FE analysis of the component combined with a number of options for calculating the elasto-plastic response to predict the stress distribution and shakedown condition for the critical component features. Two approaches have been used in the current study:

(i)   Elastic finite element stress analysis combined with approximation methods such as Neuber ($\sigma.\varepsilon$ = constant)[9] or Glinka (strain energy = constant)[10] or a modified pro-cedure to account for multiaxial stress state based on Hoffman and Seeger[11] to estimate the principal elasto-plastic stresses and strains.

(ii)   Non-linear finite element stress analysis using constitutive equations to model initial plastic response and time dependent behaviour as appropriate. Both methods are followed by application of a fatigue damage criteria, e.g., strain-life (S-N) curves defined in terms of a Coffin-Manson type equation, modified to account for the effects of particular test conditions such as temperature, frequency, dwell times, R-ratio and major-minor cycling, if necessary.

For each identified loading cycle the local stress-strain paths at critical locations are characterised by the temperature, strain amplitude and the maximum and mean stress levels. The parameters are then used in an evaluation of the extent of damage accumulation and to assign a component life time. The local strain assessment of the elasto-plastic behaviour can be represented by Ramberg–Osgood equations derived for the monotonic tensile (first cycle) and cyclic (saturated or half life) stress-strain curves for the material:

$$\varepsilon_a = \frac{\sigma_a}{E} + \left(\frac{\sigma_a}{K'}\right)^{1/n} \tag{1}$$

This equation provides initial and shakedown stress distributions. To evaluate the progres-sive stress relaxation from first cycle to shakedown a simplified isothermal version of the DERA Creep Law[12] was used:

$$\varepsilon_c = \Sigma C \sigma^\beta \, t^\kappa \qquad\qquad (2)$$

where ε_c is the creep strain, σ is the applied stress and t is time. The values for parameters C β and κ were determined from experimental data.

Figure 7 shows the predicted variation in equivalent stress ahead of the notch root after 10 750 cycles, which approaches the shakedown condition for the disc. Figure 8 shows the predicted notch root centre stress relaxation due to creep. These results indicate that creep

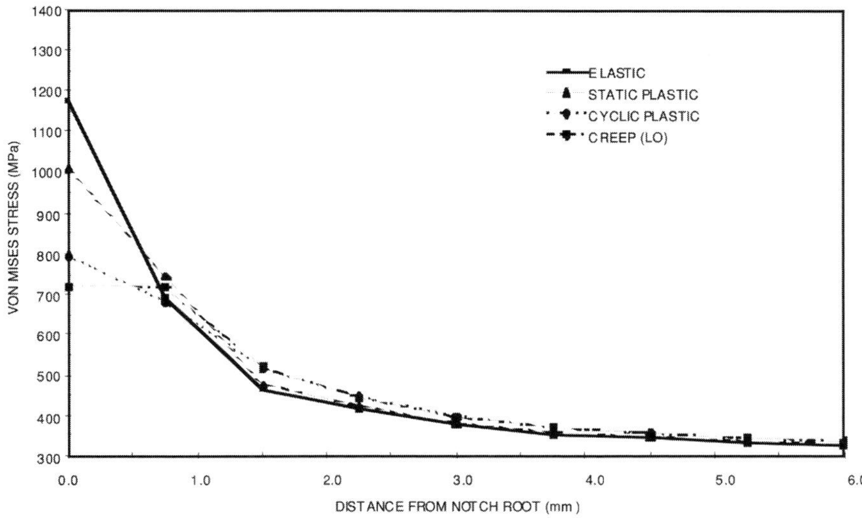

Fig. 7 3D Finite Element results showing variation in equivalent stress ahead of the notch feature in the model disc after 200 hours (10 750 test cycles).

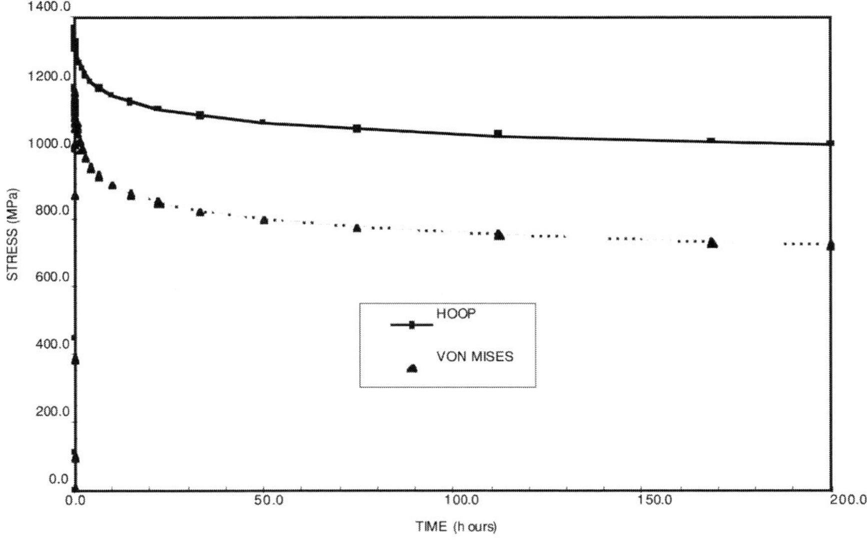

Fig. 8 3D Finite Element results showing stress relaxation due to creep at notch root centre of model disc.

relaxation is practically exhausted for the bulk of the spin test and acts to limit the stress distribution ahead of the notch during the crack initiation phase. Due to difficulties with residual stress measurements taken from the notch roots it was not possible to define the residual stresses within the FE model. However, the need to apply suitable boundary conditions to account for the initial stress distribution due to manufacturing is recognised as crucial for successful component modelling.[13]

Damage accumulation was assessed using the Smith-Watson-Topper parameter[14] as the basic S–N curve to represent materials fatigue strength. This approach combines saturated or half life maximum stress and strain amplitudes as a means of consolidating data to account for mean stress (R-ratio) effects on LCF behaviour:

$$P_{SWT} = \sqrt{\sigma_{max}\, \varepsilon_a\, E} \qquad (3)$$

To enable both LCF and HCF behaviour to be evaluated, this equation can be incorporated within the Coffin-Manson equation:

$$P_{SWT} = \sqrt{\sigma_f'^2 (2N)^{2b} + \sigma_f' \varepsilon_f' E (2N)^{b+c}} \qquad (4)$$

Constants are obtained by fitting the fatigue endurance data. In addition to the above format, fatigue life predictions were also evaluated using an equivalent alternating strain approach incorporated within a Coffin-Manson equation and an effective alternating strain approach to account for possible mean strain effects on fatigue life predictions.

3.2.2 Size Effect

Typically fatigue lives can show a scatter of times 6 in life between the shortest and the longest test results. This is attributed to the presence of material defects or 'weak-links'. Since a larger volume of stressed material has a greater probability of containing a weak link, statistically it should have a shorter fatigue life. To enable read across from plane specimen data to notch data for specimens or components an equivalent size factor can be applied. For the present study the surface qualities were considered as fatigue critical and the different notch root areas were compared with the surface area of the plane test pieces. By calculating a size factor based on the stress integral calculated for the notched specimen and a reference test piece alike, the size factor (Fg) is used to adjust the baseline S–N (i.e. P_{SWT}-life) curve in the loading direction ($Fg>1$ shifts the curve upwards).

3.2.3 Crack Growth Modelling

Fatigue crack growth prediction is dependent upon calculation of an appropriate fracture mechanics parameter (e.g. K solution) for the component geometry that relates the external loading conditions to the stress fields present at the tips of predetermined cracks. Crack growth life is declared to some cleared crack length using a parametric crack propagation model, based mainly on accurate K solution and temperature histories. Derivation of baseline da/dN versus ΔK design curves that account for the effects of retardation, crack closure and dwell-on-load is an essential step in the process.

Crack development proceeds by initiation of a microstructurally short crack that increases in size to the physically short crack and long crack regimes; the latter phase identifies the crack propagation lifing regime. Each stage may be driven by a different crack tip driving force and the parameter used in correlating crack growth behaviour will differ. The basis for much of the fatigue crack growth modelling, however, continues to be the Paris Law.

The models used either applied the Paris equation directly (after accounting for local plasticity and dwell effects) or calculated an effective ΔK to account for mean stress (*R*-ratio), crack closure and constraint effects using, for example, an Elber approach such as:

$$\Delta K_{eff} = (\alpha + \beta R)\Delta K \tag{5}$$

$$\frac{da}{dN} = C_e \Delta K_{eff}^{m_e} \tag{6}$$

Linear summation methods, where the cycle-dependent and time-dependent components and environmental factors contributing towards each increment of crack growth are summed over each loading cycle have been proposed as follows:

$$\left(\frac{da}{dN}\right) = C\Delta K^m + \int_{t=0}^{t=1/f} D_3 K^{m2} dt + E. f(loading) K_{max}^{m3} \tag{7}$$

3.3   Spin Pit Testing

Two rim featured model discs were spun at 600°C using the 17.5:30:17.5:1 waveform between 100 and 15 000 rpm. Temperature variation across the disc rim was within 5°C and the spin chamber was evacuated to 1.5 mbar to minimise wind resistance and convection losses. Periodic optical inspections, using a miniature TV camera, lens and fibre optic illumination system, were used to determine the crack initiation and propagation phases. Following problems with crack detection for the first spin test, eddy current probe, dye penetrent and acetate replica techniques were employed.

3.3.1   Test Results

Disc #1 burst after completing 23 704 cycles with little prior evidence of a crack being present. Approximately one third of the disc was found embedded in the containment wall of the spin chamber. Several segments of the rim had detached and ricocheted between the heater unit and the containment leading to severe damage. The largest remnant lay within the debris of the heater. Reconstruction and examination of disc #1 enabled the sequence of events to be established. Figure 9 shows the reassembled portions of the disc and illustrates the manner in which it failed. Notch #15 was identified as the crack initiation site. Seven segments were expelled from the disc rim and substantial distortion of the diaphragm wall and bore region was evident. The majority of the fracture surfaces were characterised by ductile overload caused when the disc disintegrated. Evidence of mixed trans and intergranular fatigue crack propagation was found on the fracture surface of notch #15 with fine scale striations evident throughout the crack growth process on transgranular regions of the fracture. The striation spacing was estimated to be approximately 1µm, regardless of crack length. i.e., the striations did not correspond to the overall crack growth rate and did not increase with crack length.

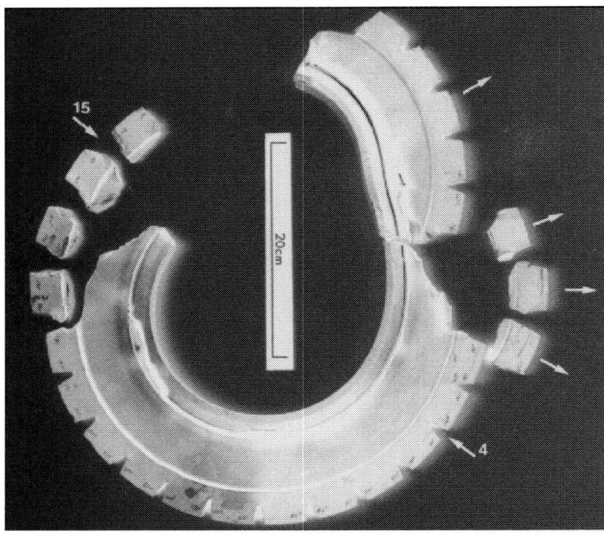

Fig. 9 Disc #1 reconstructed following sudden failure after 23 704 cycles. Following crack initiation and propagation from notch #15, a section of the disc was flung off in the direction of rotation. Three segments exploded from the opposite side of the disc rim. Drive arm side uppermost.

Examination of the remaining rim notches showed one other penetrating fatigue crack. Figure 10 shows a 2 mm deep thumb-nail shaped crack, initiated at the surface of notch #4. SEM examination found the fatigue crack growth regime was surrounded by a heat tinted ductile overload region and four distinct regions of crack growth delineated by beachmarks arising from inspection interruptions. A tendency for crack tunnelling (a/c ratio of approximately 1.4) was observed. The roots of most other notches, when broken

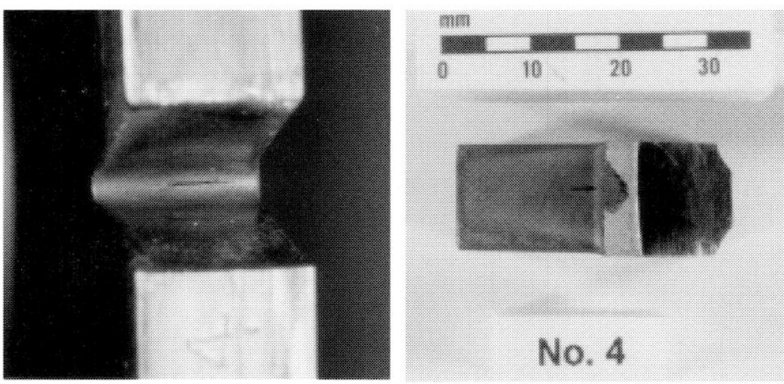

Fig. 10 Macroscopic view of Disc #1 notch #4 revealed presence of second notch root crack. After separation of the crack halves, a 3 mm deep thumb-nail shaped crack was revealed surrounded by a ductile overload region.

open, revealed heat tinting of the fracture surface to a depth of approximately 0.4 mm. SEM examination revealed that most of the heat tinted regions were due to ductile overload, however, small fatigue crack initiation sites were present as seen in Fig. 11. EDX analysis found each initiation site to be associated with either a NbC or TiC inclusion, approximately 5 to 10 μm in size.

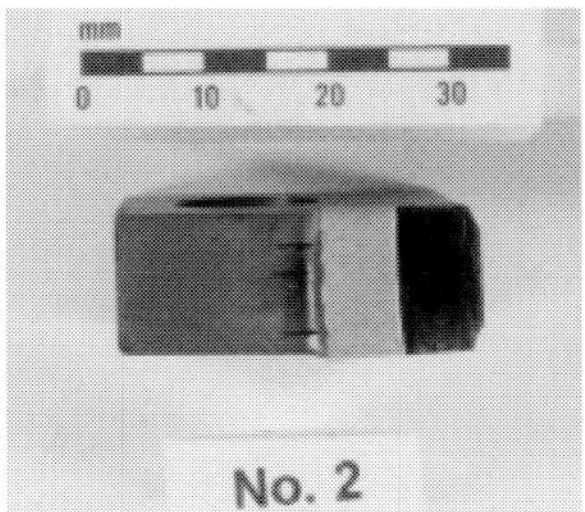
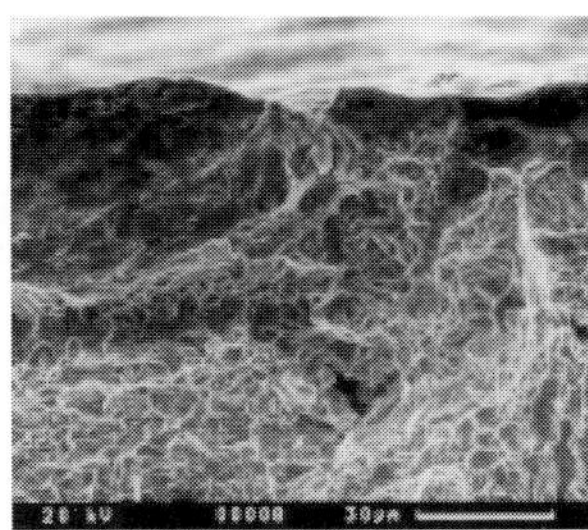

Fig. 11 After breaking open notch #2 in Disc #1 shallow heat tinted cracks were visible at the notch to a depth of approx. 0.4 mm. These were mainly due to ductile overload. Close examination revealed Nb and Ti carbides at initiation sites with evidence of fatigue crack growth.

The second disc test (Disc #2) was discontinued after 15 000 cycles due to the presence of an 8 mm deep through section crack at notch #21. SEM examination of the notch root found a single crack with an irregular region towards the drive arm side, formed by linking up of a number of neighbouring initiation sites. After separation of the crack halves a number of beachmarks were evident on the fracture surface, which enabled the crack growth rates and the development of the crack to be estimated. The initial stages of crack growth are thought to have proceeded by link-up of a number of initiation sites, prior to propagation and penetration of the rim section by the crack. Removal and examination of the remaining notches found them all to contain a significant number of fine scale initiation sites, approximately 20 to 50 μm in length, and microcracks in the 100 to 150 μm range. Typical examples of these features are shown in Fig. 12. EDX analysis found these features to be oxidised carbides.

Fig. 12 Examination of notch roots in Disc #2 reveals small scale cracks approximately 50 to 150 μm in length at oxidised surface emergent carbide particles.

Crack initiation and propagation mechanisms in both discs were found to be identical. Initiation occurred at oxidised surface emergent carbide particles along the notch root area.

A short region of transgranular propagation exists around each particle and a transition to a bimodal transgranular and intergranular crack propagation mechanism is seen as crack length increases and the individual thumb-nail shaped cracks link up (see Fig. 13).

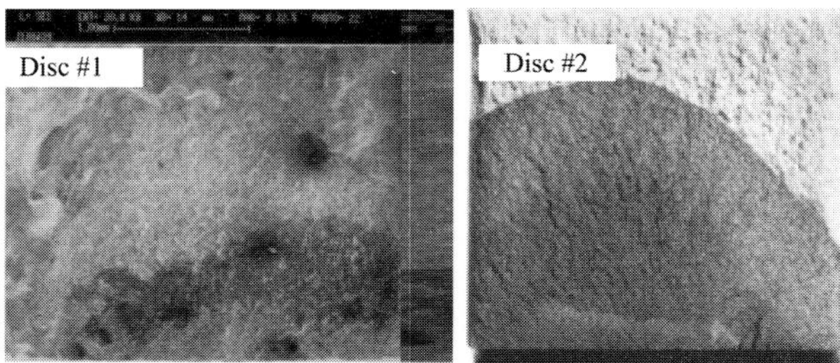

Fig. 13 Beachmark measurements enabled the crack growth response to be estimated for both discs.

3.4 Correlation of Disc Spinning Results With Plain Specimen LCF Data

The Smith Watson Topper parameter (P_{SWT}) was used as the basis for life assessment and correlation between the disc spinning results and LCF specimen data. The peak stress at the disc rim notch base obtained from FE analysis was used to calculate the equivalent cyclic stress/strain conditions from which the disc P_{SWT} parameter was calculated as 840 MPa. Failure of the first disc occurred after 23 704 cycles, whilst the second disc test was stopped after 15 000 cycles. Fractographic examination of beachmarks, as shown in Fig. 13, and correlation with the inspection and maintenance intervals provided evidence of the crack growth behaviour during spin testing. The individual notch crack measurements from both discs were combined to produce a common crack length (a) versus N curve for the number of elapsed cycles beyond 0.38 mm crack length extrapolated to a dysfunction size of 12 mm (reasonable limit of disc rim section). The crack growth analysis, described below, predicts a fatigue crack growth life of 5800 cycles from a crack initiation size of 0.38 mm. The corresponding standard crack initiation lives are 17900 and 10 300 cycles for the first and second discs, respectively.

Figure 14 shows the model disc lives plotted in terms of P_{SWT} in comparison with the scatter band for the specimen test programme. For disc life assessment the first crack size is defined as an 'engineering crack' of 0.38 mm depth and it is typically assumed that 'first crack' in the disc corresponds to typical specimen failure. The two disc spinning test results fall within the broad band of specimen failure data, however, unusually, they lie above the mean specimen results rather than towards the lower end of the specimen distribution. An explanation for this observation is that, although in general the volume of material experiencing the peak stresses is significantly greater in the component than in an individual specimen, in the current programme the critically stressed surface area in the individual specimens is slightly greater than the cumulative critical surface area for all 24 notches in the disc.

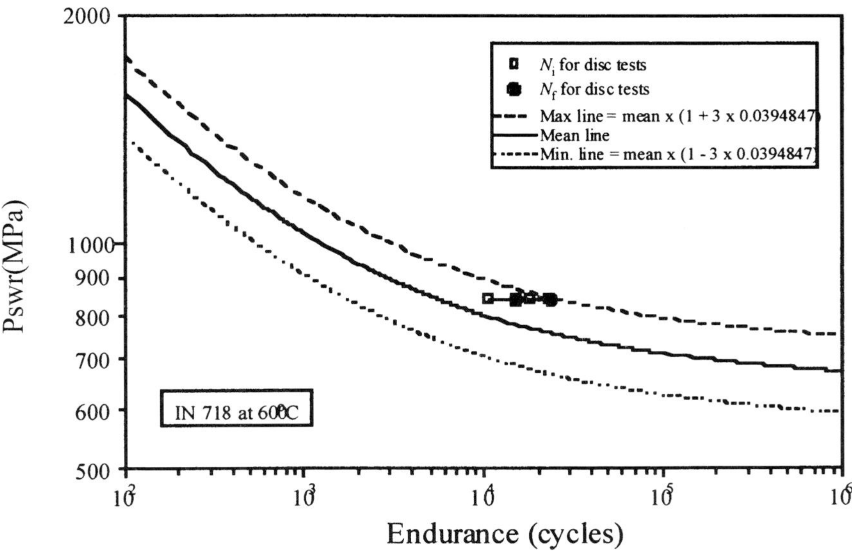

Fig. 14   Smith Watson Topper comparison of LCF test data with disc test endurance lives (N_i is life to first engineering crack of 0.38 mm).

3.5   DOUBLE EDGE NOTCH SPECIMEN FATIGUE TEST PROGRAMME

Additional tests were conducted on double edge notched specimens to generate crack growth data using the spin rig cycle and pressure. These data were compared with those produced at atmospheric pressure and under high vacuum for various fatigue cycles. The details are given elsewhere.[15] The results indicated that the crack growth mechanism is controlled by environmental pressure, dwell time at peak load, loading time, waveform, stress ratio and stress state. Reducing the pressure of air from normal atmospheric to high vacuum inhibits intergranular crack growth and encourages transgranular propagation. However for an equivalent dwell loading of 30 seconds, there was little or no influence of the reduced spin pit pressure (1.5 mbar) on the fatigue crack growth behaviour apart from an increased tendency for crack tunnelling. Crack shape was found to be controlled by the crack growth mechanism and hence the stress intensity factor used to characterise crack growth rate requires a crack shape parameter. Figure 15 compares the IN718 notch fatigue results at 600°C, 1.5 mbar for the spin pit waveform with a similar cycle having only a 1 second dwell and fully cycle dependent data generated at 2 Hz. Reducing the dwell period to 1 second provides a growth rate comparable to that found for a 2 Hz sine wave.

3.6   CHARACTERISATION OF HIGH TEMPERATURE CRACK PROPAGATION LIVES OF IN718 DISCS

The crack growth rate curve for the spin test conditions can be predicted via assessment of the da/dN specimen data for time-independent conditions and those used for the model

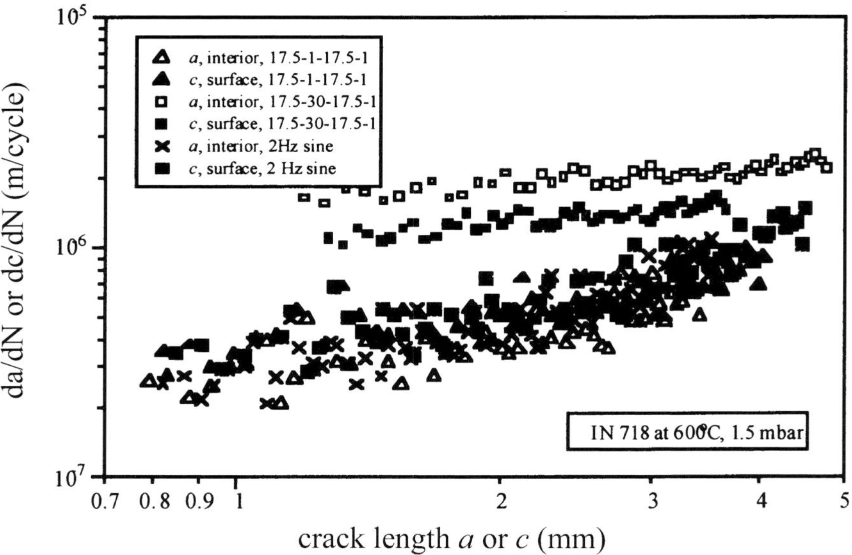

Fig. 15 Influence of dwell time at peak load and ramp time on crack growth behaviour of IN718 at a pressure of 1.5 mbar at 600°C, $R = 0.05$.

disc test coupled with estimation of the K versus 'a' relationship for a crack in the model disc notch at 15 000 rpm. Figure 15 shows that at 1.5 mbar pressure, the 17.5 second loading/1s dwell and the 2 Hz sine wave data are superimposed. The crack growth data for the model disc notch root loading conditions is a factor of 3.5 times greater and is approximately ΔK independent over the regime plotted. Therefore, the 3.5 factor can be applied over the Paris crack growth regime. The mean time independent crack growth rate behaviour represented by the 2 Hz and 5 Hz data from the CT specimen test database shown in Figure 16 can therefore be multiplied by a factor of 3.5 to provide a predicted crack growth curve for the disc.

As the crack is predominantly a non-through crack, an elastic-cyclic plastic correction to the K-solution was determined by taking the ratio of the cyclic plastic stress σ_{ep} to elastic stress σ_e for any crack depth and multiplying the K_e solution by the ratio of the corresponding two stresses. (In the cracked body, it is assumed that this ratio remains the same as in an uncracked body, due to the constraint imposed by the surrounding material at the notch. The K_{ep} solutions are applied to the midpoint crack length).

Beachmark measurements of crack depths at particular values of N were used to determine the overall crack growth rates in the disc a function of ΔK. These measurements are compared in Fig. 16 with the predicted crack growth rate curves for the disc test cycle. The 'elastic' and the corrected 'cyclic-plastic' compares the elastic and plastic K solutions as discussed above. The fatigue crack growth curves derived for the disc tests appear to show a threshold response which may be attributable to closure of the crack within the notch in plastic zone. Good agreement between the observed and predicted crack growth rate curves for the model discs is demonstrated.

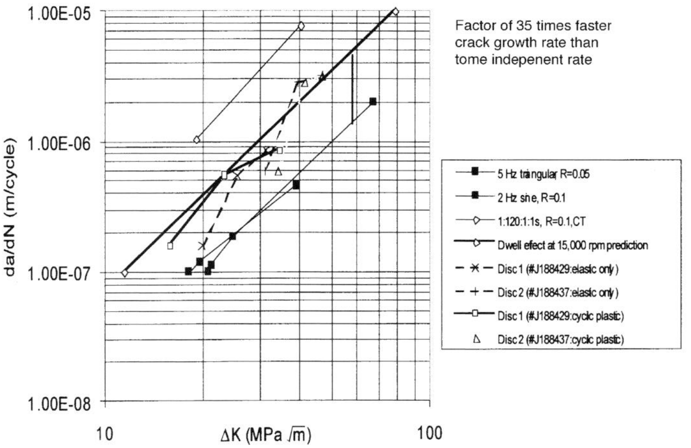

Fig. 16 Comparison of the scaled 2 Hz and 5 Hz fatigue crack growth baseline data, to account for dwell during disc spinning programme, with the observed curves for IN718 rim disc at 600°C. Experimental disc results are derived from beachmark measurements. Elastic K solution was scaled using elastic/plastic stress ratio at notch tip.

3.7 APPLICATION OF A CRACK GROWTH METHODOLOGY FOR THE PREDICTION OF MODEL DISC PROPAGATION LIVES

The predicted 'a' versus N curve for the model disc notch root can be obtained via application of a power law curve fit to the predicted da/dN versus ΔK relationship for the disc test conditions. A mean crack growth life can be calculated from a starter crack size of 0.38 mm depth to a dysfunction size of 12 mm.

$$N_p = \int_{a=0.38\text{mm}}^{a=12\text{mm}} \frac{1}{f(\Delta K(a))} \, da \qquad (8)$$

The notation $\Delta K(a)$ refers to a function that expresses ΔK in terms of crack depth 'a'. This expression gives the 'a' versus N relationship from the first 'engineering crack'. The solution is plotted in Fig. 17 in comparison with the crack growth measurements derived from the beachmarks on the fracture surfaces of the model discs. For comparison the predicted crack growth curves for a steep sided trapezoidal 120s dwell load cycle and a 2 Hz sine wave are shown on the same figure.

4 FINAL COMMENTS AND IMPLICATIONS FOR LIFING PROCEDURES

The LCF crack initiation and crack propagation life predictions for the model rim disc found good agreement with the experimental lives for the two rim featured model discs spun at 600°C, 15 000 rpm under a 17.5:30:17.5:1 waveform. These tests resulted in

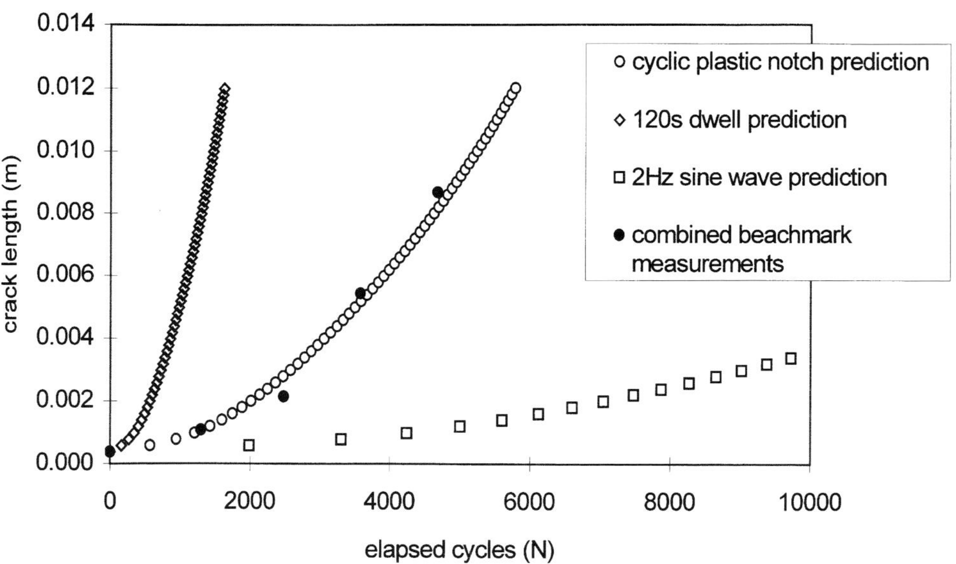

Fig. 17 Predicted crack growth behaviour on IN718 model disc at 600°C and 15 000 rpm in comparison with experimental beachmark measurements from spin testing programme.

failure lives of 23 703 for the first disc and 15 000 cycles for the second. Based on aeroengine lifing regulatory procedures, where 'first crack' is defined as a 0.38 mm deep crack, best estimates of the growth lives were 5800 cycles and 4700 cycles, respectively. The corresponding crack initiation lives are therefore 17 903 and 10 300 cycles, respectively. The difference in the crack initiation lives for the two discs is within the anticipated scatter. However, superior crack initiation is not reflected in the propagation phase. The spin test result for the second disc supports the equivalence of the life to a 0.38 mm deep crack and the '2/3 dysfunction' life. For Disc #1, however, the propagation life proportion is significantly shorter than given by this assumption. Application of a '2/3 dysfunction' criterion is, therefore, the more appropriate method for declaring a safe life as it automatically ensures a constant factor of safety.

Within the programme of research simpler local strain approximation methods for estimating the elasto-plastic response and thereby the crack initiation life were found to work just as well as more complex constitutive models, such as Chaboche. Application of the Smith Watson Topper (P_{SWT}) low cycle fatigue parameter to the model disc crack initiation data showed the disc results to fall within the estimated scatter band for the specimen test database. This demonstrates that this analysis is suitable for rationalising the effects of mean stress on high temperature crack initiation lives of IN718 specimens and discs. Crack growth life predictions were made using baseline fatigue crack growth data. More accurate predictions were achieved by application of a notch plasticity correction factor to the stress intensity K-solution and adjustment of the cycle dependent baseline data to account for the effects of dwell time at peak load. Crack growth life predictions in comparison with beachmark measurements for both discs are shown in Fig. 17. The rim te t results showed that under the temperature and loading conditions imposed, the crack propagation lives of such discs represents only a small proportion of the total dysfunction life. This has major safety implications, since application of the Def. Stan, JAR and FAA approved LTFC methodology to declare component service lives would result in such components having very little safety margin between the onset of 'first crack' and potential disc failure. In contrast to this solution, the UK '2/3 dysfunction' criterion always yields a constant safety factor between the declared service life and the potential dysfunction of the disc. Since the temperature and loading conditions used in the test programme are typical of the maximum conditions currently experienced during service usage, it is imperative that the '2/3 dysfunction' criterion be a mandatory, rather than an optional procedure, for declaring the service lives of these components. The research provides good experimental support for the DERA crack growth lifing methodology, which has been implemented within military aeroengine service life prediction procedures. Despite successful prediction of the model disc crack propagation life in this particular case, more sophisticated models are needed to assess fully the relevant input variables and loading parameters with regard to their effects on damage and crack growth rates.

High temperature crack initiation in IN718 is dominated by oxidation of niobium and titanium carbides at the notch root surfaces. Specific crack growth mechanisms were found to control the rate of elevated temperature crack growth. The crack propagation behaviour away from the notch root centres was identified as transgranular for short crack lengths and dominated by cycle dependent fatigue, followed by a transition to bimodal trans and intergranular crack propagation for longer crack lengths. Intergranular cracking produces very high rates of crack growth when compared with typical transgranular fatigue mechanisms.

5   CONCLUSIONS

1. Demands for reduced life cycle costs and increases in the flight capabilities of modern military aircraft, whilst maintaining flight safety requires a disc design and lifing strategy that encapsulates the engineering and materials technologies available. A thorough understanding of the analysis methods for specimen and component test results is required. Increased emphasis is being placed on finite element analysis and materials modelling techniques, statistical analysis of data and risk assessment procedures.

2. The functional and materials property requirements necessary to satisfy gas turbine disc applications have been reviewed. The principle property controlling turbine disc life is resistance to low cycle fatigue. However, to meet the operational parameters requires high strength-high integrity materials having a balance of key properties. A number of alloys have been designed, possessing increasing levels of high temperature strength, stiffness, dimensional stability and resistance to fatigue and creep damage and has culminated in the design of alloys such as IN718 and U720Li.

3. The LTFC, 2/3 dysfunction and fracture mechanics disc lifing philosophies have been reviewed. Each procedure aims to declare safe component lifetimes, whilst optimising materials usage without overly stringent inspection and overhaul procedures. The

methods are based wholly on crack initiation, crack growth or total life and rely on the application of suitable statistical analysis procedures and safety factors.

4. Large scale specimen testing has established a significant database for the creep and fatigue behaviour of IN718. Crack initiation life was found to be dependent on the elasto-plastic and creep response of the material, which determined the shakedown of tensile mean stresses during cyclic loading. Creep relaxation at high temperatures increased crack initiation life, whilst under certain dwell loading conditions accumulation of inelastic creep strain has a detrimental effect.

5. Crack initiation and propagation phases have been modelled independently with some success. Predictive accuracy is dependent on the ability to model stress redistribution around the fatigue critical regions and on relevant test data being derived under conditions fairly close to those of the intended service application. Further research is needed to improve the understanding of the effects of notch tip plasticity, multiaxiality, environment and dwell on the short crack growth phase and to develop lifing methods that couple together the short and long crack growth regimes.

6. High temperature notch root crack initiation in IN718 is dominated by oxidation of surface emergent carbides. All notches examined contained a number of fine-scale initiation sites, 20 to 50 μm in length, and microcracks in the 100 to 150 μm range. Crack growth rate and crack shape are dependent on dwell at peak load, loading time and waveform, stress ratio and state as well as the environment (partial pressure of oxygen). Enhanced fatigue crack growth rates are found as temperature and dwell period increases due to enhanced grain boundary oxidation embrittlement. Crack propagation was found to occur by link-up of transgranular regions with failed grain boundary ligaments that develop immediately ahead of the crack due to oxygen embrittlement.

7. The IN718 disc spinning test programme successfully validated models developed for both the crack initiation and propagation stages. However, the crack propagation lives represented only a small fraction of the total dysfunction life, which has major implications for the safety of these components when the LTFC methodology is used to declare service lives. Use of 2/3 dysfunction yields a constant safety factor between the declared service life and potential disc dysfunction life.

ACKNOWLEDGEMENTS

This research has been supported by Mechanical Sciences Sector, DERA under UK MCD funding whose support is gratefully acknowledged. Much the testing was conducted within the European collaborative programme IEPG TA31. Many thanks are given to Paul Tranter, Mike Smith and Dr. Mark Hardy of DERA and collaborative partners at ABB-Alstom, Whetstone in the UK, NLR in The Netherlands and MTU Gmbh and IABG, Munich in Germany.

REFERENCES

1. UK Military Defence Standards, Def. Stan 00–971, 'General Specification for Aircraft Gas Turbine Engines,' 1986.
2. Joint (European) Airworthiness Requirements – Engines (JAR-E), Civil Aviation Authority, 1986.
3. G.F. Harrison and A.D. Boyd-Lee, 'Life Extension Methods for Fracture Critical Aeroengine Components.' Proc. 17th NATO RTO Meeting on Qualification of Life Extension Schemes for Engine Components, Corfu, Greece, 1998, Paper 18.
4. G.F. Harrison and M.J. Weaver, 'Life Assessment of Fracture Critical Aeroengine Parts.' *Proc. Int. Conf. Engineering Against Fatigue*, J.H Benyon et al. eds., Sheffield, UK, 1997, pp. 713–720. Publisher?
5. US Federal Aviation Regulations Part 33. Airworthiness Standards: Aircraft Engines, FAA Publications, 1993.
6. US DoD Engine Structural Integrity Program, MIL STD 1783 (USAF), 1984.
7. C. Annis, A. Berrens and F. Bray, 'Proposed Mil. Std. for USAF NDE System Reliability Assessment.' Publisher?
8. J. Harris, 'Engine Component Retirement for Cause, Experimental Summary,' AFWAL-TR-87-4069.
9. H. Neuber, 'Theory of Stress Concentration for Shear-Strained Prismatical Bodies with Arbitrary Non-Linear Stress-Strain Law.' *Trans. ASME, J. Applied Mech.*, (1961), **28**, 544–550.
10. G. Glinka, 'Calculation of Inelastic Notch Tip Stress-Strain Histories under Cyclic Loading,' *Eng. Fracture Mechanics*, (1985), **22**(5), 839–854.
11. M. Hoffman and T. Seeger, 'Generalised Method for Estimating Multiaxial Elastic-Plastic Notch Stresses and Strains, Part 1: Theory, Part 2; Application and General Discussion,' *ASME Journal of Engineering Materials Technology*, (1985), **107**, 250–260.
12. T. Homewood, T.J.W. Ward, M.B. Henderson and G.F. Harrison, 'The DERA Slip System Creep Law for Modelling Face Centred Cubic Single Crystal Material Behaviour,' *Proc. Conf. Modelling of Microstructural Evolution in Creep Resistant Materials*, Institute for Materials, London, 1998.
13. G.F. Harrison and P.H. Tranter, 'Application of Materials Characterisation to Stressing and Lifing of Aeroengine Fracture Critical Components.' *Proc. of 1st Int. Conf. on Component Optimisation from Materials Properties and Simulation Software*, W.J. Evans et al eds., Swansea, UK, 1999, 251–262.
14. K.N. Smith, P. Watson and T.H. Topper, 'A Stress-Strain Function for the Fatigue of Metals,' *J. of Materials*, 1970, **5**(4), 767–778.
15. M.R. Bache, W.J. Evans and M.C. Hardy, 'The Effects of Environment and Loading Waveform on Fatigue Crack growth in Inconel 718,' *Int. J. Fatigue*, 1999, **21**, S69-S77.

Life Management Issues for Operators

BIREND RA NATH

National Power plc, Windmill Hill Business Park, Whitehill Way, Swindon SN5 6PB,
e-mail: birendra.nath@natpower.com

1   INTRODUCTION

National Power owns, operates and maintains a large and a rapidly growing fleet of advanced gas turbines worldwide. This is a challenging business particularly in the established competitive electricity market in Britain and emerging ones in some other countries. There is an unrelenting pressure on power producers to reduce the cost of generation, £/MWh, and environmental emissions. These business drivers define the following goals for new power plants.

- Low cost of EPC (Engineering, Procurement and Construction)
- Thermal efficiency and fuel price, which together determine the cost of fuel used per MWh
- Low costs of operations, including manpower
- Low maintenance costs
- High reliability and availability
- Clean technologies with low emissions

Potentially, combined cycle gas turbines (CCGTs) meet the above-mentioned requirements and have become the plant of choice for Britain in recent years. Ever increasing thermal efficiency and the competition between Original Equipment Manufacturers (OEMs) have led to increasing Turbine Inlet Temperatures (TIT), higher compression ratios and steam cooling for hot gas section components. For example, TIT (ISO definition) in commercially available turbines are as high as 1255°C and pressure ratio of up to 30 bar. Large machines with very high outputs are being developed for low EPC costs ($£/kW_{installed}$). These advanced turbines necessarily use leading edge design features and materials technologies, which have not yet been proven fully. Aero-engine experience may not be readily transferable to base-loaded stationary turbines because the later operate at the peak temperature continuously for long duration and undergo fewer cycles.

One consequence of the very rapidly developing technology is that newer models are introduced before previous ones have been proven in long-term service. This can also limit the fleet size of identical turbines. For example, there are only 5 × ABB13E1 and 5 × Siemens 94.3 turbine worldwide. Faced with design problems in such small fleets, OEMs may opt for 'quick-fix' rather than permanent redesign or re-engineering solutions, which operators would prefer.

There is a shared goal between OEMs and operators to introduce and prove new technologies for which the two parties collaborate closely. However, stronger commercial factors define the supplier-buyer relationship.

In order to understand the difficulties faced by operators, it is instructive to consider the life expectancy of Row-1 blades*. In general, blades are designed for either creep strain or rupture in 100 000 hours. The design intent for fatigue is probably in excess of 5000 cycles. However, these 'design lives' have little bearing on 'expected lives' quoted but not guaranteed by OEMs or the 'actual lives' in service. Figures given below emphasise the gulf between 'expected lives' and the interval when operators have to replace Row-1 blades.

- 'Design creep life' = 100 000 hours
- 'Expected life' = 48 to 50 000 EOH†
- 'Actual life' = 3000 to 24 000 EOH.

Much of the information (e.g. temperatures–stresses–strain, behaviour of coatings and their interaction with the substrate, definitions of critical defects etc.) required for life management of components are considered proprietary by OEMs. In general, these data are not available to operators. Consequently, operators do not always understand the reasons for 'run/repair/refurbish/replace' recommendations and they end up paying a high price for new parts. For example, a set of blades can cost around £5M and an individual row more that £1M.[1]

Life management of hot gas path (HGP) parts starts from the procurement of gas turbines with manageable risks. They have to be operated to minimise damage and maintained i.e. repaired/refurbished economically. Finally, the parts should be replaced only when life-expired and then with low-cost components, which are 'fit for the purpose'.

The purpose of this paper is to describe some tools currently available to power producers for the life management of HGP components, in particular blades and vanes, and some potential developments.

2   DESIGN AND CONSTRUCTION

The earliest opportunity for influencing through-life costs arises when purchasing new power plants. Risks associated with the GT technology influence component lives, plant Reliability, Availability and Maintainability (RAM). The risk profile acceptable to a power producer depends largely on commercial considerations.

Some of the major new technologies, which are in service now or are likely to be used in the near future, are identified below.

- Dry, low-NO_x combustion
- Sequential combustion in ABB GT24/GT26 family
- Use of directionally-solidified (DS) (e.g. GE Frame 7FA/9FA, ABB GT24/GT26) and single crystal (SX) blades (Siemens V94.3A, ABB GT24/GT26)

---

*In this document, 'vanes' and 'blades' respectively refer to stationary and rotating airfoils in the turbine.
†Equivalent Operating Hours (EOH) is defined in Section 4.1.

- Thermal barrier coating (TBC) on rotating blades (for example, ABB 13E1/13E2, Siemens V94.3 etc.)
- Film-cooling of blades and vanes
- Steam cooling in Advanced Turbine Systems being developed by GE and by Siemens-Westinghouse in USA
- Very large last stage blades cast in equiaxed or DS forms
- Possible use of Ni-alloy discs

2.1 Engineering Risk Assessment

National Power carries out engineering risk assessment of the technology offered by vendors to estimate plant availability throughout the life including the bathtub effect. This requires a detailed knowledge of GT technologies and where available, operations and maintenance experience. The Engineering Risk Assessment methodology (ERAP™), developed by the company, is continually validated by retrospective analysis of the actual availability. It may take up to 2 years for a new-technology plant to bed-in (Fig. 1) but good management delivers high availability in later years.

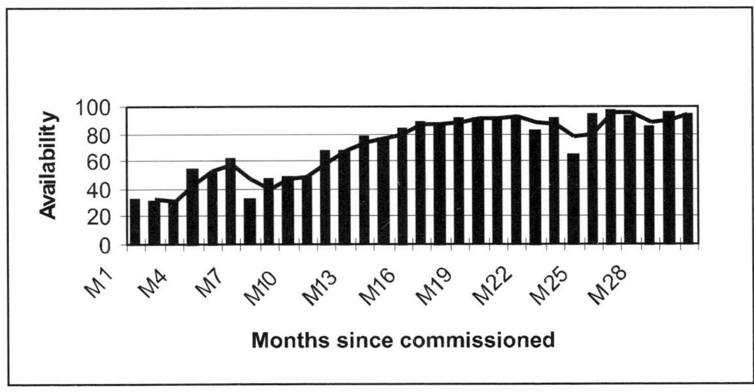

Fig. 1 Average availability of four advanced CCGT stations in National Power.[1]

Poor quality assurance of conventional components (e.g. steam turbines, circulating water pumps, low-voltage and high-voltage switchgear, condenser tubes, cooling towers, valves in the heat recovery steam generator, transformers, air-intake system etc.) has led to significant availability losses[1] but these considerations are outside the scope of the current paper.

3 OPERATION

For cost-effectiveness of HGP part lives, plant operation must control thermal, mechanical and chemical loads on components by controlling the following.

- Effective air filtration
- Fuel quality, including any treatment that may be required
- Quality of water/steam injected for cleaning and/or for power augmentation

- Combustion processes to control TIT and its circumferential distribution
- Possible combustion pulsation from dry low-NO_x burners and NO_x emissions
- Thermal ramp rates during transient operations

3.1 TURBINE INLET TEMPERATURE (TIT)

This is one of the main parameters used for controlling turbine operation. However, it is not measured directly but is calculated from measurement of the turbine exit gas temperature (TET). In a generic form,

$$TIT = f(TET, CDP, CDT, Fuel, T_{Ambient}, P_{Ambient}, P_{Exit}, RPM) \qquad (1)$$

Where,

TET	= Measured turbine exit temperature. Some OEMs use a modified figure, to take account of for example, any departure from the calculated values
CDP	= Compressor discharge pressure
CDT	= Compressor discharge temperature
Fuel	= Fuel factor
$T_{Ambient}$	= Ambient air temperature
$P_{Ambient}$	= Ambient air pressure
P_{Exit}	= Gas turbine exit pressure
RPM	= 3000 revolutions per minute for 50Hz turbines

Most OEMs do not use all of the parameters listed above for calculating TIT.

The methodology for deriving TIT from TET is considered proprietary by OEMs. From experience the calculations are subject to revisions, suggesting an empirical approach is used

National Power and some other operators, install optical pyrometers to measure blade metal temperatures. In future it may be more sensible to use these measured values for controlling turbines.

3.2 VIBRATION MONITORING

On-line vibration monitoring is used to detect abnormal behaviour of gas turbines and failures of components. For example, the system used by National Power has successfully identified cracked tie bolts and discs in large gas turbine rotors, enabling safe shutdown and avoiding major damage. In another incidence, the company rejected a rotor due to unsatisfactory vibration characteristics, and the supplier found a major crack in it but only after a similar rotor operated by another user failed.

4 MAINTENANCE

The cost of maintaining HGP components is the largest single element in the maintenance cost of a CCGT power plant. In order to maintain HGP parts in the most cost-effective manner, an operator needs to 'run' parts in the engine until their condition warrants either 'repair/refurbishment' or 'replacement'. The tools currently available for these decisions and those under development are discussed below.

4.1 Equivalent Operating Hours (EOH)

Decisions on inspection, repair, refurbishment and replacements of HGP components are currently based on EOH and in the case of General Electric designs also on Equivalent Starts (ES). OEMs typically recommend a maintenance regime covering inspection, repair/refurbishment and replacement (Table 1).

Table 1 Typical Inspection, Repair/Refurbishment Regime for Blades and Vanes in Stationary Gas Turbines.

	EOH	Scope of work
'A' or Combustion Inspection	8000	Visual inspection of first stage blades and vanes
'B'	16 000	Visual inspection of first stage blades and vanes
'C' or Hot Gas Path Inspection	24 000	Strip GT, refurbish high temperature vanes and blades
Major Overhaul	48 000	Replace blades and vanes in first and in most models, second stages

Conceptually, EOH is a method for summing creep, corrosion, and cyclic damage. In general terms, it can be expressed as follows. [For detailed relationships the reader is referred to UNIPEDE[2]]

$$EOH = t_{gas}^{normal} + a.t_{gas}^{peak} + b.t_{fuel}^{other} + c.t_{injection}^{water} + d.N_{starts}^{normal} + e.N_{trips}^{low}$$

$$+ f.N_{trips}^{low} + g.N_{load\text{-}shed}^{preventative} \qquad (2)$$

Where

t	= Operating Hours
N	= Number of Cycles
High trips	= Trips from above a TIT or MW threshold
Low trips	= Trips from or below the TIT or MW threshold
b	= Fuel factor, typically 1 to 2.5, depending on the fuel grade
a, c to g	= Constants, some of which may be set to zero by some OEMs

The EOH formulation, used by ABB, Siemens, and Siemens–Westinghouse, represents linear summation of creep and cyclic damage, the gradient of which varies with the manufacturer. General Electric separates the cyclic damage terms from Eq. 2, to define the parameter, ES, and employs non-linear damage summation.

Different OEMs adopt different methods of calculating and summating creep and fatigue damage in identical materials. Consequently, the life cycle of HGP components can vary significantly.

From an operator's perspective, some other limitations of EOH are listed below.

- It does not take account of the site (for example, coastal sites are more likely to suffer from corrosion), how the turbine is operated etc.

- Refurbishment and replacement lives of components as diverse as rotors, compressor blades and turbine blades, may be all measured in terms of the same EOH.
- The formulations penalise increased TIT at peak loads (MW above rated capacity) but give no 'credit' for a reduction in TIT at low loads (MW below rated capacities).

OEMs can also introduce other empirical criteria. For example, one company faced with thermo-mechanical fatigue (TMF) cracking of a Row-1 blade design introduced an 'equivalent cycle' term to prescribe an inspection regime.

4.2   NON-DESTRUCTIVE TESTING

Unless other major damage occurs (e.g. critical cracks, excessive creep, significant impact damage etc.), the recommended interval for the C-inspection (24,000 EOH) is largely based on coating life exhaustion. The life of the coating can be considered exhausted if the aluminide phase, β or γ, is fully consumed or if the Al-content of the coating falls below say 4 at%. Laboratory evaluation suggests that magnetic permeability may be one possible method for monitoring Al-depletion.[3] Another potential technique under development is the Swept Frequency Eddy Current,[4] which utilises an empirical correlation between electrical resistivity and width of the Al-depleted zone.

4.3   MODELS FOR COATING LIFE

Some research programmes are trying to develop models for estimating lives of oxidation and corrosion resistant coatings[5] as well as TBCs.[6] When validated models become available, the operators will be able to use them to determine the refurbishment intervals.

4.4   USERS' GROUP

Service experience is widely used by OEMs, independent contractors and operators for 'run/repair/refurbish/replace' decisions. OEMs have the advantage of fleet-wide experience and the genealogical features of their design but others are better able to cross-read experience with different makes of turbines. One of the most effective methods available to operators for life management of HGP parts is pooling of service information by Users' Groups. Quantitative data on service experience can help define limits on defect acceptance and repairs, methods of repairs, suitability of non-proprietary coatings etc.

5   REMAINING LIFE ASSESSMENT

5.1   METAL TEMPERATURES

Most mechanisms of damage on HGP parts (*viz.* creep, TMF, oxidation-corrosion, embrittlement etc.) are dependent on the metal temperature. This is one of the most important parameters required for life management. However, metal temperatures of HGP parts are u ually not available to operators. Even OEMs may have only limited measurements, which may not cover different operational regimes. For this reason, some operators have installed o tical pyrometers to measure the temperature of each blade. Further complexities arise on the case of TBC coated blades because the ceramic, at least when new, is transparent to infrared. Developments are in hand to interpret the data gathered on such parts.[7]

It is also possible to estimate an 'equivalent temperature' from microstructures and this can be defined as follows: if the component had been exposed to this notional temperature for the entire service duration then the material would have registered changes observed. In principle, any time-temperature dependent parameter can be used for this purpose and some that have been, are listed below.

- γ'-size is a widely used parameter.

$$(r_t)^3 - (r_0)^3 = Kt \tag{3}$$

Where $K = k^3 = \dfrac{c_1}{T} \exp\left(-Q/_{RT}\right)$,

r_t and r_0 are the radii of γ' at time = t and = 0, K, k (used by some investigators), and c_1 are constants, Q is the activation energy for coarsening = 269 kJ mol^{-1}.[7]

Considerable scatter is observed in the data due to a variety of reasons, some of which are given below.

(i)   Growth by particle impingement and sintering rather than Ostwald ripening
(ii)   Micro-heterogeneity, e.g. eutectic regions
(iii) Irregular shapes as particles loose coherency during coarsening
(iv) Possible rafting, particularly in DS and SX alloys
(v)   Whether the etchant attacks the γ-matrix or the γ'-particles
(vi) Subjective judgements in image analysis etc.

The accuracy of the method is probably ±25°C,[8] which is large for a very meaningful assessment of creep life. For practical purposes, its use is probably limited to above 700°C because of the very slow kinetics at lower temperatures.[9]

- Width of the interdiffusion zone between the coating and the base material has also been used.[10,11,12] Ellison, Daleo and Boone[12] report good agreement between temperature estimates from interdiffusion and γ'-coarsening.

$$\xi^2 = k_p t \tag{4}$$

Where $k_p = c_2 \exp\left(-Q/_{RT}\right)$

ξ =   Width of interdiffusion zone below the original interface in μm
Q =   366 kJ mol^{-1} for GT29+/DS GTD-111 system[12]
c =   constant

- The metal temperature of the cooling passage has been estimated from the oxide thickness on the bare metal,[13] using a relationship similar to Eqn (4).

5.2   Destructive Assessment of Remaining Life

Blades/vanes from a set can be destructively examined to determine the types of damage that occur in service, the locations and the magnitude of such damage and their effect on the remaining life. In the absence of other information, the sample is selected on the basis of a visual inspection. However, if pyrometer data is available then the hottest blades might be chosen for a conservative assessment.

It may be tempting to use the γ'-particles in ex-service parts as a measure of the remaining life but the validity of such an approach is questionable for the following reasons.

(i)   The γ'-size measurement is likely to suffer from significant scatter due to the reasons given in Section 5.1.

(ii)   Even very marked degradation of the γ' may not lead to a significant reduction in long-term creep properties.[14]

(iii)   The knowledge of the remaining life in ex-service condition is of little value to an operator who wants to know what life the component will have if it is refurbished. The heat treatment, which necessarily forms a part of the coating process during refurbishment, will very largely restore the γ'-particles to near virgin state. The extent to which it also restores the life is less certain.

Creep/stress-rupture tests[13] and stress relaxation tests at constant-strains[14] are two potential techniques for estimating remaining lives. Considerable care should be exercised in applying measured creep parameters to predict the remaining life of components for the following reasons.

(i)   The sampled blade may not have the shortest remnant life.

(ii)   Test samples often span regions with different thermal and mechanical loads.

(iii)   Section size effects: samples extracted from a thin stock is markedly weaker in creep than identical specimens from a thicker stock.[15] Small samples also creep significantly faster than larger test-pieces from the same material.[14]

(iv)   Limited numbers of test pieces make it difficult to verify statistical scatter.

(v)   Rupture tests for residual life are usually carried out at high stresses and they show significant reduction in life after service exposure. However, at low stresses and long term tests, rupture life of service exposed blades is not very different from identical new blades.[14] Long term service exposure of up to 50 000 hours has remarkably little effect on creep rates measured in stress relaxation tests.[15]

One consideration before destructive testing is the availability of replacement parts to make-up the complete set.

5.3   Aerothermal Analysis

Full aerothermal analysis and lifing of blades is an extremely difficult task even for OEMs who have a detailed knowledge of design parameters. This is evident from the fact that in our experience no first version of Row-1 blade design in recent large frame engines has yet attained the expected life. Nevertheless, publications from some operators suggest a developing capability to carry out aerothermal analysis for estimating damage in GT blades.[16,17] It is not clear if any operator is actually making 'run/repair/refurbish/replace' decisions solely on the basis of such analysis. Nevertheless, the analysis can provide a basis for informed judgement when looking at options, operational changes and design improvements.

6   REPAIR AND REFURBISHMENT

HGP components are repaired and refurbished at periodic intervals, as outlined above. The scope of the usual work is summarised below.

- Inspection – usually visual and by liquid penetrant, but depending on the condition of the part may include ultrasonic, eddy current and radiography
- Removal of coatings – generally by chemical leaching and/or mechanical means e.g. grit-blasting, grinding
- Some parts (e.g. blades after long term service) may be Hot Isostatically Pressed (HIPed) to rejuvenate the material properties
- Blending – grinding of minor damage, small indentation etc. to a smooth profile
- Brazing to build-up wall thickness and to repair cracks
- Welding to restore blade tips and seal fins, repair cracks, or replace damaged segments
- Coating with oxidation-corrosion resistant material and/or TBC
- Heat treatment to restore the microstructure
- Restoration of cooling holes
- Renewal of honeycomb seals

Operators can have the parts repaired and refurbished either by OEMs or by independent contractors, many of whom also act as sub-contractors to OEMs. The two options offer different advantages and disadvantages.

6.1   COATINGS

Independent contractors can not offer coating systems, which are proprietary to OEMs without their approval. Therefore, operators generally have to accept equivalent or other non-proprietary coating systems, which offer opportunities and risks. It may be possible to use a coating system, which is better suited to the particular duty cycle of the GT, taking account of the fuel, TIT, loading pattern, geographical location, prevailing ambient conditions etc. However, this opportunity has to be balanced against the risk of coating-related failures unless either engine experience is available for comparable duty or sound metallurgical assessment can be made. As the gas and metal temperatures increase and the operators look to extend refurbishment intervals, the oxidation/corrosion resistance of the coating must also increase and the TBC integrity must improve.

6.2   RAINBOW TESTING

Operators can follow the route often taken by OEMs and conduct rainbow trials to evaluate available coating systems. Indeed, such trials may be undertaken by OEMs on

GTs for which they have maintenance contracts, sometimes apparently without the knowledge of the operator.

6.3   INSPECTION LIMITS, ACCEPTANCE CRITERIA

Independent contractors will not have 'engine-proven' specifications on repair and refurbishment of HGP parts of new and advanced models because such turbines have not been operating for sufficient duration. Therefore, inspection and acceptance criteria, limits of blending and other repairs etc. will often be based on the experience with mature turbines. The operator has to consider if adoption of these specifications will result in 'fit-for-purpose' parts and assess associated risks. This assumes particular importance in the case of DS and SX parts.

7   REPLACEMENT PARTS

7.1   SPARES HOLDING

Operators need spares to replace parts after breakdowns and for planned maintenance. Some of the options available to an individual operator are (i) to buy, own and probably store spares on site (ii) enter into a pool owned by operators (iii) spares pool operated by OEMs (iv) buy parts just-in-time and possibly, pay a premium for urgent supply. Which spares holding strategy suits an operator depends on several factors, some of which are identified below.

- Cost of spares, which for a complete set of HGP parts run into several millions of pound sterling.
- Likelihood of the design becoming obsolete.
- Probability of a type-fault in the design.
- Reliability of the component, which can be defined by mean-time-to-failure and mean-time-to-repair. In the absence of quantitative values for relatively newer models, the operator may have to assess engineering risks.
- Lead time for the supply of engine-ready parts. Typical figures for blades and vanes could be 12–18 months.
- Repair and refurbishment interval.
- Likely replacement interval.
- Turn-around time from repair/refurbishment.
- Cost of repair/refurbishment.
- Scrap rate.
- Size of the fleet.
- Exchangeability of some parts with other engines in the fleet.
- Anticipated duty cycle.
- Commercial factors e.g. contractual agreements with investors and customers of electricity, agreements with OEMs and/or other suppliers.

National Power has developed a model, which can analyse spares strategies for very complex systems.

7.2   SOURCE OF NEW PARTS

Parts for newly introduced models are available exclusively from OEMs. However, for mature engines, independent suppliers begin to offer alternatives, which generally lead to significant reductions in prices of parts. Operators' commitment to use parts from independent sources is a necessary prerequisite for such suppliers to make the investment necessary to develop suitable parts.

One of the factors, which may influence the decision, is the goodwill between the OEM and the operator.

Under certain circumstances, parts from third party can offer opportunities for improving the design and/or materials. For example, Arizona Public Service Company successfully replaced Row-1 blades in forged Udimet-520 with those in cast In-738LC.[17]

8   BREAKDOWN SUPPORT

In the author's experience, if major failures or breakdowns occur then OEMs generally offer good support to operators for repairing the damage and restoring the turbine to an operational state.

9   SUMMARY

- Operators can assess and manage technological risks when buying power plants.
- To optimise costs, they need information and tools for 'run/refurbish/repair/replace' decisions, but much of this is proprietary to OEMs.
- Successful developments in the following areas would assist operators considerably in managing the life of HGP parts.

  (i)   Optical pyrometers to measure blade metal temperatures
  (ii)   New and improved NDT methods
  (iii)   Models for estimating lives of oxidation/corrosion resistant coatings and TBCs
  (iv)   Pooling of Users' experience
  (v)   Estimating remaining lives either by destructive or by analytical methods
  (vi)   Novel repair and refurbishment technologies for welding CC blades, coating of blades with shaped film-cooling holes, DS and SX components
  (vii)   Developing acceptance criteria and extending limits of repair, particularly for DS and SX parts
  (viii) A competitive market for HGP parts

- Notwithstanding commercial considerations, technical co-operation between OEMs and operators is essential for low life-cycle cost of advanced gas turbines.

10 ACKNOWLEDGEMENTS

This paper contains the author's personal opinions based on his knowledge and experience of National Power and other power producers but does not necessarily represent the Company views. The author is grateful to many colleagues for numerous helpful discussions. The paper is published with the permission of the Company. © National Power plc 2000.

REFERENCES

1. N.R.D. Orchard, 'Advanced Design CCGTs – Asset or Liability', IMechE Conf., 1999.
2. UNIPEDE, 'Combustion Turbines in UNIPEDE Countries: Design and Performance Characteristics, Maintenance and Operation Experience', Report No. 02006Ren9763, 1997.
3. N. Czech, F. Kirchner and W. Stamm, 'Life Assessment of Service Stressed High Temperature Coatings by Permeability Measurements', *Elevated Temperature Coatings: Science and Technology II*, N.B. Dahotre and J.M. Hampikian eds, TMS, 1996, 361–371
4. R. Viswanathan, R. Frischmouth and D. Gandy, High Tech Turbines Require High Tech Maintenance, *Power Engineering*, January 1999.
5. K.S. Chan, N.S. Cheruvu and G.R. Leverant, *Predicting Coating Degradation under Variable Peak Temperatures*, IGTI, 1999.
6. K.S. Chan, Y.-D. Lee, G.R. Leverant, T.J. Fitzgerald and J.G. Goedjen, *Prediction of Damage and Failure in Thermal Barrier Coatings, Damage and Failure of Interfaces*, Rossmanith ed , Balkema, 1997.
7. S.D. Alaruri, L. Bianchini and A.J. Brewington, 'Emissivity Measurements for YSZ Thermal Barrier Coating at High Temperatures using a 1.6 µm Single-Wavelength Pyrometer', *Optical Engineering*, 1998, **37**(2), 687.
8. J.M. Aurrecoechea and W.D. Brentnall, 'Operating Temperature Estimation and Life Assessment of Turbine Blade Airfoils', *Proc. Conf. Life Assessment & Repair Technology for Combustion Turbine Hot Section Components*, EPRI/ASM, 1990.
9. G.M. McColvin, Technical Report 3296 for COST-50/II Project UK 6, 1980, Henry Wiggin referred to in Summary of Physical and Mechanical Property Data of Ni-Base Superalloy In-738, COST Report 81-UW-COST-B1 by R Stickler, 1981.
10. V. Srinivasan, N.S. Cheruvu, T.J. Carr and M. O'Brien, 1995, 'Degradation of MCrAlY Coating and Substrate Superalloy during Long-Term Thermal Exposure', *Materials and Manufacturing Processes*, 1995, **10**, 955–969.
11. N.S. Cheruvu and G.R. Leverant, 'Influence of Metal Temperature on Base Materials and Coating Degradation of GTD-111 Bucket', IGTI Gas Turbine Expo, June, Stockholm, Sweden, 1996.
12. K.A. Ellison, J.A. Daleo and D.H. Boone, Metallurgical Temperature Estimates Based on Interdiffusion Between CoCrAlY Overlay Coatings and a Directionally Solidified Nickel-Base Superalloy Substrate, *Proc. 6th Liége Conference on Materials for Advanced Power Engineering*, 1998, Liège, Belgium, 1998
13. V.P. Swaminathan, J.M. Allen and G.L. Touchton, 'Temperature Estimation and Life Prediction of Turbine Blades using Post-Service Oxidation Measurements', *Trans ASME*, 1997, **119**, 922–929 [presented at IGTE, June 1996, Birmingham, UK].
14. R. Castillo and A. K. Koul, 'Factors Influencing the Residual Creep Life Prediction of Service Exposed Ni-base Superalloy Components', *Proc. Conf. Life Assessment & Repair Technology for Combustion Turbine Hot Section Components*, EPRI/ASM, 1990.
15. D.A. Woodford, 'The Design for Performance Concept Applied to Life Management of Gas Turbine Blades', *BALTICA III Int. Conf Plant Condition & Life Management*, June 1995, Helsinki-Stockholm, Vol. 1, 319–331.

16. R. Frischmouth, 'Blade Life Management System for GE Frame 6B Gas Turbines', EPRI Reports, TR-109196-V1 and TR-109196-V2, 1998.
17. J.K. Hepworth, 1999, 'Accurately Assessing the Remaining Life of Hot Gas Path Components to Enable Informed Operational Decisions and Independent Appraisal of OEM Recommendations', CCGT Generation Conf., IIR Ltd., London.
18. J.B. Lovelace and J.C. Hendelman, 'Assessment of Reliability and Life Improvement Utilising Alternate Materials for Gas Turbine Blades', Steam and Combustion Turbine Blading Conference, January, EPRI, Orlando, USA, 1992.

Microstructural Design of IN706-type Disc Materials for Improved Creep Crack Growth Resistance

STEFFEN MÜLLER* and JOACHIM RÖSLER

Institut für Werkstoffe, D-38016 Braunschweig, Germany

1 INTRODUCTION

The demand for high efficiency gas and steam turbines in power generation requires, amongst others, new material solutions for turbine rotors. A candidate material to replace current ferritic steels in applications where the component temperature exceeds 600°C is the Ni-Fe-base superalloy IN706 as it combines superior creep strength with proven manufacturability of large components.[1] However, one design limitation of wrought Ni-base superalloys such as IN706 in this temperature regime is accelerated creep crack growth due to embrittling action of the environment, known as Stress Accelerated Grain Boundary Oxidation (SAGBO).[2,3]

In this article, possible concepts to improve resistance against creep crack propagation by microstructural design of the material are outlined. Heat treatment modifications are discussed in the first part. By stabilisation of grain boundaries with discontinuous η (Ni_3Ti) [2], which forms between 750°C and 975°C,[4] it is possible to improve creep crack growth resistance by several orders of magnitude. It is believed that η alters the segregation behaviour of deleterious elements at the grain boundary[5] or traps oxygen atoms.[6] Another mechanism of reducing susceptibility to SAGBO may be crack tip blunting due to deformation of the γ'/γ'' denuded zone around η plates.[7] Surprisingly, in our study it was found that conventional stabilising heat treatment at 850°C for 3 h[1] does not necessarily lead to acceptable results and that slight heat treatment modifications can have dramatic effects on the alloy's creep crack growth resistance. It is demonstrated that creep crack growth resistance is also critically dependent on cooling history which is interpreted in terms of the η-γ'/γ'' precipitation sequence.

In the second part, a modified boronising treatment is discussed which allows to alter the grain boundary chemistry close to the component surface and, by that, protects against SAGBO. It is well established that small amounts of boron are sufficient to suppress SAGBO.[8] Boron enrichment at grain boundaries is achieved by formation of a diffusion zone after powder pack boronising of IN706. Constant strain rate tests at 600°C are used to quantify the effect of boronising on mechanical performance. A change from intergranular to transgranular fracture and a four times longer lifetime compared to unmodified material indicate successful avoidance of SAGBO in IN706.

---

*New at Alstom Power Ltd, Dept GTDB, CH-5452 Baden, Switzerland.

2 EXPERIMENTAL PROCEDURE

2.1 MATERIAL

For this study two different batches of IN706 were used. First, a section of a triple melt (vacuum induction melting, electroslag remelting, vacuum arc melting) and forged turbine disc supplied by ABB Power Generation Ltd. Baden (Switzerland) was used for creep crack growth tests and heat treatment modifications. The chemical composition is given in Table 1. The material had an ASTM 5 grain size.

Table 1   Chemical Composition of IN706 (wt. %) – disc material.

Ni	Fe	Cr	Nb	Al	Ti	C	Si	B
41.96	36.97	16.02	3.02	0.20	1.55	0.01	0.08	0.003
Mn	**Cu**	**S**	**Mg**	**Co**	**Ta**	**P**	**O**	**N**
0.07	0.02	0.0006	<0.001	0.05	0.01	0.008	0.002	0.004

Secondly, boronising tests were performed on hot rolled sheet material (12 × 120 mm cross section) of IN706 supplied by INCO Alloys International Inc. An ASTM 3 grain size was measured after heat treatment. The chemical composition of this material is given in Table 2.

Table 2   Chemical Composition of IN706 (wt. %) – sheet material.

Ni	Fe	Cr	Nb	Al	Ti	C	Si
40.98	Rem.	16.50	2.90	0.225	1.68	0.017	0.04
Mn	**Cu**	**S**	**Mg**	**Co**	**Ta**	**P**	**B**
0.08	<0.03	0.0002	<0.001	0.03	<0.03	<0.005	0.003

2.2 HEAT TREATMENT

For different fields of application, two different heat treatments are proposed by the alloy manufacturer:[1]

1. Two-step heat treatment resulting in high tensile strength at moderate temperatures,
2. Three-step heat treatment for high rupture strength at temperatures up to 700°C.

The time–temperature profile of both heat treatments is given in Fig. 1. The second heat treatment uses an additional stabilising step (850°C for 3 h) for η precipitation at grain boundaries. In order to simulate the cooling characteristics of a large component, a cooling

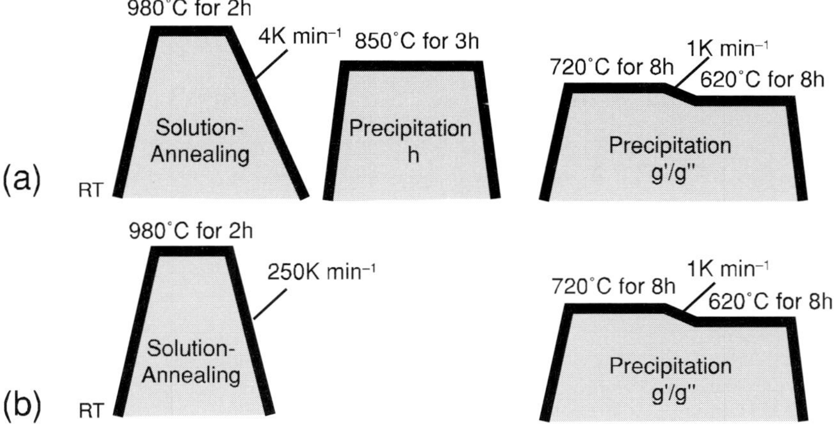

Fig. 1 Time–temperature profile of three-step (a) and two-step (b) heat treatment cycles for IN706.

ra e of 4 K min⁻¹ was selected for the three-step heat treatment, whereas air cooling (~250 K min⁻¹) was used for the two-step heat treatment.

2.3 Creep Crack Growth Experiments

Fci creep crack growth measurement, compact tension (CT) specimen with 25 mm th ckness and 2.5 mm deep side grooves according to ASTM E 1457–92 were used. The fracture surface was in the tangential-radial plane of the forging with crack propagation in ra ial direction. All specimen were precracked under fatigue loading at room temperature. C eep crack growth tests were performed using electromechancial testing machines with chamber furnaces. The temperature was controlled within ±3°C using thermocouples which were attached directly to the specimen. Crack propagation was monitored using the pctential drop technique as described in ASTM E1457–92. Crack length was calculated from the measured potential signal using Johnson's formula (see e.g. ASTM E1457–92). Because brittle fracture was observed in all tests, linear-elastic fracture mechanics was used.

2.4 Boron Deposition and Constant Strain Rate Tests

IN706 was boronised using a boron containing powder pack. This method is widely used for production of wear resistant coatings on steels.[9] Here, a recently developed powder formulation for nickel based superalloys[10] is used. Finished tensile test specimen (gauge length 25 mm, diameter 8 mm, specimen axis perpendicular to rolling direction) were completely covered with powder and heat treated at 850°C for 3 h in argon atmosphere. During heat treatment, boron atoms are adsorbed at the surface via chemical vapor deposition. The specimen were diffusion heat treated at 1000°C for 0.5 h after removal from the powder pack. Because of the solutioning effect of this heat treatment, two step heat treatment was applied again.

Since it was shown that constant strain rate (CSR) testing with strain rates of $d\varepsilon/dt=10^{-3}$ ... 10^{-4} h^{-1} is an excellent method to establish the resistance of nickel base superalloys against SAGBO,[2] this method was also used here. CSR tests were performed in a MTS servohydraulic testing machine at 600°C using induction heating. A strain rate of $d\varepsilon/dt=5.10^{-4}$ h^{-1} was selected.

2.5 METALLOGRAPHY

For metallographic investigations specimen were ground and polished mechanically and etched with a mixture of 100 ml HCl, 10 ml HNO$_3$, 0.3 ml Sparbeize (trade name of Wirtz-Buehler GmbH Düsseldorf, Germany) and 100 ml H$_2$O at a temperature of 40°C. For transmission electron microscopy (TEM), a Philips CM12 was used. Specimen for TEM investigations were cut, punched and jet-polished at –15°C using 30 ml ethyleneglycol monobutyl ether, 63 ml Ethanol and 7 ml HClO$_4$ (6%). Boron concentration profiles were measured quantitatively by glow discharge optical spectroscopy (GDOS). For that purpose, surface layers of approx. 20 µm thickness were removed by grinding after each measurement to gain adequate depth resolution.

3 RESULTS

3.1 MICROSTRUCTURE

Typical microstructures after different standard heat treatments are shown in Fig. 2. Primary, blocky carbides and nitrides (see arrow 1 in Fig. 2a) as well as γ' and γ'' are visible independent of heat treatment cycle. Two-step heat treatment produces 'clean', η-free grain boundaries, whereas η precipitates are evident at grain boundaries (arrow 2 in Fig. 2b) after stabilisation at 850°C for 3 h. However, not all grain boundaries are decorated

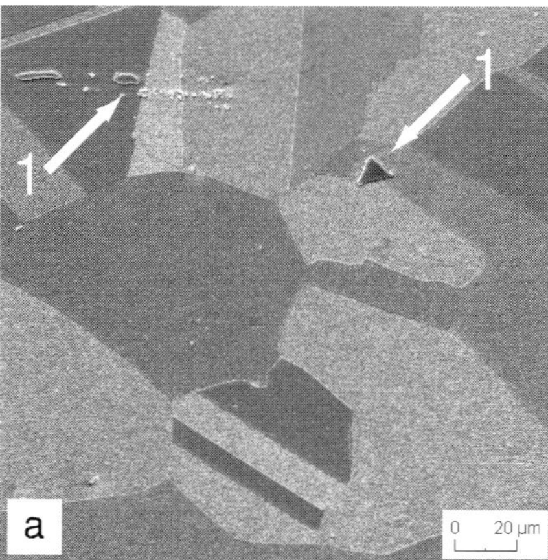
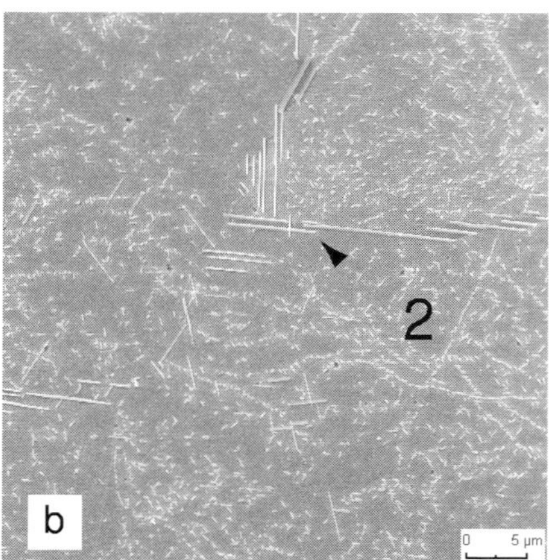

Fig. 2 Microstructure of IN706 after two-step (a) and three-step (b) heat treatment. Arrow 1 indicates primary carbides and nitrides, arrow 2 indicates η-phase.

with η and, sometimes, film like precipitation is observed. In contrast to literature results,[4] these films are identified as η and not as carbides.

3.2   CREEP CRACK GROWTH OF IN706 AFTER STANDARD HEAT TREATMENT

All creep crack growth tests showed immediate crack propagation after loading without incubation time. Results for both heat treatments are given in Fig. 3. The two-step heat treated material exhibited very high crack growth rates as reported in the literature[2] and fractograhic investigation revealed fully intergranular fracture with many secondary cracks (Fig. 4).

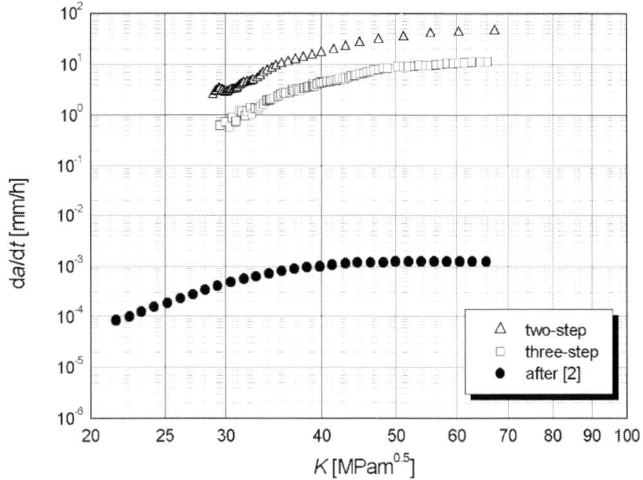

Fig. 3   Creep crack growth rates of IN706 at 600°C for standard heat treatments according to Fig. 1. For comparison, data from Ref. 2 are included.

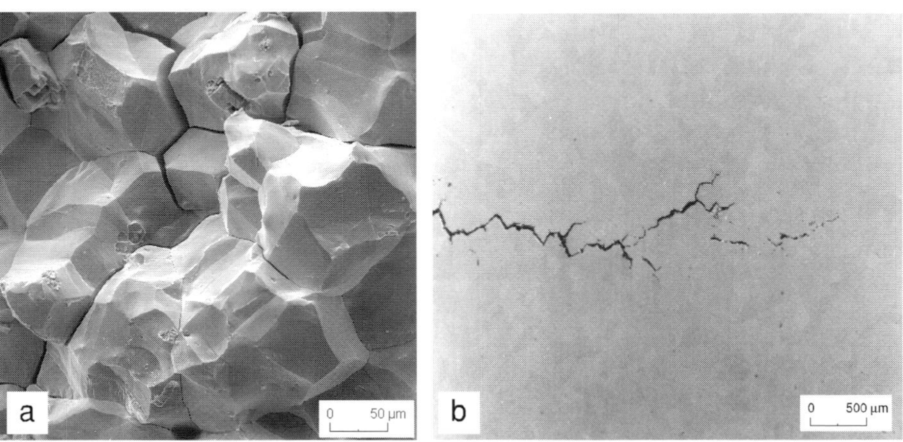

Fig. 4   Fracture surface of two step heat treated IN706 (a: fracture surface, b: cross section of fracture path) after creep crack growth at 600°C.

According to literature data,[2] three-step heat treatment should improve creep crack growth resistance by several orders of magnitude. However, only a marginal benefit was found (Fig. 3) and fracture was again intergranular. Yet, a slight difference is noticeable presence of η after three-step heat treatment leads to teeth-like serrations on former grain boundaries (Fig. 5). These can be attributed to pull-out of η-platelets upon fracture, explaining the slightly improved creep crack growth resistance compared to two-step heat treated material.

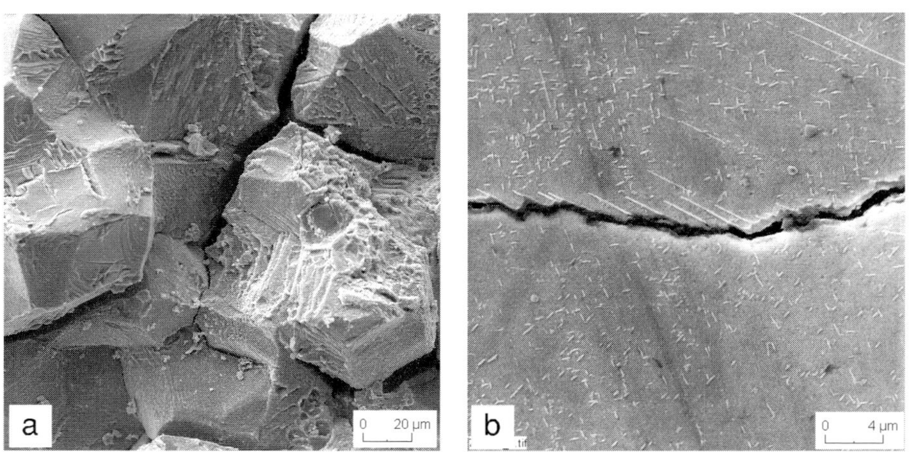

Fig. 5 Fracture surface of three step heat treated IN706 (a: fracture surface, b: cross section of fracture path) after creep crack growth at 600°C.

Following this argument it appears necessary to increase the amount of η by further heat treatment modification in order to (i) improve grain boundary cohesion and (ii) avoid fast fracture along grain boundaries not decorated with η platelets. Potential measures to achieve the above objective are discussed in the following.

3.3   THREE-STEP HEAT TREATMENT MODIFICATIONS

As η precipitation occurs during the stabilising step at 850°C, the stabilising duration was extended from 3 h to 10 h with the goal to precipitate more η at grain boundaries. However, as shown in Fig. 6, the effect on creep crack growth resistance is marginal and metallographic examination revealed little increase in η volume fraction at grain boundaries. Further variation in time and temperature of solutioning and stabilising heat treatment did not change the situation.

Surprisingly, a heat treatment cycle with direct cooling from solution annealing (3 h @ 980°C) to stabilisation (10 h @ 820°C) at 4 K min⁻¹ cooling rate (see Fig. 7) showed better microstructure than all trials before. Subsequent heat treatment steps were unchanged. This cycle type, referred to as 'direct ageing' in the following, was originally proposed by Shibata et al.[11] for heat treatment of large IN706 components to save time and costs.

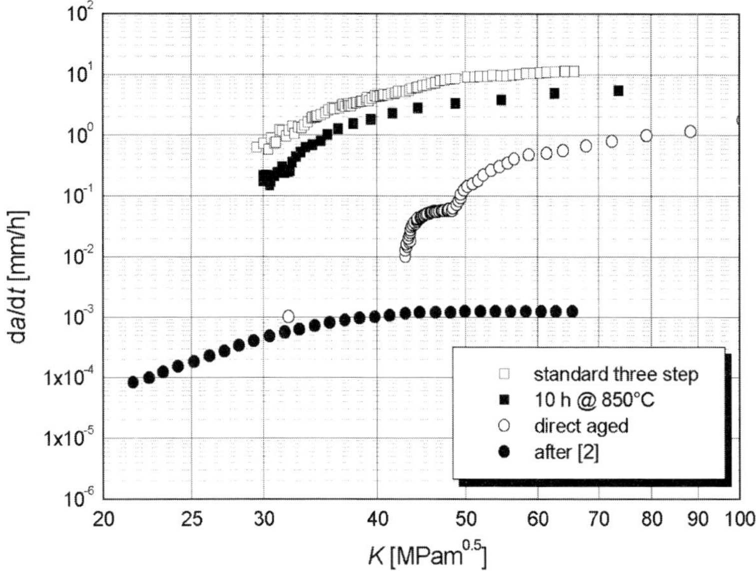

Fig. 6   Creep crack growth of IN706 for different stabilising heat treatments (for explanations see text). Data from [2] are included for comparison.

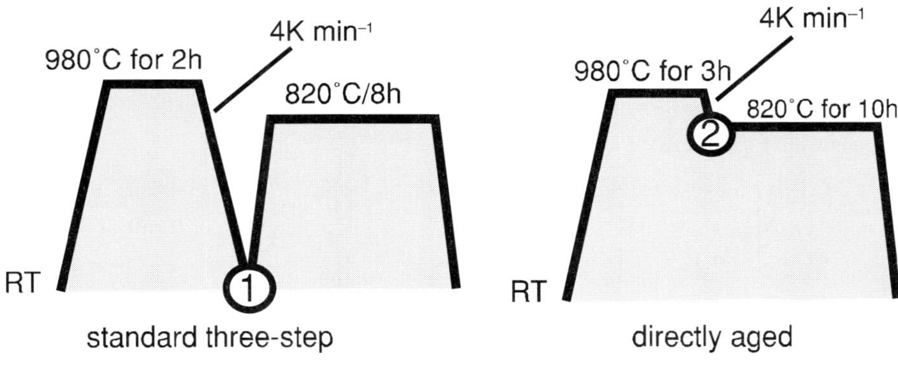

Fig. 7   First part of the cycles for 'standard three-step' and 'directly aged' heat treatment. Heat treatment was interrupted at points '1' and '2' for microstructural investigations.

Creep crack growth data show a strong improvement of the directly aged over the standard material (Fig. 6). At low stress intensities, the results are comparable with data from,[2] while at higher K values a discrepancy remains. The fracture surface is similar to that shown in Fig. 5.

In order to clarify the observed effect of direct aging on microstructure and mechanical behaviour, heat treatment cycles were interrupted by rapid water quenching at points given in Fig. 7 and the microstructure was investigated subsequently. Results of these investigations are discussed in relation to the η precipitation sequence in the next chapter.

3.4 η Precipitation Sequence

According to the time–temperature precipitation diagram of IN706,[4] annealing at 980°C dissolves all precipitates except for primary carbides and nitrides. Upon slow cooling at K min^{-1} to room temperature during standard heat treatment, formation of intragranular γ'/γ'' precipitates is expected because incubation time is short [4]. Indeed, TEM investigation of materials state '1' (see Fig. 7a) shows fully developed γ'/γ'' particles with a size of approximately 10 nm (Fig. 8). The hardness increases to 333 HV10 compared to 139 HV10 after rapid quenching from solutioning (2h @ 980°C/water quenched). In contrast, γ'/γ'' precipitates were hardly detectable after quenching from material state '2' (see Fig. 7). The particle size was less than 1nm and hardness increased only to 156 HV10. Lack of γ'/γ'' precipitation is understandable, noting that the γ'/γ'' 'nose' of the time-temperature-precipitation diagram is significantly below 820°C.

Fig. 8 TEM image of material state '1' according to Fig. 7, showing γ'/γ'' matrix precipitates and η plates at a grain boundary.

Heterogeneous η nucleation at grain boundaries occurred in both material states (Fig. 8) seemingly without precursor phase. This contrasts previous findings in austenitic alloys, suggesting that the strain field of γ' particles in the matrix is needed as driving force for η formation [12]. Also [11] assumes γ' as precursor phase of η.

Upon further heat treatment at 820°C, the situation with respect to η growth is, however, quite different for the two material states. Upon standard heat treatment a significant amount of Ti and Nb is captured by γ'/γ'' precipitates prior to aging at 820°C so that η growth is hindered. In contrast, direct aging produces η prior to γ'/γ'' so that competition for Ti and Nb is won by η. In consequence, a higher volume fraction of grain boundary phase results, leading to superior creep crack growth resistance as observed experimentally.

4 BORONISING

4.1 MICROSTRUCTURE AFTER BORON DEPOSITION

After boronising and subsequent diffusion heat treatment, a boron enriched matrix zone is formed with precipitation of (Ni,Fe)-rich borides in the matrix and at grain boundaries. Additionally, an approx. 35 µm thick boride layer is observed (Fig. 9). This layer consists of different boride types, details are given elsewhere.[13]

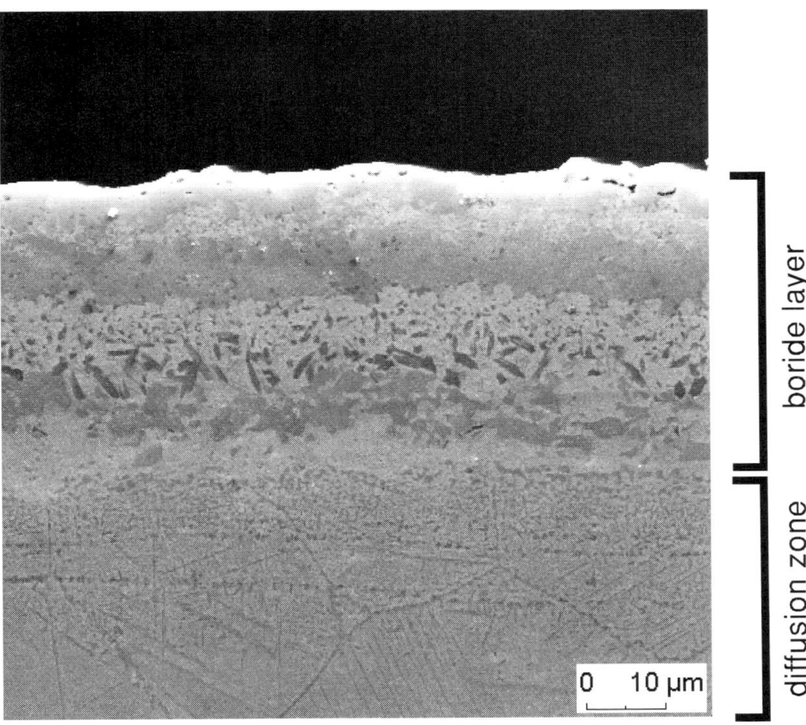

Fig. 9 SEM micrograph showing the boride layer and the boron enriched diffusion zone in IN706.

Because boron is the key element in preventing SAGBO, the concentration profile after diffusion heat treatment was examined by GDOS. Beginning with a maximum boron concentration of approx. 8 wt% at the specimen surface, the boron content drops sharply. For the diffusion zone, which is the region of prime interest, the concentration profile is given in Fig. 10. Considering an initial boron content of 30 ppm in IN706, boron enrichment is found to a depth of approx. 200 µm.

4.2 MECHANICAL BEHAVIOUR

Results from CSR tests of boronised and unboronised specimens are given in Fig. 11. In the case of unmodified IN706, fast, intergranular fracture occurred after approx. 20 h and

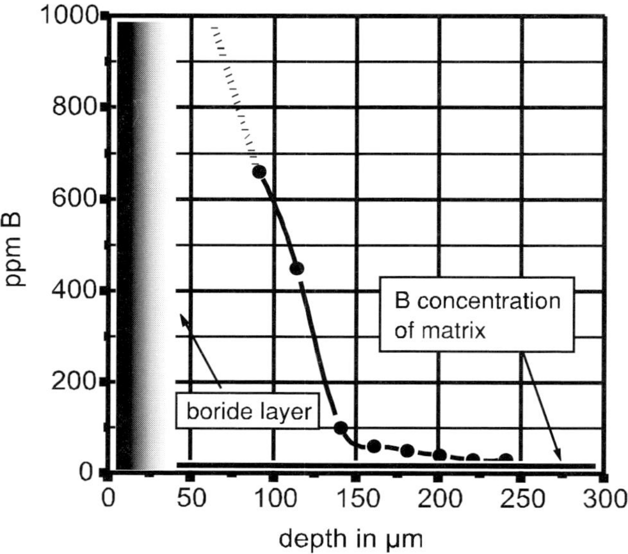

Fig. 10 Boron concentration profile of the interdiffusion zone after heat treatment at 1000°C for 0.5 h as measured by glow discharge optical spectroscopy.

0.5% plastic elongation. Before complete fracture occurred, the stress decreased continuously from 800 to 200 MPa within 1 h which can be attributed to stable crack propagation. Due to the action of SAGBO, the fracture morphology was identical to that obtained after creep crack growth testing (see Fig. 4).

For boronised specimen, however, fracture strain and lifetime increased drastically to about 3.5% and 78 h, respectively. Fractographic investigations showed a markedly different type of fracture (Fig. 12). Within a zone of approx. 150 μm, a ductile, transgranular

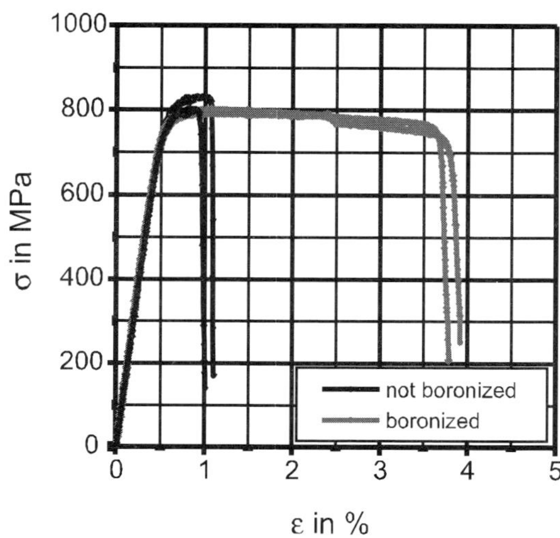

Fig. 11 Stress-strain diagram of the constant strain rate tests performed at 600°C.

fracture is apparent which changes to intergranular fracture as observed in unmodified material. According to Fig. 10, the depth of the boron enriched zone corresponds well with the depth of transgranular fracture. Thus, a boron concentration of about 100 ppm seems to be sufficient to protect IN706 against SAGBO.

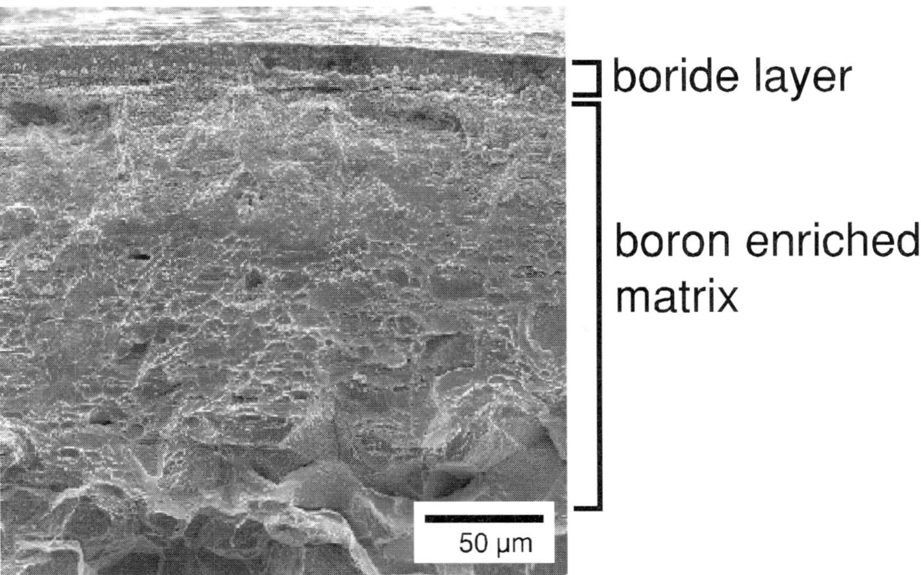

Fig. 12   SEM micrograph of the fracture surface after constant strain rate testing of a boronised sample at 600°C. Brittle fracture of the outer boride layer and the transition zone with transgranular fracture is visible.

Additionally, brittle fracture of the outer boride layer is visible in Fig. 12 which is believed to fracture immediately upon loading. It means that this layer is not necessary to protect the alloy against oxygen attack. Instead, only the diffusion zone protects against SAGBO because it 'seals' grain boundaries against oxygen penetration.

5   CONCLUSIONS

The environmental sensitivity of IN706 at 600°C, known as SAGBO, was investigated by measurement of creep crack growth behaviour. Whereas two-step heat treatment leads to fast fracture, three-step heat treatment with stabilised grain boundaries may prevent SAGBO. But, dependent on heat treatment cycle results show that standard three-step procedures recommended in the literature may not lead to optimal creep crack growth resistance. With respect to very large turbine components an optimised heat treatment is proposed. Basis for understanding this improvement is the analysis of the γ'/γ'' and η precipitation sequence. The modified heat treatment maximises the content of η grain boundary phase and, by that, reduces the creep crack growth rate by up to two orders of magnitude.

Another possibility to increase resistance of unstabilised IN706 (two-step heat treatment) against SAGBO is local enrichment of boron in a diffusion zone at the component surface. As shown by CSR tests, the failure mechanism may be altered completely by the presence of about 100 ppm boron in the matrix. Brittle, intergranular fracture due to SAGBO as observed after creep crack growth is suppressed and a retarded, transgranular fracture occurs leading to significant extension of lifetime and creep ductility.

6   ACKNOWLEDGEMENTS

The authors would like to thank the Deutsche Forschungsgemeinschaft, Germany, for financial support and Dr. H.-J. Hunger from Bortech GmbH Grefrath (Germany) for boronising the samples. Material supply by ABB Power Generation Ltd. Baden (Switzerland) and INCO Alloys International Inc. Huntington (USA), is greatly appreciated.

REFERENCES

1. *Inconel 706,* brochure from Huntington Alloys, Inc., 1974.
2. G. Härkegard, W. Balbach, K. Stärk and J. Rösler, in: *Superalloys 718, 625, 706 and Various Derivatives,* E.A. Loria ed., TMS, 1997, 425–430.
3. G. Härkegard and J.Y. Guedou, in *Materials for Advanced Power Engineering,* J. Lecomte-Beckers *et al.* eds, Jülich, Forschungszentrum Jülich 1998, 913–931.
4. K.A. Heck, in *Superalloys 718, 625, 706 and Various Derivatives,* E.A. Loria ed., TMS, 1994, 393–404.
5. S.P. Lynch, T.C. Radtke, B.J. Wicks and R.T. Byrnes, *Int. Met. Reviews,* 1974, **19**, 199–222.
6. S.D. Antolovich, in *Superalloy 718 – Metallurgy and Applications,* E.A. Loria ed., TMS, 1989, 647–653.
7. R. Molins, J.-C. Chassaigne and E. Andrieu, in *Superalloys 718, 625, 706 and Various Derivatives,* E.A. Loria ed., TMS 1997, 655–664.
8. M. Nazmy, M. Staubli, C. Noseda, in *Superalloys 718, 625, 706 and Various Derivatives,* E.A. Loria ed., TMS 1997, 419–424.
9. H.-J. Hunger and G. Trute, *Heat Treat. Metals,* 1994, **21**, 31.
10. H.-J. Hunger, *Mater. Sci. Forum,* 1994, **163**, 341.
11. T. Shibata, T. Takahashi, Y. Shudo and M. Kushuhashi, in *Superalloys 718, 625, 706 and Various Derivatives,* E.A. Loria ed., TMS, 1997, 379–388.
12. B.R. Clark and F.B. Pickering, *JISI,* 1967, 205, 70–84.
13. J. Rösler and S. Müller, *Scripta Mat.,* 1999, **40**, 257–263

Fatigue in Modern Nickel-Base Alloys for Gas Turbine Applications

R.R. SEELEY and V.R. ISHWAR;[a] P.K. LIAW and V.H. HE[b]

[a]Haynes International, Inc., P.O. Box 9013, 1020 West Park Avenue, Kokomo, Indiana 46904-9013, USA
[b]University of Tennessee, Knoxville, Tennessee 37996, USA

1   INTRODUCTION

In today's modern gas turbines, there is increasing emphasis being placed towards improving the efficiency and decreasing the NOx emissions. Material selection of thc hot gas path components plays a significant role in meeting these objectives through use of higher operating temperatures and reduced cooling. As the turbine entry temperatures approach 1200°C or higher, alloys with higher creep-rupture strength, higher fatigue resistance, and better environmental resistance are required.

Hot gas path components in modern gas turbines experience cyclic conditions of stress or strain by virtue of operation of the gas turbine. Cyclic strains can develop because of temperature changes during operation or due to start-ups and shut-downs. Cyclic stresses can occur by vibration or simply by cyclic mechanical loads. Therefore, fatigue stresses and strains are an operational reality for hot gas path components. The fatigue properties of the materials used in these components are important for the optimum design and operation of the component.

The fatigue properties of high-temperature alloys are influenced by a number of factors operating sometimes singly and sometimes in combination. The most prominent factor is the cyclic stress or strain that is imposed on the structural component or on fatigue test specimens.

High cyclic stresses or strains result in short lives. Temperature is probably the next most influential effect where higher temperatures generally result in shorter fatigue lives. Other influences include applied mean stresses or strains and hold times occurring at peak stresses or strains. High mean stresses have been shown to result in lower fatigue lives. Hold times at peak stresses or strains may influence the fatigue life of a metal depending on where in the loading cycle these hold times are imposed. Hold times in the compressive portion of the cycle may or may not have a significant impact on fatigue life, while hold times in the tension portion of the cycle generally reduce the fatigue life. Metallurgical factors, such as grain size, may also effect the fatigue life of an alloy. Larger grain sizes are known to reduce the fatigue life. He and Liaw[1] have noted the presence of secondary-phase particles and deformation substrucures in broken LCF specimens of the HAYNES® HR-120®* alloy. They conclude that these microstructural features have an influence on the fatigue life of this material.

---

*230, HR-120 and 242 are trade marks for material produced by Haynes International, Inc.
617 is a trade mark for material produced by Huntington Alloys.

This paper illustrates the fatigue properties of six high-temperature alloys, four of which are currently used for hot gas path components. These alloys are the HAYNES 188, 230®*, HR-120, 242™*, HASTELLOY® X and 617* alloys. The 188 alloy is a cobalt-base alloy, while the other five materials are nickel-base alloys. The influence of cyclic strains, temperature and mean strain on the low cycle fatigue (LCF) life are discussed for these alloys. Some high-temperature, high cycle fatigue (HCF) data are discussed for the 230 alloy. Table 1 contains the product forms and sizes for the materials discussed and the details of the fatigue tests.

Table 1   Material Product Form and Fatigue Test Parameters.

Alloy	Product form and size	Test temperature °C	R-ratio
188	19 mm diameter bar	20, 425, 650, 870	−1
X	4.8, 12.5 mm plate	20, 540, 650, 870	0, −1.0
230	15.9, 19 mm plate	760, 870, 980	+0.05, −1.0
617	16 mm plate	760, 870, 980	−1
HR-120	4.8, 19 mm plate	20, 650, 760, 870	0, −1.0
242	19 mm plate	425, 540, 650, 760	−1

2   MATERIALS

The X alloy is a nickel-base alloy containing principally 22% Cr, 18% Fe and 9% Mo. Compositions are expressed in weight percent in all cases. It possesses an exceptional combination of oxidation resistance, high-temperature strength and fabricability. Developed in the 1950s, this alloy has been the 'work horse' of the gas turbine industry as a preferred combustor material.[2]

The 188 alloy, a cobalt-base alloy containing approximately 22% Ni, 22% Cr and 14% W, was developed in the 1960s for military jet engines to provide higher temperature capability than the X alloy. This alloy combines excellent high-temperature strength with good oxidation and excellent resistance to hot sulphate corrosion up to 1095°C.[3]

Alloy 230 is a nickel-based alloy containing approximately 22% Cr, 14% W, 2% Mo.[4] While discussions of the physical and environmental resistance characteristics of this alloy have been presented extensively,[5] it may be worthy to note that the 230 alloy was developed to have the best combination of high temperature creep-rupture strength, LCF resistance, environmental resistance, resistance to grain coarsening, and thermal stability. The high temperature strength of alloy 230 is significantly superior to alloy X and almost equal to that of cobalt-based a; by 188. Alloy 230 is a solid solution strengthened alloy and

---

*230, HR-120 and 242 are trade marks for material produced by Haynes International, Inc.
617 is a trade mark for material produced by Huntington Alloys.

is easily fabricable and repairable. The fabricability of an alloy is an important consideration due to the complex nature of some of the components in the gas turbines. The 230 alloy is currently used for flying and land-based gas turbines as combustor or liner material, as transition pieces or ducts, as afterburner flame holders, and other components.

The 617 alloy is a nickel-base alloy containing 22% Cr, 12.5% Co, 9% Mo and exhibits an excellent combination of high-temperature strength and oxidation resistance up to 980°C. This alloy is an attractive material for hot gas path components, such as combustion cans, ducting and transition liners.[6]

The HR-120 alloy is a nickel-base alloy system containing 37% Ni, 33% Fe, 25% Cr and small but potent additions of 0.7% Nb and 0.2%N.[7] The presence of niobium (columbium) and nitrogen in combination with carbon provides both nitride- and carbide-strengthening. HR-120 alloy has creep-rupture strength significantly superior to stainless steels such as 310 SS, and equivalent or superior to some nickel-based alloys such as alloy X, alloy 601 and others. This alloy has good environmental (oxidation, carburisation and sulphidation) resistance characteristics.

The unique combination of relatively low cost, good high temperature strength and good environmental resistance makes HR-120 alloy an excellent candidate as an upgrade to stainless steels, and as a cost effective alternate to alloys, such as alloy X in certain applications. The HR-120 alloy is currently being used as cooling shroud material by a major producer of land-based gas turbines who had to upgrade from 310 SS. For this application, large diameter (up to almost 3 metres GD) seamless rings of HR-120 alloy are currently being produced. In another application, HR-120 alloy is being used as exhaust domes on small land-based gas turbines, also as an upgrade to stainless steel. HR-120 alloy is currently being evaluated, by a major gas turbine producer, for the exhaust diffuser section of land-based gas turbines as a cost effective alternate to alloy X.

The 242 alloy is a Ni-8Cr–25Mo alloy that is strengthened by a long-range ordered phase Ni2(Cr,Mo) which is developed by ageing at 650°C.[8] The presence of 25% Mo gives the alloy its low thermal expansion characteristics, while the presence of 8% Cr provides the alloy with oxidation resistance equivalent or superior to other alloys designed specifically for low thermal expansion. Again, a detailed discussion on the physical metallurgy of this alloy has been presented elsewhere.[9] The combination of its long-range strengthening, low thermal expansion and good oxidation resistance characteristics makes alloy 242 uniquely suitable as seal rings and nozzle/turbine casings in both flying and land-based gas turbines.

The nominal compositions for the alloys are contained in Table 2.

3 TEST METHODS

The fatigue properties reported in this paper were determined for five nickel-base alloys and one cobalt-base alloy. The 188, X, 230, 617 and HR-120 alloys were tested in the solution annealed condition. The 242 alloy was tested in the solution annealed and aged (650°C, 24h) condition. These are the typical conditions applications using the alloys discussed here.

Table 2 Nominal Composition of High Temperature Alloys (wt%).

Element	188	X	230	617	242	HR-120
Ni	20–24	Bal	Bal	Bal	Bal	Bal
Cr	21–23	22.00	22.00	22.00	7–9	25.00
W	13–15	0.60	14.00	—	—	2.5*
Mo	—	9.00	2.00	9.00	24–26	2.5*
Fe	3*	18.00	3*	1.50	2*	33.00
Co	Bal	1.50	5*	12.50	2.5*	3*
Mn	1.25*	1*	0.50	0.50	0.80*	0.70
Si	0.20–0.50	1*	0.40	0.50	0.80*	0.60
Al	—	—	0.30	1.20	0.50*	0.10
C	0.05–0.15	0.10	0.10	0.07	0.03*	0.05
La	0.02–0.12	—	0.02	—	—	—
B	0.015*	0.008*	0.015*	—	0.006*	0.004*
Cb	—	—	—	—	—	0.70
N	—	—	—	—	—	0.20
Cu	—	—	—	0.20	0.50*	—
S	—	—	—	0.00	—	—
Ti	—	—	—	0.30	—	—

* Maximum

Low cycle fatigue (LCF) testing was accomplished using cylindrical specimens with gauge diameters of 4 mm to 6.35 mm and gage length-to-diameter ratios of 4 or less. The total strain range experienced by the specimens was controlled throughout the test. The test speeds for LCF testing were either 0.33–1 Hz or a strain rate of $4 \times 10^3 \text{s}^{-1}$ The elevated temperatures for these tests ranged from 425°C to 980°C.

Where fatigue lives are expected to be of the order of 10^5 cycles or less, strain is the most appropriate mode of control. During these relatively shorter fatigue tests, a considerable amount of plasticity occurs in the test specimen which has a strong influence on the resultant fatigue life.

Figure 1 exhibits the elastic and plastic components of a strain-controlled LCF test of the HR-120 alloy for two test temperatures. At lower fatigue lives, the plastic strain component is much greater than the elastic component, and thus, controls the overall fatigue life. The point at which the plastic strain component dominates appears to be a function of the test temperature. At 24°C, the transition point (plastic strain range–elastic

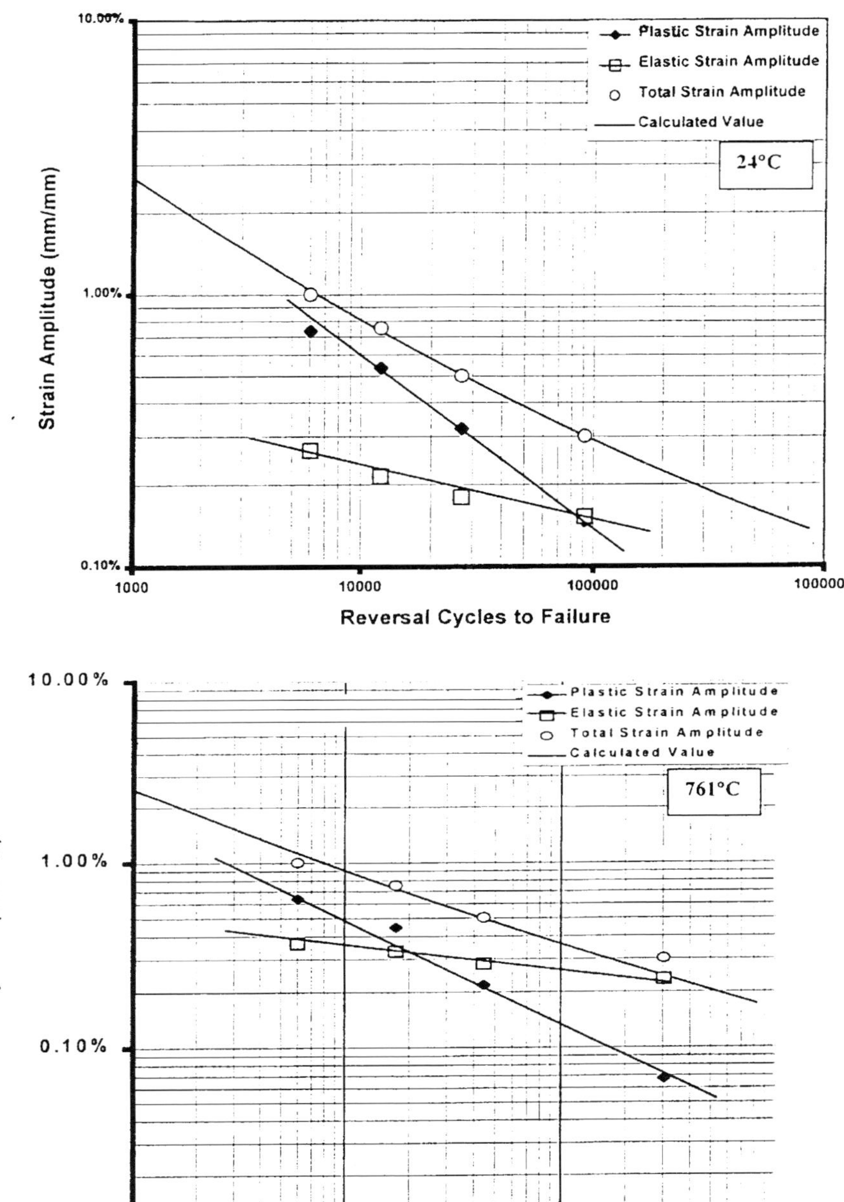

Fig. 1 Elastic and plastic strain amplitudes for Alloy HR-120 during LCF.

strain range) occurs at about 8500 cycles, while at 761°C the transition point is about 2000 cycles. Therefore, in order to obtain representative fatigue data of the pedigree to allow comparisons among materials, the total strain range must be controlled during the LCF test. The LCF properties, i.e. cycles to failure, discussed in this paper are presented in terms of **total strain range** in percent.

Some high cycle fatigue (HCF) properties for the 230 alloy are also discussed. In the case of HCF testing, the stress on the specimen is controlled throughout the test. The HCF properties discussed here are presented in terms of **total stress range**. During these longer fatigue tests the stresses on the specimen are relatively low, and thus, the strains are low and primarily elastic (Fig. 1). In this case the elastic strains control the overall fatigue life. High cycle fatigue data may be converted from total stress range to total strain range using the elastic modulus, at the temperature of the fatigue test, since these tests are essentially elastic. The HCF test specimens had a gauge diameter of 6.35mm and a gauge length of 10mm. The HCF tests were run at a frequency of 59 Hz.

4 FATIGUE PROPERTIES

4.1 ALLOY X

The LCF properties of Alloy X are illustrated in Fig. 2 for temperatures of 20°C–870°C and an *R*-ratio (min. stress or strain of the cycle/max stress or strain of the cycle) of –1.0 and total strain ranges up to about 4%. Increasing temperatures reduced the fatigue life of this material but the influence of temperature tends to decrease at higher total strain ranges. The solid curve in Fig. 2, and in other figures, is drawn through the LCF data point at 870°C.

The fatigue properties of alloy X at 650°C for *R*-ratios of 0 and –1.0 are illustrated in Fig. 3. There is no influence of mean strain for alloy X under these conditions.

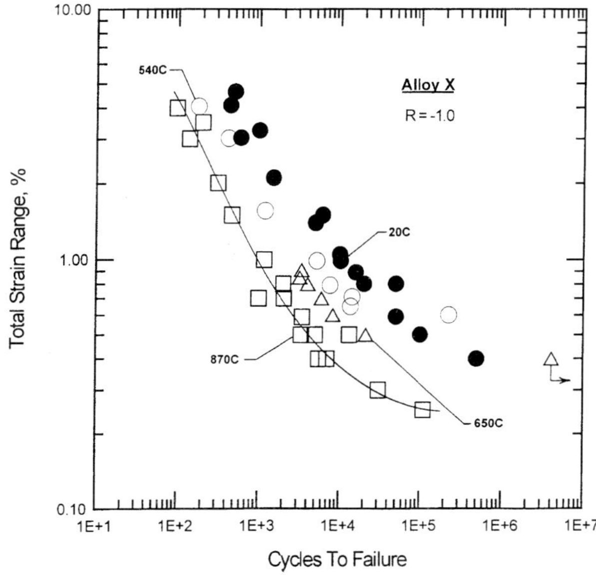

Fig. 2 LCF properties of Alloy X.

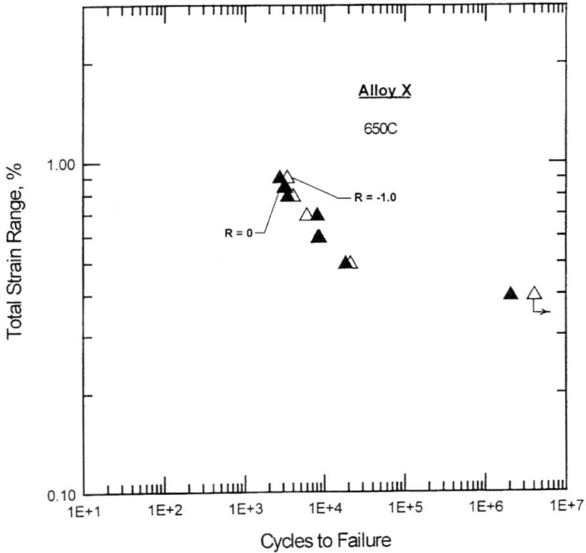

Fig. 3 Effect of mean strain on the LCF properties of Alloy X.

4.2 ALLOY 188

The LCF properties of the 188 alloy at temperatures of 20°C–870°C and total strain ranges of 3% are illustrated in Fig. 4. The *R*-ratio for these tests was –1.0. Temperatures cf 650°C and 870°C appear to have a significant effect on the fatigue life of this material. These fatigue properties also suggest the temperature influence diminishes at higher total strain ranges.

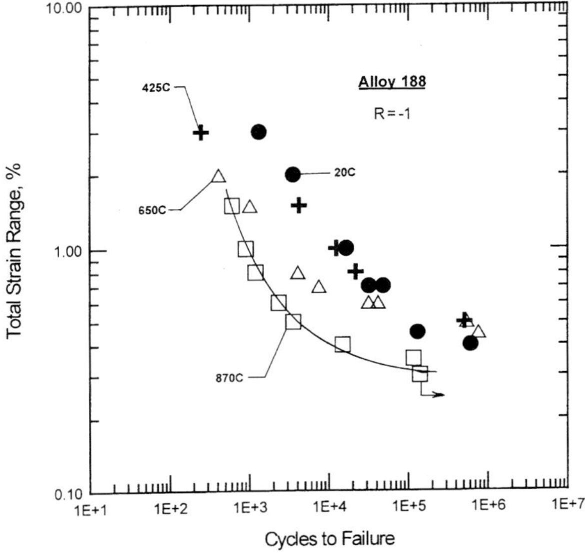

Fig. 4 LCF properties of Alloy 188.

4.3   ALLOY 230

Figure 5 illustrates the LCF properties for the 230 alloy at temperatures of 760°C–980°C and at total strain ranges up to 1%. The temperature dependence of the LCF properties appears to diminish at total strain ranges of 1% for the 760°C and 980°C test temperatures.

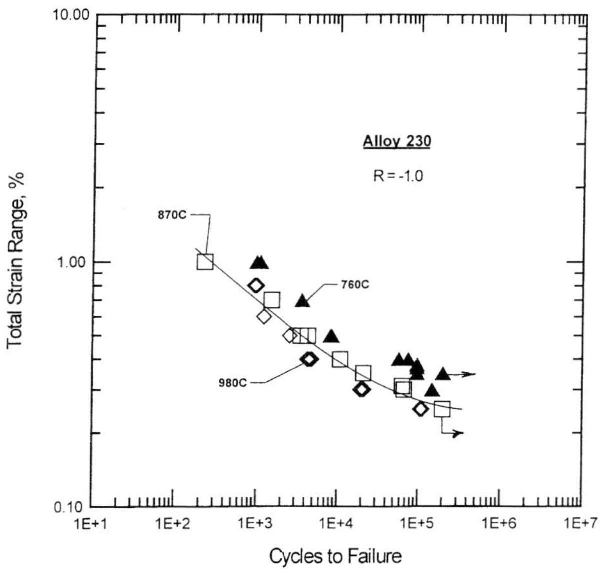

Fig. 5   LCF properties of Alloy 230.

The high cycle fatigue properties of the 230 alloy are illustrated in Fig. 6 for temperatures of 760°C and 980°C. When the HCF stress ranges (at an *R*-ratio of C 0.05) are converted to strain ranges and plotted with the LCF data at the two common test temperatures, there appears to be little or no influence of mean stress on the fatigue properties of the 230 alloy at these temperatures, as shown in Fig. 7.

4.4   ALLOY 617

Figure 8 illustrates the LCF properties of alloy 617 at a number of temperatures from 540°C to 980°C and an *R*-ratio of –1.0. The temperature dependence of the fatigue properties of the 617 alloy seems to be similar to that for the 230 alloy in the temperature range of 760°C–980°C. This temperature dependence appears to diminish considerably at total strain ranges greater than 1%.

4.5   ALLOY HR-120

Figure 9 illustrates the LCF properties of the HR- 120 alloy at temperatures of 20°C–980°C and an *R*-ratio of –1.0. The LCF properties appear to be similar at 200°C and

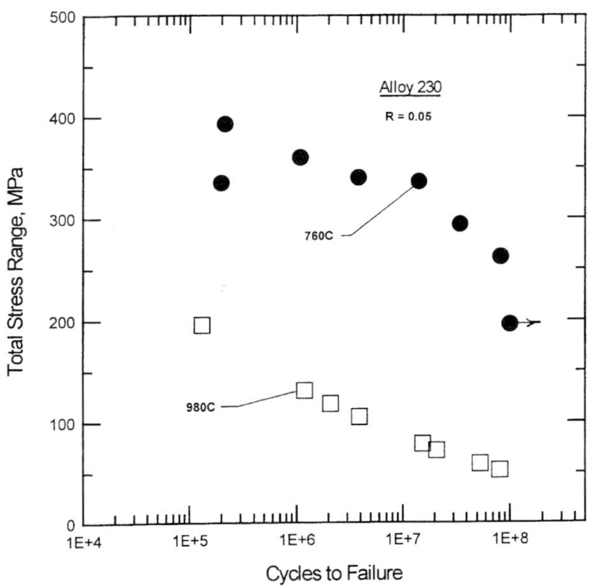

Fig. 6   High cycle fatigue properties of Alloy 230.

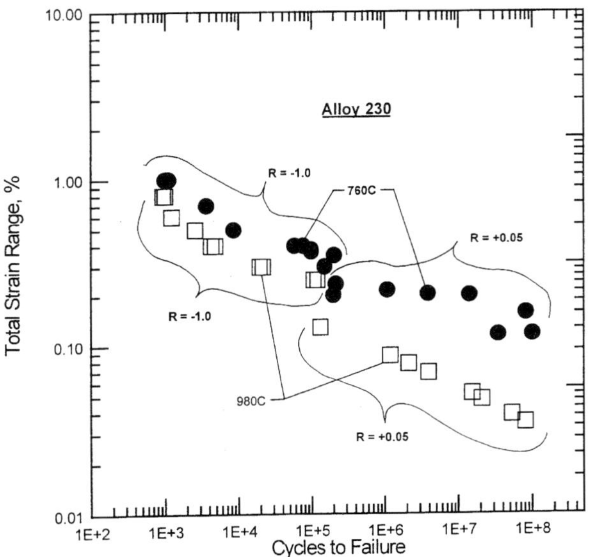

Fig. 7   Effect of *R*-ratio on the fatigue properties of Alloy 230.

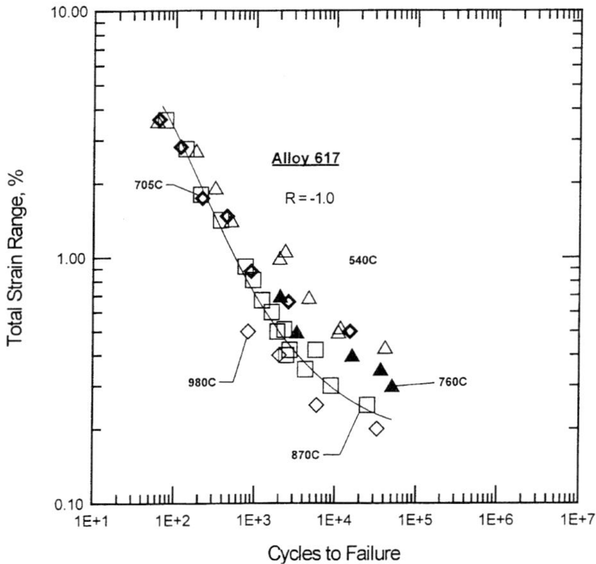

Fig. 8   LCF properties of Alloy 617.

650°C but at higher temperatures the fatigue lives are lower especially at higher total strain ranges. The temperature dependence at temperatures of 760°C–980°C begins to diminish at total strain ranges of 1%.

The cyclic hardening/softening behaviour of the HR-120 alloy is shown in Fig. 10. The peak hardening stress always occurred at 760°C regardless of the total strain range

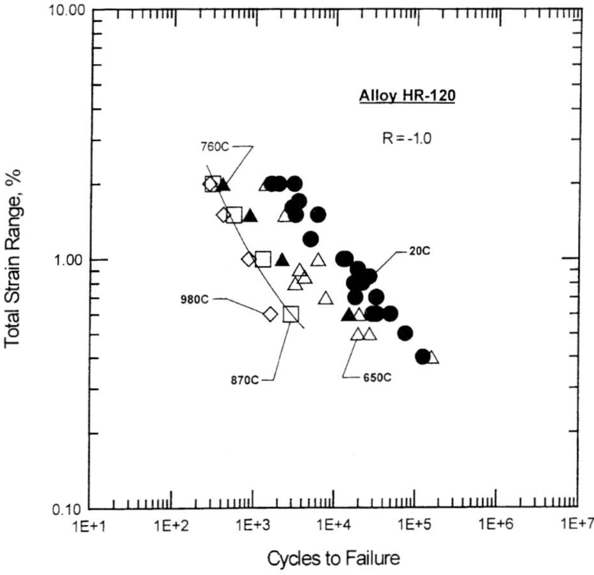

Fig. 9   LCF properties of Alloy HR-120.

He and Liaw have examined the microstructures of fatigued specimens of the alloy and attribute the high peak hardening stress at 760°C to the precipitation of secondary-phase particles and formation of slip bands.[3]

The 650°C LCF properties of the HR-120 alloy at *R*-ratios of 0 and −1.0 are illustrated in Fig. 11. As for the X alloy, there is no effect of mean strain on the LCF properties for HR-120 under these conditions.

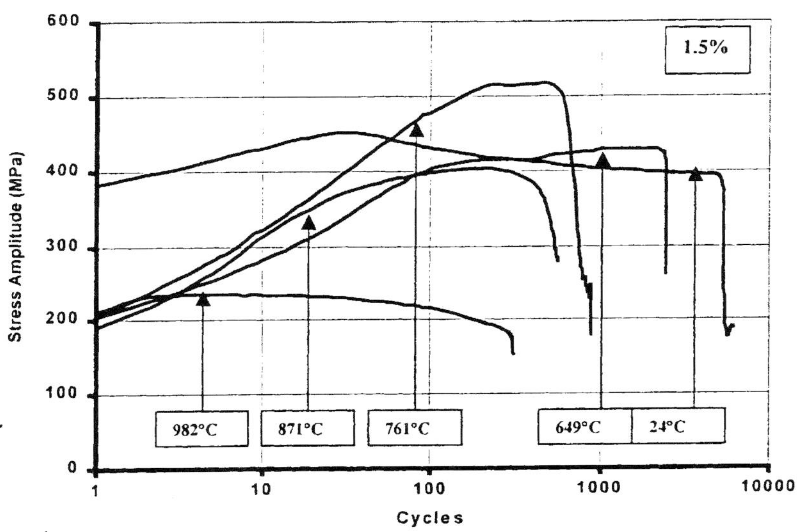

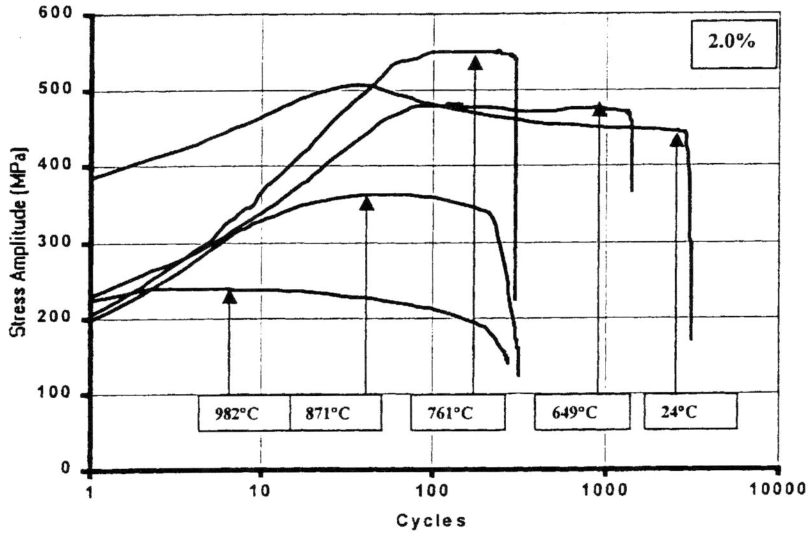

Fig. 10 Cyclic hardening behavior for Alloy HR-120.

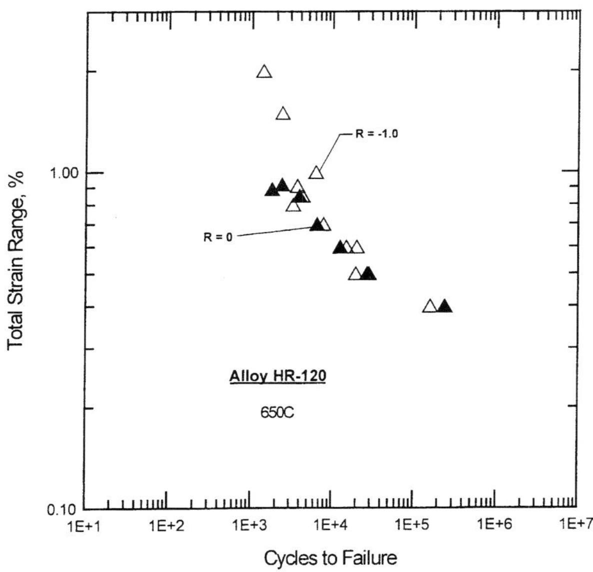

Fig. 11   Effect of mean strain on the LCF properties of Alloy HR-120.

4.6 ALLOY 242

The LCF properties of the 242 alloy are illustrated in Fig. 12 for temperatures of 425°C–760°C. The 425°C and 540°C properties are similar, and those properties at 650°C are close to the lower temperature properties for lives of 10 000 cycles and higher. The properties at 760°C are the lowest in these tests. It should be stated that this alloy is strengthened by ageing at 650°C. The strengthening mechanism for this alloy is long range ordering. At temperatures of 760°C and higher, the long range order is destroyed with an accompanying loss of tensile properties. Thus, the lowering of the LCF properties at 760°C is a due to the loss of the strengthening mechanism as well as the high temperature. The temperature dependence of the LCF properties at 650°C and 760°C appears to diminish at total strain ranges of 1%, similar to the phenomenon demonstrated by the 188, X, 617 and the HR-120 alloys.

Figure 13 illustrates a comparison of LCF properties at 650°C for the X, 617, 230, 188 and HR-120 alloys. The 188, X and HR-120 alloys exhibit similar properties. The 230 alloy exhibits properties similar to those of the 188, X and HR-120 at total strain ranges below 0.10%. The 617 material reported here exhibited lower LCF properties due to a much larger grain size ASTM 0 to 2 compared to ASTM 4 to 5 for the other alloys. Larger grain sizes promote better stress rupture resistance and the 617 was heat treated to achieve this end.

The tabular fatigue data for these alloys are contained in Appendix A.

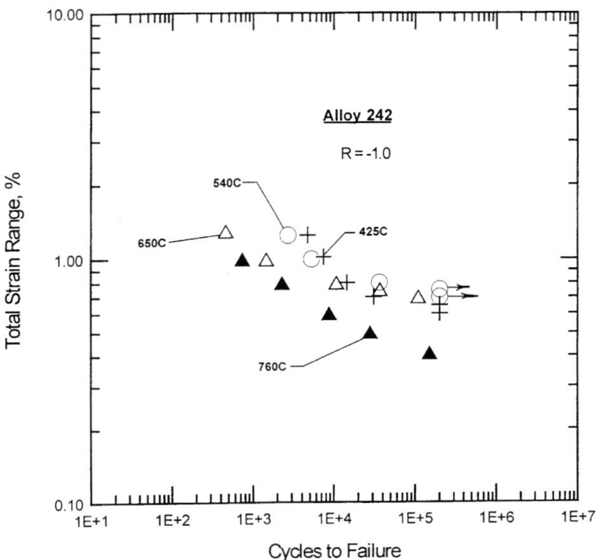

Fig. 12 LCF properties of Alloy 242.

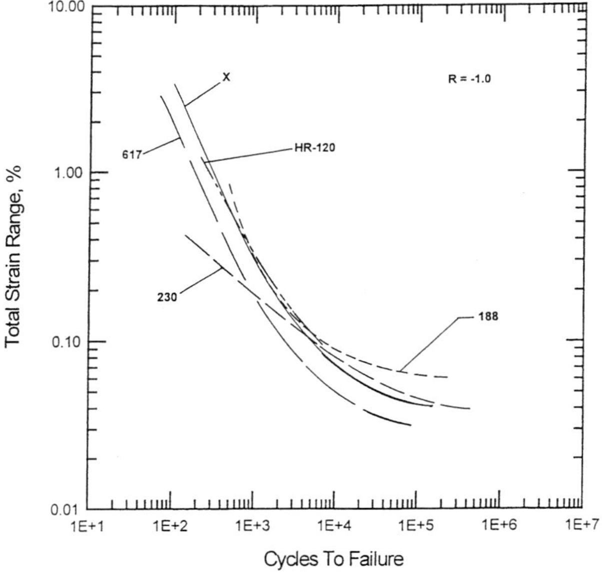

Fig. 13 Comparison of 870°C LCF properties.

5 CONCLUSIONS

1. The LCF properties of six high-temperature alloys used in hot gas path components of gas turbines are illustrated, and comparisons are made at several temperatures.
2. Some HCF properties for the 230 alloy are shown at temperatures of 760°C and 980°C.
3. The 188, X, 230 and HR-120 exhibit similar LCF properties at 650°C and $R = -1.0$.
4. The temperature dependence of LCF properties appear to diminish at higher total strain ranges.
5. There appears to be no influence of mean strain on the LCF properties for a range of test temperatures.

ACKNOWLEDGEMENTS

The authors gratefully acknowledge the work of Drs Y.H. He and P.K. Liaw for their work on the low cycle fatigue of HR-120.

The authors are grateful for the financial supports provided by Haynes International Inc. and the National Science Foundation (DMI-9724476 and EEC-9527527 with Dr D Durham and Ms M. Poasts as contract monitors, respectively).

REFERENCES

1. Y.H. He, P.K. Liaw, R.R. Seeley and D.L. Klarstrom, 'Low-Cycle Fatigue Behavior of HAYNES® HR-120R Alloy', to be published in the *International Journal of Fatigue*.
2. Haynes International Technical Bulletin H-3009A, HASTELLOY® X Alloy.
3. Haynes International Technical Bulletin H-3001A, HAYNES® 188 Alloy.
4. Haynes International Technical Bulletin H-3000F, HAYNES® 230® Alloy.
5. D.L. Klarstrom, H.M. Tawancy, D.B. Fluck, and M.F. Rothman, 'A New Gas Turbine Combustor Alloy', International Gas Turbines Conference, ASME Paper 84-GT-70, New York, NY, ASME, 1984.
6. INCONEL alloy 617, Huntington Alloys Technical Bulletin, 1979.
7. Haynes International Technical Bulletin H-3 125A, HAYNES® HR-120® Alloy.
8. Haynes International Technical Bulletin H-3079D, HAYNES® 242™ Alloy.
9. S.K. Srivastava and GY. Lai, 'A New Low-Thermal-Expansion, High-Strength Alloy for Gas Turbines', International Gas Turbines Conference, ASME Paper 89-GT-329, Toronto, Canada, ASME, 1989.
10. J.P. Strizak, C.R. Brinkman, M.K. Booker, and P.L. Rittenhouse, 'The Influence of Temperature, Environment and Thermal Ageing on the Continuous Cycle Fatigue Behaviour of HASTELLOY X and Inconel 617', ORNL Report TM-8130, April, 1982.
11. K.S. Vechio, M.D. Fitzpatrick and D.L. Klarstrom, 'Influence of Subsolvus Thermomechanical Processing on the Low-Cycle Fatigue Properties of HAYNES 230 Alloy', *Metallurgical and Materials Transactions A*, March 1995, **26A**, 673–689.
12. S.K. Srivastava and D.L. Klarstrom, 'The LCF Behavior of Several Solid Solution Strengthened Alloys used in Gas Turbine Engines', Gas Turbine and Aeroengine Congress and Exposition ASME Paper 90-GT-80, Brussels, Belgium, 1990.

APPENDIX A
Fatigue Properties of High-Temperature Alloys

Table A1   Low Cycle Fatigue Properties of Alloy X (R = −1.0).

Temperature, °C	Total strain range, %	Cycles to failure, N_f	Temperature, °C	Total strain range, %	Cycles to failure, N_f
20	0.4	500 000	20[10]	0.80	49 644
	0.59	100 000		0.89	15 796
	0.8	20 000		0.99	10 245
	1.05	10 000		1.50	6101
	1.4	500		2.11	1494
	3.25	1000		3.04	591
	4.65	500		4.10	439
20	0.4	169 636	540[10]	0.60	225 000
	0.5	49 132		0.65	13 675
	0.6	33 436		0.71	14 240
	0.7	24 106		0.79	7596
	0.8	15 888		0.99	5182
	0.85	16 522		1.56	1181
	0.91	13 279		3.02	418
650	0.4	4 074 352 R		4.06	177
	0.5	21 238	870[10]	0.50	13 234
	0.6	8313		0.59	3433
	0.7	5964		0.80	2002
	0.8	4025		1.00	1156
	0.85	3208		1.50	466
	0.91	3380		2.01	312
870	0.25[12]	111 155		3.01	138
	0.3[12]	30 698		4.00	96
	0.4[12]	5505			
	0.5[12]	3361			
	0.5[10]	7000			
	0.6[10]	5000			
	0.7[12]	2040			
	0.7[10]	1000			

R – Test was stopped

Table A2   Low Cycle Fatigue Properties of Alloy X (R = 0).

Temperature, °C	Total strain range, %	Cycles to failure, N_f
20	0.40	135 812
	0.50	57 345
	0.60	31 596
	0.70	23 669
	0.80	17 081
	0.85	16 564
	0.91	14 001
650	0.40	2 079 468
	0.50	18 502
	0.60	8699
	0.70	8130
	0.80	3408
	0.85	3059
	0.91	2718

Table A3   Low Cycle Fatigue Properties of Alloy 188[3] (R = −1.0).

Temperature, °C	Total strain range, %	Cycles to failure, N_f
26	0.40	608 640
	0.45	130 860
	0.70	48 258
	0.70	31 989
	1.00	16 630
	2.00	3523
	3.00	1312
425	0.50	510 440
	0.80	21 808
	1.00	12 647
	1.50	4185
	3.00	246
650	0.45	761 440
	0.50	546 639
	0.60	41 727
	0.60	31 796
	0.70	7535
	0.80	4035
	1.50	1015
	2.00	409
870	0.30	139 750 R
	0.35	116 400
	0.40	14 924
	0.50	3535
	0.60	2325
	0.80	1189
	1.00	904
	1.50	604

R – Test was stopped.

Table A4 Low Cycle Fatigue Properties of Alloy 230 (R = −1.0).

Temperature, °C	Total strain range, %	Cycles to failure, N_f
760[12]	0.30	150 000 R
	0.35	97 612
	0.40	75 470
	0.50	8490
	0.70	3622
	1.00	1097
760[11]	0.35	202 920 R
	0.37	97 612
	0.38	96 844
	0.40	57 819
	0.50	8153
	1.00	990
870[12]	0.25	200 770 R
	0.30	66 926
	0.35	20 964
	0.40	10 781
	0.50	4299
	0.70	1504
870[11]	0.31	63 253
	0.40	10 837
	0.50	3473
	1.00	228
980[12]	0.25	112 957
	0.30	20 286
	0.40	4504
	0.50	2562
	0.60	1218
	0.80	991
980[11]	0.25	106 020
	0.30	19 200
	0.30	21 311
	0.40	4784
	0.40	4223
	0.80	933

R – Test was stopped.

Table A5 High Cycle Fatigue Properties of Alloy 230 (R = 0.05).

Temperature, °C	Total stress range, MPa	Cycles to failure	Temperature, °C	Total stress range, MPa	Cycles to failure, N_f
760	196	100 424 536 R	980	196	131 518
	335	198 268		131	1 187 515
	294	34 492 907		118	2 098 832
	336	14 167 912		105	3 918 360
	340	3 830 880		78.5	15 449 294
	360	1 079 078		58.9	53 364 009
	393	214 446		52.4	80 884 125
	262	82 252 642		72	20 844 436

R – Test was stopped

Table A6   Low Cycle Fatigue Properties of Alloy 617 (R = –1.0).

Temperature, °C	Total strain range, %	Cycles to failure, N_f
760[12]	0.30	50 360
	0.35	36 333
	0.40	16 160
	0.50	3321
	0.70	2054
870[12]	0.25	20 505
	0.30	8882
	0.35	4260
	0.40	2461
	0.50	1894
980[12]	0.20	33 110
	0.25	5924
	0.40	2000
	0.50	821
540[10]	0.43	41 180
	0.50	10 929
	0.52	11 520
	0.69	4672
	1.00	2009
	1.07	2357
	1.42	498
	1.93	316
	2.73	181
	3.60	60
705[10]	0.50	15 142
	0.66	2600
	0.88	886
	1.47	441
	1.74	218
	2.81	116
	3.64	63

Table A7 Low Cycle Fatigue Properties of Alloy HR-120[1] (R = −1.0).

Temperature, °C	Total strain range, %	Cycles to failure, N_f	Temperature, °C	Total strain range, %	Cycles to failure, N_f
20	0.60	48 580	650	0.40	161 062
	0.60	29 250		0.50	27 077
	0.70	18 000		0.50	19 567
	0.80	17 300		0.60	14 963
	1.00	12 711		0.70	7773
	1.00	13 546		0.80	3273
	1.20	5000		0.85	4260
	1.50	6104		0.91	3673
	1.50	3246		0.60	20 100
	1.60	2950		1.00	6212
	1.70	3536		1.50	2380
	2.00	2000		2.00	1377
	2.00	3112	760	0.60	15 187
	0.40	125 615		1.00	2220
	0.50	74 901		1.50	876
	0.60	32 821		2.00	400
	0.70	32 602	870	0.60	2900
	0.80	21 830		1.00	1284
	0.85	26 534		1.50	556
	0.91	19 167		2.00	300
			980	0.60	1600
				1.00	859
				1.50	410
				2.00	276

Table A8 Low Cycle Fatigue Properties of Alloy HR-120 (R = 0).

Temperature, °C	Total strain range, %	Cycles to failure, N_f
20	0.4	130 711
	0.5	73 583
	0.6	37 133
	0.7	24 568
	0.8	22 562
	0.85	19 637
	0.91	19 314
650	0.4	240 853
	0.5	28 471
	0.6	12 541
	0.7	6461
	0.85	3903
	0.89	1776
	0.92	2369

Table A9 Low Cycle Fatigue Properties of Alloy 242[8] (R = −1.0).

Temperature, °C	Total strain range, %	Cycles to failure, N_f
425	0.60	200 000 R
	0.65	200 000 R
	0.70	31 037
	0.80	14 540
	1.02	7482
	1.25	4681
540	0.70	200 000 R
	0.75	200 000 R
	0.80	36 490
	1.00	5237
	1.25	2700
650	0.70	109 695
	0.75	36 922
	0.80	10 837
	1.00	1455
	1.30	459
760	0.41	151 614
	0.50	28 021
	0.60	8793
	0.80	2281
	1.00	735

(a) Aged at 650°C for 24 hr.
R – Test was stopped

Influence of Creep and Oxidation on the Elevated Temperature Crack Growth Behaviour of a Near-α Titanium Alloy

M.C. HARDY,[a] M.R. BACHE,[b] G KÖNIG[c] and M.B. HENDERSON[d]

[a]*Rolls-Royce plc, PO Box 31, Derby DE24 8BJ*
[b]*.R.C. in Materials for High Performance Applications, University of Wales, Singleton Park, Swansea SA2 8PP*
[c]*MTU München GmbH, Post fach 50 06 40, D-8000 München 50, Germany*
[d]*Defence Evaluation Research Agency, Ively Road, Farnborough, Hampshire GU14 0LS*

1 INTRODUCTION

The drive to improve the efficiency and thrust-to-weight ratio of aircraft gas turbine engines is exceptionally demanding for fatigue-limited, rotating components, such as compressor discs, since they are required to provide longer services lives at higher stresses and temperatures. At temperatures above 550°C, the fatigue performance of near-α titanium alloys, in particular, may be compromised by oxidation and time dependent deformation. To ensure against service failure requires, therefore, an understanding of crack nucleation and growth under these conditions.

Over the past two decades,[1–8] considerable attention has been given to the investigation of elevated temperature crack growth in compressor and turbine disc materials. This work has examined the effects of dwell time at peak stress, cycle frequency, stress ratio and cycle shape. It has been found that these factors can influence the mode and rate of crack growth, especially in oxygen or water vapour containing atmospheres. Slow fatigue cycles, notably those with a dwell period at peak stress, can result in relatively fast, intergranular crack growth. In contrast, transgranular crack growth, and much lower growth rates, are produced from high frequency cycles.

The propensity for intergranular cracking has been attributed to the combined effects of oxidation and creep deformation. During elevated temperature exposure in air, it has been proposed[9] that oxygen diffusion to the boundaries causes boundary pinning, which prevents deformation via slip or grain boundary sliding. Once this occurs, wedge cracks and cavities may nucleate at triple points and along boundaries. These features concentrate stress and result in cracks that link up around grains and lead to intergranular cracking. Another theory,[10] specific to nickel alloy, Inconel 718, proposes that intergranular cracking results from the fracture of a brittle niobium oxide (Nb_2O_5) film that forms from elemental segregation at grain boundaries. It is possible, therefore, that the oxidation of grain boundary carbides and precipitates may produce similar effects in other nickel alloys.

Although intergranular cracking has been observed widely in nickel disc alloys under creep-fatigue conditions, it occurs less frequently in near-α titanium alloys. In both β and α plus β heat treated microstructures, fatigue cracks develop and propagate as a result of intense planar slip bands on the basal plane of the α phase. This mechanism, which is exclusive at room and ambient temperatures,[11-12] is retained at temperatures as high as 600°C in air, even for low frequency fatigue cycles.[6-8] However, intergranular cracking has been observed along prior β grain boundaries, at high stress intensity values, in Ti-1100 at 593°C from fatigue cycles that include a 300 second dwell period at peak load.[7]

Trends in the crack growth behaviour of nickel alloy, Inconel 718, have been identified according to the frequency of cycling.[2] Three regimes, namely, fully time dependent crack growth, fully cycle dependent crack growth, and a mixed mode regime between these two, were defined. Where time dependent cracking occurred, crack growth was found to be inversely proportional to frequency and, consequently, crack extension is controlled by time and is independent of the number of cycles elapsed.

Crack growth, as a result of high frequency fatigue cycles, occurs so quickly that there is insufficient time for oxygen or water vapour to reach the damage zone ahead of the crack tip. So much so that crack growth in air is essentially cycle dependent. Full cycle dependency, however, is achieved at very low air pressures in high vacuum. Under these conditions, crack growth is transgranular and results from the cracking of intense planar slip bands (faceted crack growth) or from two slip systems intersecting on the crack plane (striations). Between the time and cycle dependent regimes, it has been proposed[13] that environment alone controls the rate of crack growth, so that the dependence to frequency is given by a power of minus one-fifth to minus one-third.[5,8] In nickel alloys, both intergranular and transgranular crack growth are observed in this regime, giving rise to the term, mixed mode.

The investigation of Inconel 718 by Nicholas and Ashbaugh[3] provides a clear illustration of the combined effects of stress ratio (R) and frequency on crack growth behaviour. Their data showed that an increase in stress ratio brings about an extension of the time dependent and mixed mode regimes to faster cyclic frequencies. Clearly, this has serious implications for components that suffer from vibration or high stress ratio fluctuations. The behaviour is not unreasonable since, as R is increased for a given frequency, the crack experiences high loads for longer periods. Consequently, there is more opportunity for environment enhanced crack growth, particularly if the crack is open for the majority of the cycle.

This paper examines the influence of oxidation and creep on the crack growth behaviour of $\alpha+\beta$ heat treated, near-α titanium alloy, TIMETAL 834, at 630°C. To extend the understanding gained from earlier work, this study has focused on near threshold or short fatigue crack growth since this regime accounts for the majority of the crack propagation life. The experimental programme also includes tests conducted in vacuum and at low air pressures, typical of those encountered in component rig testing.

2 MATERIAL

TIMETAL 834 (Ti–5.8Al–4Sn–3.5Zr–0.7Nb–0.5Mo–0.35Si–0.06C) is a near-α titanium alloy used in the compressor stages of aero-engines at temperatures up to 550°C. It was designed, however, to tolerate short term, low stress excursions to 600°C.

Isothermal forgings suitable for disc manufacture were supplied by MTU. These were solution treated in the α plus β phase field (α content of 12–15%) at 1028°C for two hours and then oil quenched. Subsequently, they were aged at 700°C for two hours and finally, air cooled. This heat treatment produces a bimodal microstructure consisting of primary α grains less than 50 μm in size, surrounded by larger (50–100 μm) transformed β grains showing the basket-weave α morphology. As a result, the alloy offers improved strength and fatigue resistance over β heat treated systems with no loss in creep performance or fracture toughness.[14]

3   EXPERIMENTAL PROCEDURE

The crack growth tests defined in Table 1 were performed in tension at 630°C on 10 × 10 mm, square section, corner crack test pieces. A schematic is provided in Fig. 1 to aid in the definition of the fatigue cycle components.

Table 1   Test schedule (with the exception of 0.1/0.1 cycle, a 1 second (s) dwell period was applied at the minimum load, P_{min}).

Air pressure (Pa)	P_{max}	R ratio	Load time (s)	Dwell time at P_{max} (s)	Comments
10^5, 150, 10^{-4}	$0.7\sigma_y$	0.05	1	1	Effect of air pressure
10^5	$0.4\sigma_y$	0.05	1	1, 120	Effect of max load, dwell time at P_{max}
10^5	$0.7\sigma_y$	0.05	1	1, 120	
10^5	$0.7\sigma_y$	0.05	18	120	Effect of load time & dwell time at P_{max}
10^5	$0.7\sigma_y$	−0.5, 0.2, 0.4, 0.7, 1	1	1	Effect of R ratio (2 tests at $R = 0.2$)
10^5	$0.7\sigma_y$	0.05	0.1, 10, 18, 30	0 for load time of 0.1, otherwise 1	Effect of load time (2 tests at load times of 0.1 & 30)

Crack propagation was monitored by an automated pulsed d.c. electrical potential drop system. Prior to elevated temperature testing, precracks were grown from the 0.25 mm starter slit to at least 0.5 mm at room temperature. To avoid load history effects, the maximum load (P_{max}) intended for elevated temperature testing was applied. Tests were stopped after growing a crack of approximately 4 mm.

In all cases, a conventional radiant furnace was used to attain the test temperature. In those tests conducted in laboratory air, displacements across the crack mouth were measured using a 10 mm gauge length, axial extensometer fitted with two V-chisel quartz arms. Those tests that required a reduced air pressure were conducted in an environmental chamber, the details of which are provided elsewhere.[15]

Crack lengths (a) were determined on the basis of a linear relationship between measured voltage and crack size.[16] Crack growth rate data were produced from 50 a versus cycles (N) data pairs using the 3 point secant method. Stress intensity factor solutions, derived by Pickard,[17] were used to calculate ΔK.

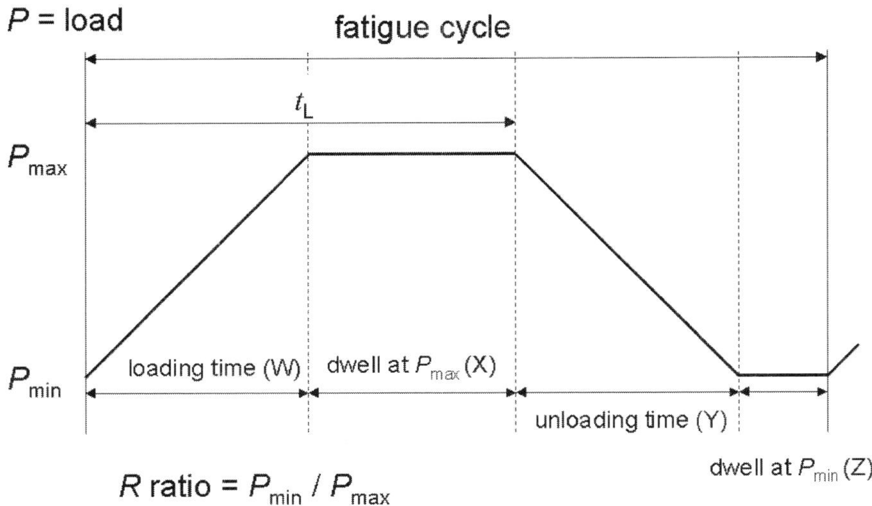

Fig. 1   Schematic showing the wave shape (W/X/Y/Z) variables of a fatigue cycle.

4   RESULTS

The data in Fig. 2 show that the rate of crack growth at 630°C, in TIMETAL 834, is significantly greater in laboratory air (10^5 Pa) or under the reduced air pressure of 150 Pa than in high vacuum (10^{-4} Pa). The cause of this increase in growth rate is not immediately apparent from the fracture surface micrographs in Fig. 3. Facets indicative of crystallographic crack growth and fatigue striations are common to both fracture surfaces. However, fracture surfaces produced from crack growth in air are distinguished by more angular facets and by the presence of oxide debris.

Lines representing growth rate data for various values of stress ratio are also presented in Fig. 2 in terms of ΔK. The behaviour appears to be typical of fatigue crack growth in metals, in that an increase in stress ratio from 0 to 0.7 has produced an increase in growth rate, presumably due to crack closure reducing the effective driving force of the applied stress range.[18] It should be noted that stress intensities were calculated using the full stress range. This accounts for the difference in growth rates shown by the –0.5 and 0.05 data.

The influence of loading time on crack growth is shown in Fig. 4. The lines indicate that crack growth rates increase with loading time.

Growth rate versus ΔK data, also plotted in Fig. 4 for 1/1/1/1 fatigue cycles, suggest that increasing the maximum stress from 40 to 70% of the 0.2% proof stress accelerates crack growth by a factor of about two. From this evidence, it is difficult to speculate whether this acceleration is a real effect or just an indication of scatter in crack growth. There is an influence of maximum stress on strain accumulated across the crack mouth during a dwell period of 120 seconds (Fig. 5). For a maximum load equivalent to 70% of the 0.2% proof stress, the inelastic strains produced at a crack length of approximately 2.5 mm were in excess of 15%. As a result, the test piece showed a considerable reduction in section size

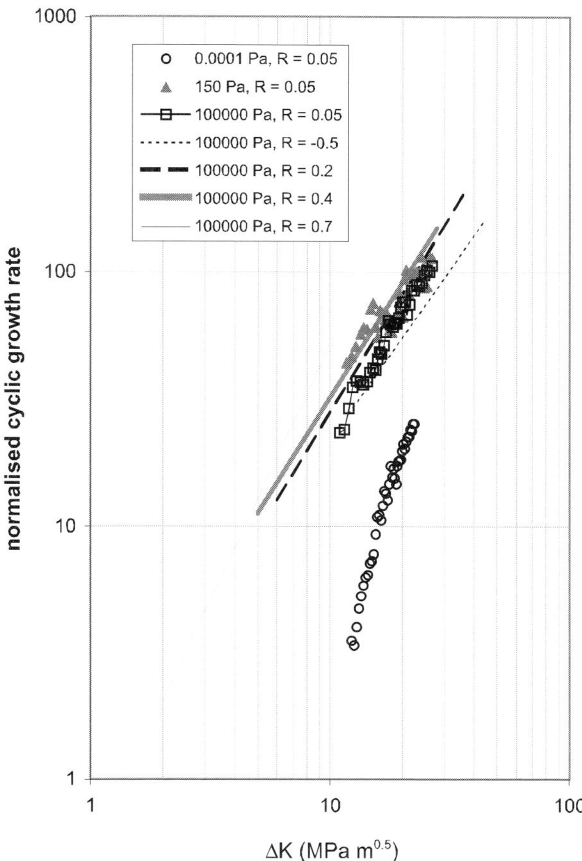

Fig. 2 Crack growth rate data for TIMETAL 834 at 630°C, generated from 1/1/1/1 fatigue cycles, to a maximum load equivalent to 70% of the 0.2% proof stress. The graph shows the influence of air pressure and stress ratio on crack growth behaviour. ΔK values were calculated using the full stress range.

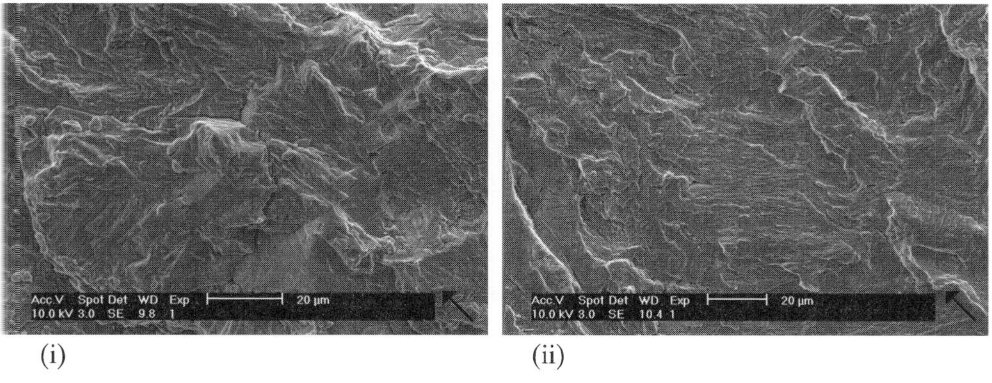

(i)   (ii)

Fig. 3 Fracture surface micrographs taken at a ΔK of 15 MPa√m, after 1/1/1/1 fatigue cycles applied at a stress ratio of 0.05 in (i) laboratory air (10^5 Pa), and (ii) high vacuum (10^{-4} Pa). Arrow indicates the direction of crack growth.

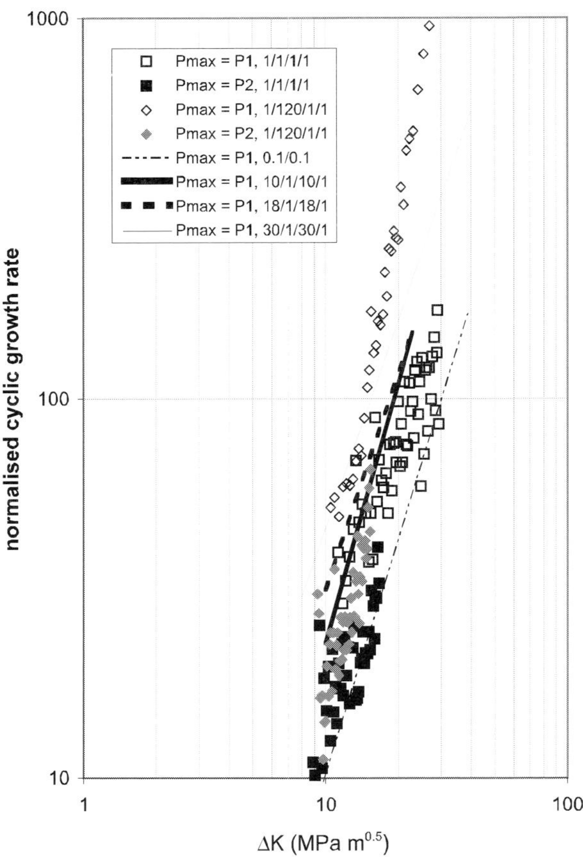

Fig. 4 Crack growth rate data for TIMETAL 834 at 630°C, generated in laboratory air (10^5 Pa) at a stress ratio of 0.05. The graph shows the effects of maximum load, dwell time and loading time on crack growth behaviour. P1 and P2 are equivalent to 70 and 40% of the 0.2% proof stress respectively.

and cracking of the surface oxygen-rich layer. In contrast, an inelastic strain of 1% was measured at a similar crack size from a maximum load equivalent to 40% of the 0.2% proof stress. A reduced dwell period of 1 second at the high stress level accumulated only 0.13% inelastic strain at a crack length of 3.1 mm. Despite the accumulation of inelastic strain during the 120 second dwell period at maximum load, the mode of cracking remained transgranular, as illustrated in Fig. 6.

Assuming constant volume, the inelastic strains measured during a 120 second dwell period were used to determine the cross sectional area at a given instant and, therefore, the stress range. To enable a comparison with growth rate data generated for a 1 second dwell, these stress range values were used to calculate ΔK for the 120 second dwell data in Fig. 4. Clearly, though, the strains encountered during a 120 second dwell at maximum load would violate the small scale yielding criterion, necessary for the use of linear elastic fracture mechanics. Nevertheless, the data indicate that the addition of a significant dwell period at maximum load does increase the growth rate, even for a maximum load

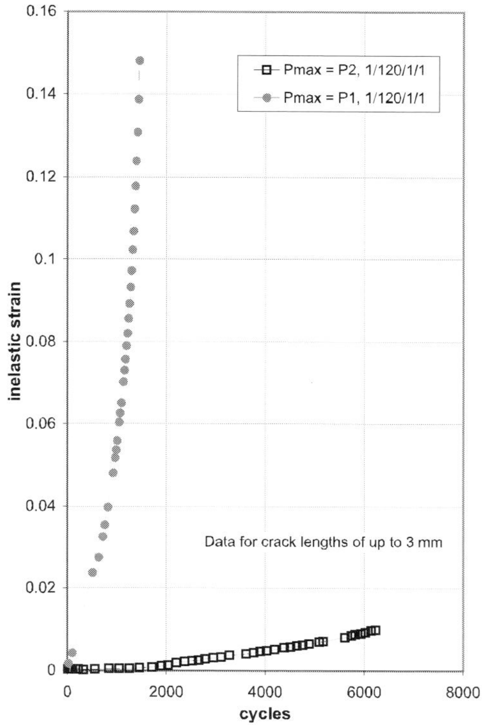

Fig. 5   Strain accumulates across the crack mouth during crack growth tests that were conducted in laboratory air (10^5 Pa) at 630°C from 1/120/1/1 fatigue cycles, applied at a stress ratio of 0.05. P1 and P2 are equivalent to 70 and 40% of the 0.2% proof stress respectively.

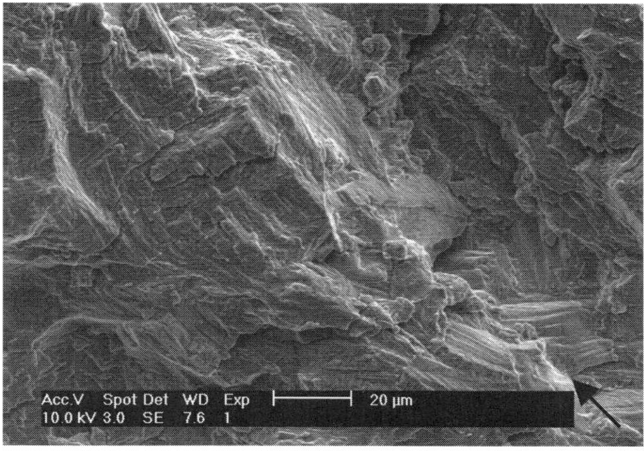

Fig. 6   Fracture survace micrography taken at a ΔK of 15 MPa√m, from fatigue 18/120/1/1 cycles, applied at a stress ratio of 0.05, in laboratory air (10^5 Pa). Arrow indicates the direction of crack growth.

equivalent to 40% of the 0.2% proof stress, which produces modest amounts of strain accumulation (Fig. 5).

Very little crack growth occurred from static loads although the potential values measured were typical for cracks of 3–4 mm in length. Instead, there was significant strain accumulation (>20%) and necking as a result of creep deformation. This was evident from intense cavitation and microstructural deformation close to the tip of the fatigue crack (Fig. 7). Post test fatigue loading quickly brought about failure from cavity linkage, which resulted in intergranular cracking. On reflection, the occurrence of crack blunting may be expected as the material exhibits a low stress exponent (4–8) at 630°C for the stresses examined.

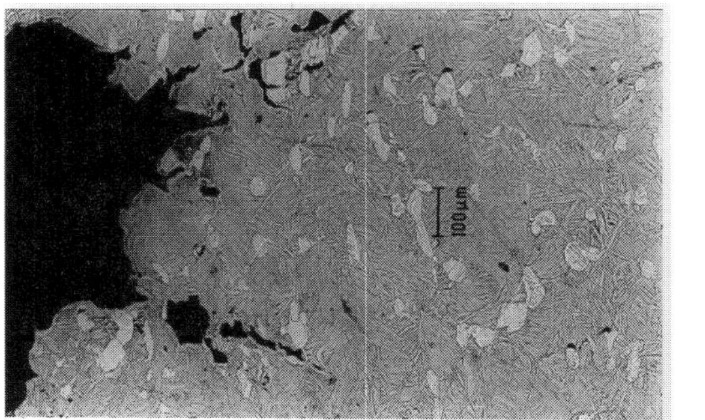

Fig. 7 Cavitation and microstructural deformation found in test piece after 45 hours at 630°C and a stress of 374 MPa. Arrow indicates the loading axis.

5 DISCUSSION

The evidence in Fig. 2 suggests that crack growth in TIMETAL 834 at 630°C is highly sensitive to the presence of oxygen or another aggressive species like water vapour. At this temperature, the rate of oxygen diffusion into a near-α titanium alloy[19] during the 4 second, 1/1/1/1 fatigue cycle can produce a saturated oxygen (33 at.% O_2) layer which is very similar in size to the intrinsic crack extension per cycle measured in high vacuum. Since crystallographic facets are formed in air as well as in vacuum, it is proposed that the presence of oxygen or oxide debris must weaken, wedge open and unzip basal planes, resulting in the rapid crack growth observed in air.

The influence of air pressure on crack growth has been established for nickel alloys by Andrieu et al.[4] Their work showed that the rate of crack growth in Inconel 718 and N18 begins to increase from the high vacuum level at 650°C for oxygen partial pressures in excess of 10^{-3} mbar (0.1 Pa) but then tends towards a maximum in laboratory air (10^5 Pa). Despite the material differences, the current study shows very similar trends in that there is a significant increase in growth rate from vacuum to laboratory air (10^5 Pa). There is also a small difference in the crack growth data generated under 10^5 and 150 Pa for the 1/1/1/1 fatigue cycle considered.

In common with other forms of corrosion fatigue, wave shape has a profound effect on the rate of oxidation accelerated crack growth. In a recent study,[5] the crack growth responses of nickel alloy, Waspaloy, from triangular, asymmetric and dwell cycles were correlated successfully using the loading and dwell time components of the fatigue cycle, called $t_{opening}$, or t_L in this current work (Fig. 1). This correlation, like others since the original work of Weerasooriya,[2] is based on a mechanism map but in this case the abscissa is the logarithm of t_L. Data generated in the current investigation are presented on a mechanism map in Fig. 8. Since the data, for a given value of ΔK, can be described by a straight line of slope α, an expression of the form;

$$\frac{da}{dN} = C\Delta K^n t_L{}^\alpha \tag{1}$$

can be used to characterise crack growth rate. Here, the values of C and n are those attributed to a fatigue cycle where t_L is 1.

The ability of the model to describe the observed crack growth behaviour is demonstrated in Fig. 9 for different loading times and for a cycle with a long dwell period at

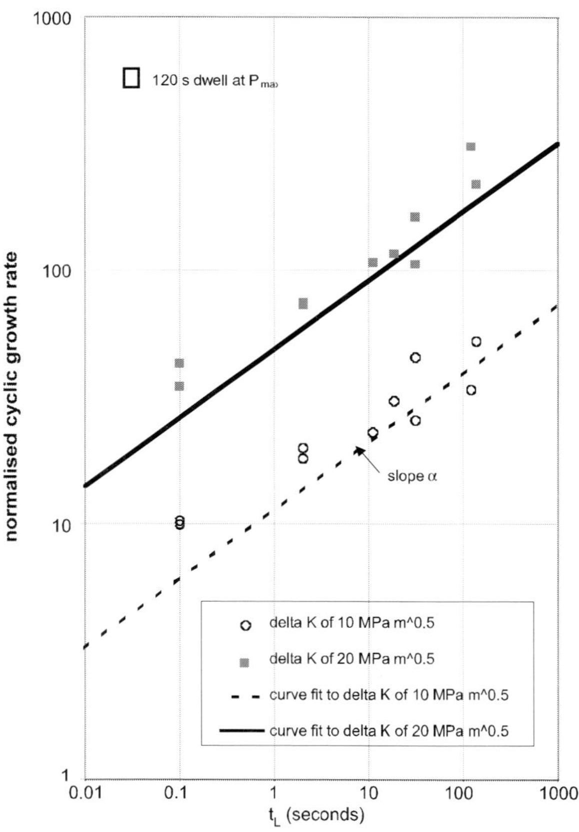

Fig. 8 Mechanism map for TIMETAL 834 at 630°C. Data generated in laboratory air (10⁵ Pa), at a stress ratio of 0.05 and to a maximum load equivalent to 70% of the 0.2% proof stress.

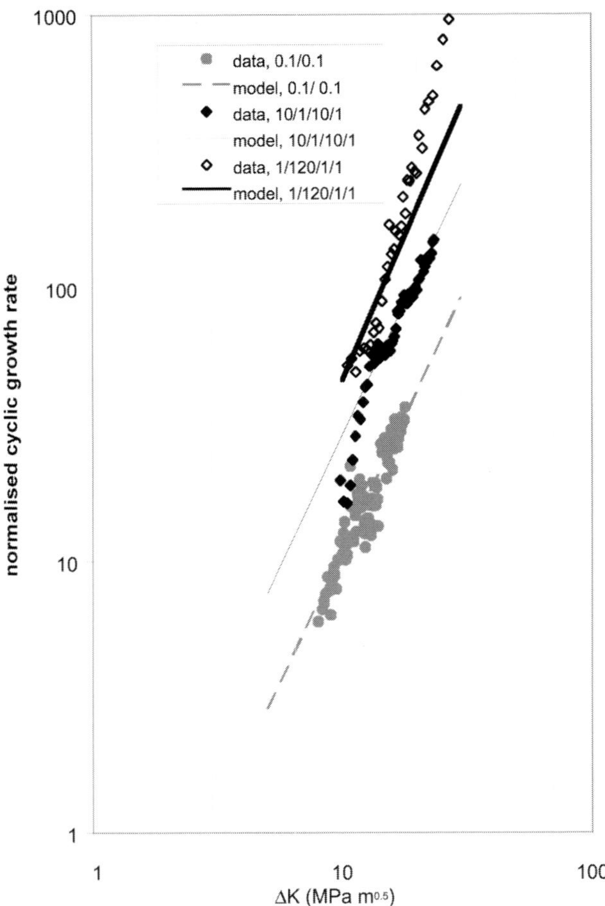

Fig. 9 Predicting the effects of loading time and dwell period on the crack growth response of TIMETAL 834 in laboratory air (10⁵ Pa) at 630°C. Data were generated at a stress ratio of 0.05 and to a maximum load equivalent to 70% of the 0.2% proof stress.

maximum stress. As expected, within the ΔK and t_L values captured in the mechanism map, a good correlation is produced. Extrapolation to higher stress intensities for the 120 second dwell underestimates the measured growth rates.

However, due to test conditions and the test piece design, there was a significant accumulation of inelastic strain at these ΔK levels, which raises serious doubts over the applicability of linear elastic fracture mechanics. More importantly, it is likely that for longer t_L periods, crack growth will become fully time dependent and thus α will equal 1.

Future investigations to examine higher t_L values in TIMETAL 834 above 600°C should consider using a compact tension test piece with side grooves.[20] In such cases, provided that sufficient constraint is imposed, the use of K as a characterising parameter should remain appropriate. This includes the case of cracks in discs where local creep strain accumulation can be redistributed to surrounding elastic material.

Since oxidation accelerated crack growth is dependent on the length of time with the crack open and subject to tensile stresses, it is vital that a model should incorporate the effects of stress ratio. It has not been possible, however, to characterise the combined effects of stress ratio and cycle frequency as the required experimental programme is beyond the scope of the current study.

Instead, an expression for U (where $U = \Delta K_{eff}/\Delta K$) as a function of stress ratio has been established for a specific fatigue cycle (1/1/1/1) in air, first by the determination of the crack opening stress intensity (K_{op}). This has been achieved by assuming that cracks remain fully open during R of 0.7 fatigue cycles so that $\Delta K = \Delta K_{eff}$. Consequently, a crack growth rate versus ΔK curve generated at R of 0.7 can be used to correct other curves produced at lower stress ratios.[21] The crack opening stress intensity factor can then be calculated via:

$$K_{op} = K_{max} - \Delta K_{eff} \tag{2}$$

provided that the same value of maximum stress is used to generate crack growth rate data.

The resulting U versus R data produced are shown in Fig. 10 and can be described by the following second-order polynomial:

$$U = a + bR + cR^2 \tag{3}$$

where a, b and c are experimentally determined parameters.

6 CONCLUSIONS

This study has examined the influence of oxidation and creep on the short crack growth behaviour of TIMETAL 834 at 630°C. The rate of crack growth has been shown to be a function of air pressure, loading time, dwell period at peak stress and stress ratio. Whilst significant increases in growth rate were produced in laboratory air from long loading times and dwell periods at peak load, crack growth was found to be transgranular under the conditions explored. The test piece used did not permit the acquisition of creep crack growth data at the high temperatures and stresses considered, due to crack blunting and excessive strain accumulation caused by a lack of constraint.

A mechanism map has been constructed for data generated in laboratory air at a stress ratio of 0.05. This was used to establish a model, based on the stress intensity factor range, to predict the effects of loading time and dwell time at peak load. Although, the model accounts for the effects of stress ratio in air for a specific fatigue cycle, it does not extend to the prediction of crack growth behaviour from the combined effects of stress ratio and wave shape. However, model predictions were shown to give good agreement with actual data, within the ΔK and cycle times examined in the study.

7 ACKNOWLEDGEMENTS

This research contributed towards Brite-Euram Project 6021. As such, the authors are grateful for the support of the CEC and those organisations involved in the collaboration,

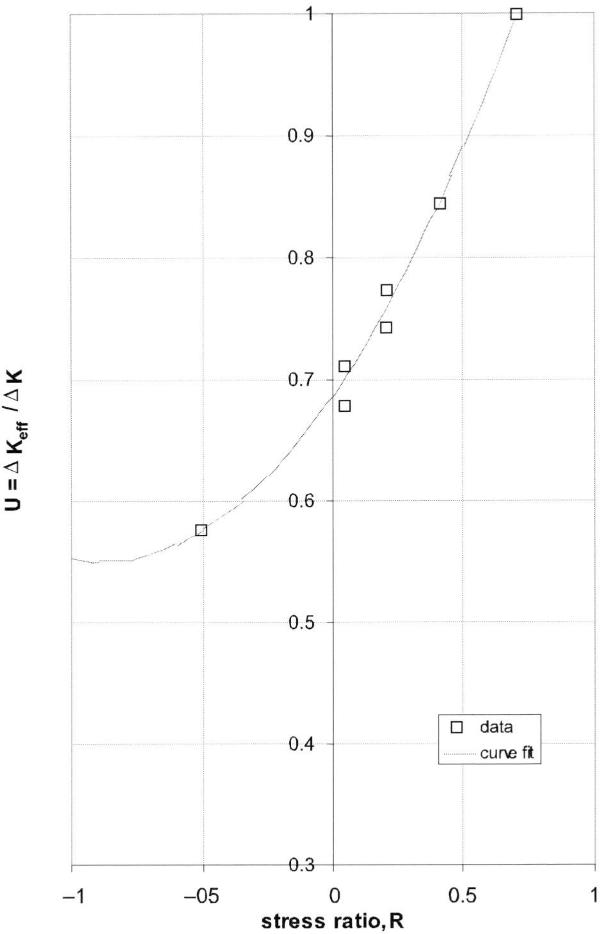

Fig. 10 *U* versus *R* data for TIMETAL 834 at 630°C, generated from 1/1/1/1 fatigue cycles in laboratory air (10^5 Pa) and to a maximum load equivalent to 70% of 0.2% proof stress.

in particular, Rolls-Royce plc, Derby, UK, Defence Evaluation Research Agency, Farnborough, UK and Motoren- und Turbinen-Union, Munich, Germany. The UK contributors would also like to acknowledge the financial support provided by the Department of Trade and Industry.

REFERENCES

1. M. Clavel and A. Pineau, 'Frequency and Waveform Effects on the Fatigue Crack Growth Behaviour of Alloy 718 at 298 K and 823 K', *Met. Trans.*, 1978, **9A**, 471–80.
2. T. Weerasooriya, 'Effect of Frequency on Fatigue Crack Growth Rate of Inconel 718 at High Temperature', *Fracture Mechanics: 19th Symposium*, T. Cruse ed., ASTM STP 969, American Society for Testing and Materials, 1988, 907–923.
3. T. Nicholas and N.E. Ashbaugh, 'Fatigue Crack Growth at High Load Ratios in the Time-Dependent Regime', *Fracture Mechanics: 19th Symposium*, T. Cruse ed., ASTM STP 969, American Society for Testing and Materials, 1988, 800–817.

4. E. Andrieu, G. Hochstetter, R. Molins and A. Pineau, 'Oxidation and Intergranular Cracking Behaviour of Two High Strength Ni-Base Superalloys', *Corrosion Deformation Interactions*, CDI '92, T. Magnin and J.M. Gras eds, Editions de Physique, 1992, 461–475.

5. V. Hodkinson and J. Byrne, 'The Effect of Waveshape on Fatigue Crack Growth Behaviour in Waspaloy at 650°C', *Engineering Against Fatigue*, J.H. Beynon, M.W. Brown, R.A. Smith, T.C. Lindley and B. Tomkins eds, A.A. Balkema, 1999, 591–598.

6. H. Ghonem and R. Foerch, 'Frequency Effects on Fatigue Crack Growth Behaviour in a Near-α Titanium Alloy', *Materials Science and Engineering*, 1991, **A138**, 69–81.

7. R. Foerch, A. Madsen and H. Ghonem, 'Environmental Interactions in High Temperature Fatigue Crack Growth of Ti-1100', *Met. Trans*, 1993, **24A**, 1321–1332.

8. M.C. Hardy and A. J. Leggett, 'Effect of Loading Rate on the Elevated Temperature Fatigue Crack Growth Behaviour of a Near-α Titanium Alloy', *Titanium '95: Science and Technology*, P.A. Blenkinsop, W.J. Evans and H.M. Flower eds, The Institute of Materials, 1996, 1195–1202.

9. R.H. Bricknell and D.A. Woodford, 'The Embrittlement of Nickel Following High Temperature Air Exposure', *Met. Trans.*, 1981, **12A**, 425–433.

10. M. Gao, D.J. Dwyer and R.P. Wei, 'Chemical and Microstructural Aspects of Creep Crack Growth in Inconel 718', *Superalloys 718, 625 and 706 and Various Derivatives*, E.A. Loria ed., TMS, 1994, 581–592.

11. D. Eylon and J.A. Hall, 'Fatigue Behavior of Beta Processed Titanium Alloy IMI 685', *Met. Trans.*, 1977, **8A**, 981–990.

12. D. Eylon, J.A. Hall, C.M. Pierce and D.L Ruckle, 'Microstructure and Mechanical Property Relationships in the Ti-11 Alloy at Room and Elevated Temperatures', *Met. Trans.*, 1976, **7A**, 1817–1826.

13. L.F. Coffin, 'Overview of Temperature and Environmental Effects on Fatigue of Structural Metals, *Fatigue, Environment and Temperature Effects*, J.J. Burke ed., Plenum, 1983, 1–40.

14. D.F. Neal, 'Development and Evaluation of High Temperature Titanium Alloy IMI 834', *Sixth World Conference on Titanium*, P. Lacombe, R. Tricot and G. Béranger eds, Les Editions de Physique, 1988, 253–258.

15. W.J. Evans, M.R. Bache, M. McElhone and L. Grabowski, 'Environmental Interactions with Fatigue Crack Growth in Alpha/Beta Titanium Alloys', *Int. J. Fatigue*, 1997, **19**, 177–182

16. M.A. Hicks and A.C. Pickard, 'A Comparison of Theoretical and Experimental Methods of Calibrating the Electrical Potential Drop Technique for Crack Length Determination', *Int. Journ. of Fracture*, 1982, **20**, 91–101.

17. A.C. Pickard, 'Stress Intensity Factors for Cracks with Circular and Elliptical Fronts Determined by 3D Finite Element Methods, *Numerical Methods in Fracture Mechanics*, D.R.J. Owens and A.R. Luxmore eds, Pineridge Press, 1980, 599–614.

18. W. Elber, 'The Significance of Fatigue Crack Closure', *Damage Tolerance in Aircraft Structures*, ASTM STP 486, American Society for Testing and Materials, 1971, 230–242.

19. T. A. Wallace, 'The Effect of Oxidation Exposure on the Mechanical Properties of Timetal-1100 (Ti–6Al–2.75Sn–4Zr–0.4Mo–0.07O_2–0.02Fe wt%)', *Titanium '95: Science and Technology*, P.A. Blenkinsop, W.J. Evans and H.M. Flower eds., The Institute of Materials, 1996, 1943–1950.

20. ASTM E1457, *Annual Book of ASTM Standards*, Volume 03.01, American Society for Testing and Materials, 1990, 648.

21. S. Zhang, R. Marissen, K. Schulte, K.K. Trautmann, N. Nowack and J. Schijve, 'Crack Propagation Studies on A1 7475 on the Basis of Constant Amplitude and Selective Variable Amplitude Loading Histories', *Fatigue Fract. Engng. Mater. Struct.*, 1987, **10**(4), 315–332.

Effects of Grain and Precipitate Size Variation on Creep-Fatigue Behaviour of Udimet 720Li

N.J. HIDE,[a] M.B. HENDERSON[b] and P.A.S. REED[a]

[a]*Materials Research Group, School of Engineering Sciences, University of Southampton, SO17 1BJ, UK*
[b]*Mechanical Sciences Sector, DERA Farnborough, Farnborough GU14 0LS, UK*

ABSTRACT

The high temperature fatigue characteristics of two base-line alloy variants of Udimet 720 (high Cr U720 and low Cr U720Li) have been compared over the temperature range 600°C to 725°C under imposed dwell times (at maximum load) of 1–120 seconds in vacuum and air conditions. The effect of varying grain size and coherent precipitate size of U720Li has also been studied in relation to its performance under vacuum conditions only.

For the base-line materials the low Cr U720Li gave the best resistance to high temperature fatigue crack growth and creep/oxidation-fatigue crack growth. Testing in air resulted in oxidation dominated intergranular crack growth at all temperatures and dwell times, with the slope (*m*-values) of the crack growth rate curves remaining constant. By contrast, in vacuum crack growth rates were much lower than in air and a purely cyclic dependent regime was evident at 600/650°C. As temperature and dwell time at maximum load were increased, m-values increased and were accompanied by a change in crack growth mechanism from transgranular to intergranular type cracking. This indicated the presence of true, time-dependent, creep-fatigue processes.

The U720Li microstructural variants all demonstrated both cyclic and time dependent fatigue regimes under vacuum conditions (c.f. the base-line U720Li). The large grain variant of the U720Li showed little advantage in crack growth rates within the cyclic dependent regime, but did show a significant increase in resistance to crack growth in the time dependent (creep-fatigue) regime. The effect of the large precipitate variant was to give similar or worse crack growth resistance than the baseline U720Li at temperatures up to 725°C (1 second dwell) but improved crack growth resistance when creep processes predominated at 725°C with an imposed 20 second dwell.

INTRODUCTION

In order to improve aero-engine efficiency, engine operating temperatures are being pushed ever higher and higher with target temperatures for nickel-base disc materials such as U720 and U720Li now approaching 725°C. Service conditions for these alloys are thus extremely severe due to temperature, oxidation and the fluctuating loads imposed during engine start-up and shut-down cycles as well as loads imposed during manoeuvres. As such these conditions pose problems in terms of high temperature fatigue, oxidation assisted fatigue and (when sub-surface defects occur which grow in an effective vacuum) creep-fatigue.

Optimisation of turbine materials so that operating temperatures can be maximised thus requires a fundamental understanding of the mechanisms involved in high temperature fatigue and oxidation/creep-fatigue, as well as the influencing factors.

In terms of alloy chemistry, higher Cr levels in U720 have been associated with lower fatigue crack growth rates.[1] However higher Cr levels have also been associated with an increased propensity to form the topologically closed packed sigma phase which tends to delineate and embrittle grain boundaries.[2] Studies of high temperature fatigue crack growth in other nickel-base superalloys have shown that increased Cr levels may limit intergranular oxygen diffusion ahead of the crack tip through the formation of protective Cr_2O_3, thus reducing high temperature oxidation assisted fatigue cracking.[3] Hence the two Cr alloy variants, U720 and U720Li have been studied in this project.

Nickel-base superalloys are largely comprised of an austenitic fcc matrix (γ phase) and coherent nickel rich γ' precipitates, from which the material derives most of its strength and creep-rupture properties. As such the characteristics of the γ grains and γ' precipitates can significantly alter the fatigue and oxidation/creep-fatigue properties of the material. Increased grain sizes have been associated with reducing the intrinsic driving force for fatigue crack growth by promoting slip reversibility when plasticity is contained within a grain. Damage accumulation may then be reduced to higher crack tip stress levels compared with that for smaller grain sizes, Krueger.[4] Larger grains are also associated with a reduction in time dependent crack growth rate through increased resistance to grain boundary motion (creep-fatigue) and fewer paths for oxygen diffusion (oxidation-fatigue), Ghonem et al.[3] Variations in the coherent γ' precipitates have also been linked to reduced high temperature fatigue and time dependent crack growth rates from modification of the dislocation-precipitate interactions. Larger precipitates are thought to increase slip homogeneity within the material leading to lower stresses occurring at grain boundaries which improve creep-rupture strength.[5] Whilst in an oxidising environment more intense, heterogeneous slip from finer precipitates is considered to provide a preferential path for oxidising species to embrittle grain boundaries.[3,6]

In this study a fundamental examination of the effects of microstructure and Cr on the high temperature fatigue and oxidation/creep fatigue of U720 has been carried out. The aim is to optimise the alloy's microstructural characteristics for maximum resistance to crack growth under these conditions.

EXPERIMENTAL

Four material variants were studied, namely standard as heat-treated forms of U720 and U720Li (base-line materials) as well as a large grain and precipitate variant of U720Li. The U720Li microstructural variants were produced by increasing the solution temperature (to increase grain size) and by slowing the cooling rate from the solution temperature (to increase the coherent precipitate size). The heat treatment routes used for each material are listed in Table 1.

The materials were characterised by etching with Fry's reagent and subsequent image analysis to determine the grain size, incoherent γ' size and γ' volume fraction. Coherent γ' characterisation was performed using TEM carbon extraction replicas of each material so that size and size range distributions of the secondary and tertiary γ' could be found. Vickers hardness tests were also performed on each of the materials so that the effect of heat treatment could be assessed in terms of mechanical strength (flow stress variations).

Table 1 Heat treatment conditions of alloy variants. All alloys received a subsequent two stage ageing treatment of 24hr @ 650°C (air cool) followed by 16hr @ 760°C (air cool).

Alloy variant	Solution treatment	Cooling medium
U720	4hr @ 1105°C	Oil quench
U720Li	4hr @ 1105°C	Oil quench
U720Li (LG)	4hr @ 1135°C	Oil quench
U720Li (LP)	4hr @ 1105°C	Insulated air cool

High temperature fatigue testing was conducted in both air and vacuum between 600°C and 725°C at an *R* ratio = 0.1 using a trapezoidal 1–X–1–1 waveform, where X was the dwell time at maximum load which varied between 1 and 120 seconds. All vacuum testing was conducted on an Instron 8501 servohydraulic testing machine fitted with an ESH vacuum furnace attachment, using single edged notch bend (SENB) specimens. Air environment testing was conducted at DERA Farnborough using compact tension (CT) specimens. Standard d-c potential drop techniques were used to monitor crack growth over the ΔK range 15–40MPam$^{1/2}$. Constant ΔK tests were also conducted at a ΔK of 20 (±2) MPam$^{1/2}$ with cracks grown for approximately 2mm under constant or test conditions. Fracture surface and etched fracture surface sections were observed under the SEM to determine the crack propagation mode.

RESULTS AND DISCUSSION

MICROSTRUCTURAL CHARACTERISATION

The microstructural characteristics and Vickers hardness results of the alloy variants used in this study are presented in Table 2. The compositions of U720 and U720Li were also analysed by wet chemical analysis and found to be very similar bar the Cr levels which are higher in the U720 material (17.9wt% and 15.9wt% respectively).

Table 2 Microstructural characteristics of the alloy variants.

Alloy variant	Grain size (μm)	Primary γ′ (μm)	Primary γ′ V_f(%)	Secondary γ′ (nm)	Tertiary γ′ (nm)	Vickers hardness
U720	8.3	2.9	15.4	100	21	462
U720Li	5.9	2.6	18.2	102	16	459
U720Li (LG)	17.4	2.9	6.9	149	25	445
U720Li (LP)	10.1	2.9	23.2	200	22	425

From Table 2 it is apparent that the grain size of the base-line U720Li is slightly smaller than the U720, with consequent variations in the volume fraction of primary γ'.

Slight differences are also apparent in the secondary and tertiary γ' sizes. The variation in solutionising temperature and cooling rates of U720Li have produced significant variations in grain size and coherent γ'. The large grain variant has a grain size three times that of the as-received alloy whilst coherent γ' characteristics have been maintained. The slower cooling rate has produced coherent secondary precipitates that are twice the size of the as-received alloy. However, some grain growth has occurred as well as an increase in the volume fraction of primary γ'. Hardness testing indicates that increasing γ' size significantly decreases flow stress, whilst increasing grain size has little effect (with similar implications for yield strength variations).

Fatigue Testing of the As Received Alloys (U720 and U720Li)

The fatigue crack propagation (f.c.p.) rates obtained under vacuum for the base-line materials (U720 and U720Li) are shown in Figure 1 for testing temperatures of 600 and 650°C (1 second dwell). Both U720 and U720Li are seen to exhibit very similar f.c.p. rates

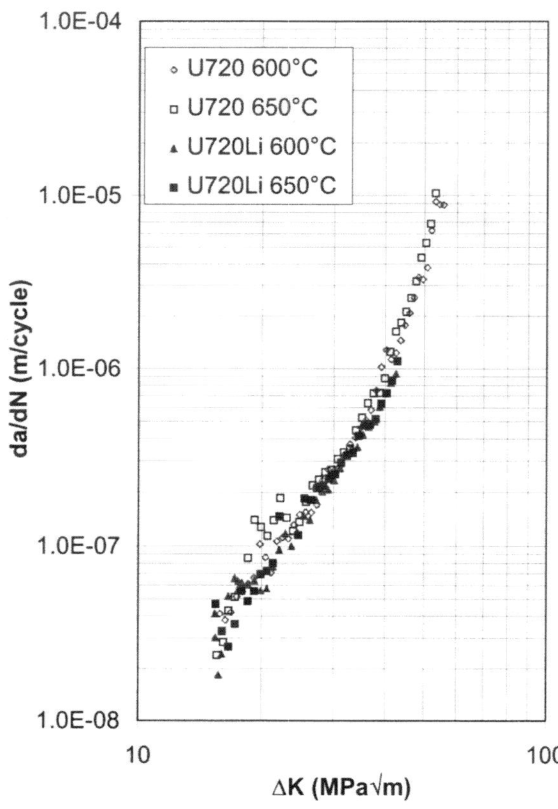

Fig. 1   F.c.p. rates of U720/U720Li at 600/650°C (1–1–1–1) in vacuum.

and m-values. SEM examination of the fracture surfaces and etched fracture surface sections obtained from growth out and constant ΔK tests (Fig. 2a–b) showed flat, featureless cracking that was highly transgranular in nature. The similarity between the f.c.p. rates of the two alloys at 600°C and 650°C and the smooth, transgranular crack growth mode indicate that crack growth under these conditions may be of a purely cyclic nature.

Fig. 2 (a) Fracture surface section of U720, 650°C (1–1–1–1), showing typical example of transgranular cracking. (b) Sectioned constant ΔK test: U720Li, 650°C (1–1–1–1), $\Delta K\approx20$MPa, crack tip region.

Figure 3 shows the f.c.p. rates obtained under vacuum for U720 and U720Li tested at 725°C with varying dwell times (testing conducted at 600/650°C is indicated by the single C.D.R. line in Fig. 3). The f.c.p. rates and m-values for both U720 and U720Li are seen to increase significantly as test temperature and dwell time at maximum load are increased. For the U720 material an increase in temperature to 725°C produces the most dramatic increase in f.c.p. rate with further, less pronounced increases in f.c.p. rate as the dwell time is increased over 1 second. U720Li shows a similar trend but with a less dramatic effect when temperature is increased to 725°C and a more significant increase in f.c.p. rate and m-value when the dwell time is increased. In terms of an alloy chemistry effect, the U720Li material (lower Cr content) exhibits an increased resistance to crack growth at 725°C compared with the higher Cr content U720.

SEM observations of tests conducted at 725°C (for both alloys) indicated much rougher fracture surfaces (c.f. lower temperature tests) with the failure modes becoming increasingly intergranular in nature as dwell times were increased (Fig. 4(a)). Crack tunnelling was also evident for tests conducted under these conditions. The change in crack growth mode coupled with the increasing f.c.p. rates and m-values noted may indicate the presence of time dependent (creep-fatigue) behaviour, with the lower Cr U720Li exhibiting a higher resistance to crack growth within this regime of high temperature fatigue and creep-fatigue.

Analysis of constant ΔK tests using backscattered SEM showed that at 725°C voids were appearing ahead of the crack tip (Fig. 4(b)) and were seen to be more prevalent as dwell time was increased. This further indicates the presence of static creep modes under these conditions.

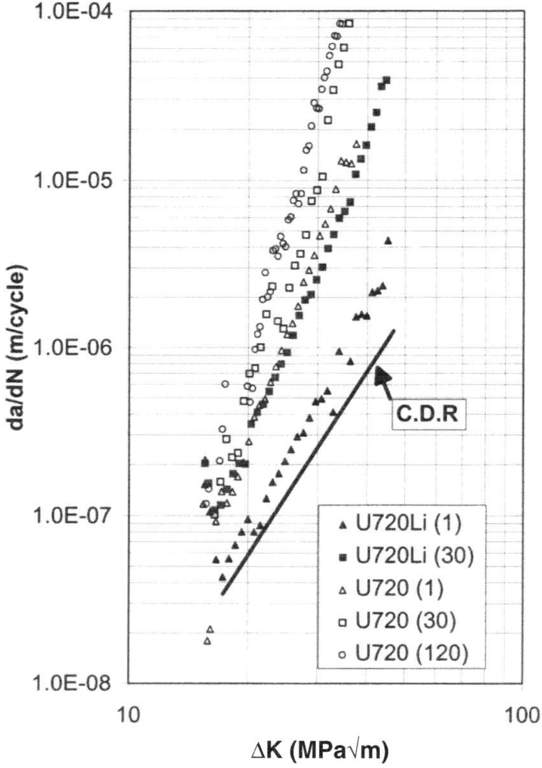

Fig. 3   F.c.p. rates of U720/U720Li at 725°C with varying dwell time.

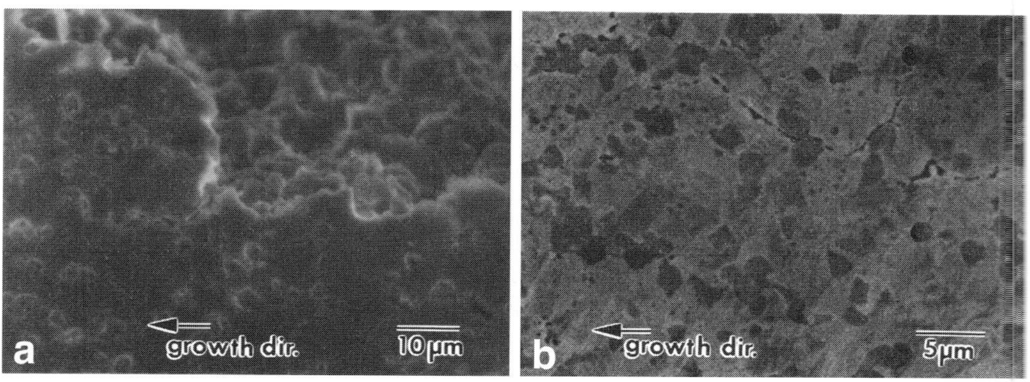

Fig. 4   (a) Fracture surface section of U720, 725°C (1–120–1–1), showing typical intergranular cracking. (b) Sectioned constant ΔK test: U720Li, 725°C (1–30–1–1), $\Delta K \approx 20$MPa, crack tip region.

The effect of environment is shown in Figs 5 and 6 where the vacuum data is compared with CT tests carried out in air at DERA Farnborough. In an air environment, crack growth rates are much higher under all testing conditions with increases noted as temperature and dwell times are increased. The same effect of alloy chemistry is apparent in air conditions (c.f. vacuum testing) with the lower Cr U720Li showing an increased resistance to crack growth compared with the Higher Cr U720. Of particular significance are the slopes of the crack growth curves (*m*-values) in Figs 5 and 6, which remain relatively constant over all the testing conditions. Intergranular cracking modes were also noted to predominate under all testing temperatures and dwell times, indicating that the mechanism of crack growth remains constant throughout. This is in stark contrast with the vacuum data where m-values increased and the crack growth mode changed from transgranular to intergranular as test temperatures and dwell times were increased.

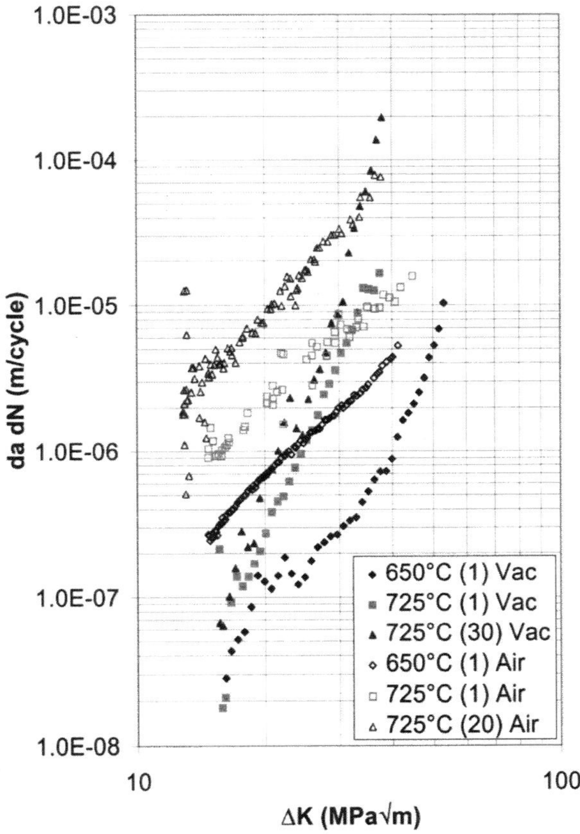

Fig. 5   Comparison of f.c.p rates in air and vacuum for U720.

In an air environment the results indicate that oxidation embrittlement along grain boundaries dominates the nature of crack growth under all testing conditions. This would explain the constant m-values and the consistently intergranular nature of the crack growth observed.

However, in vacuum the effects of oxidation are all but removed and at high temperatures with lower frequencies (increased dwell), the change in *m*-values and crack growth mode would indicate the onset of true creep processes. It is unlikely that any true creep

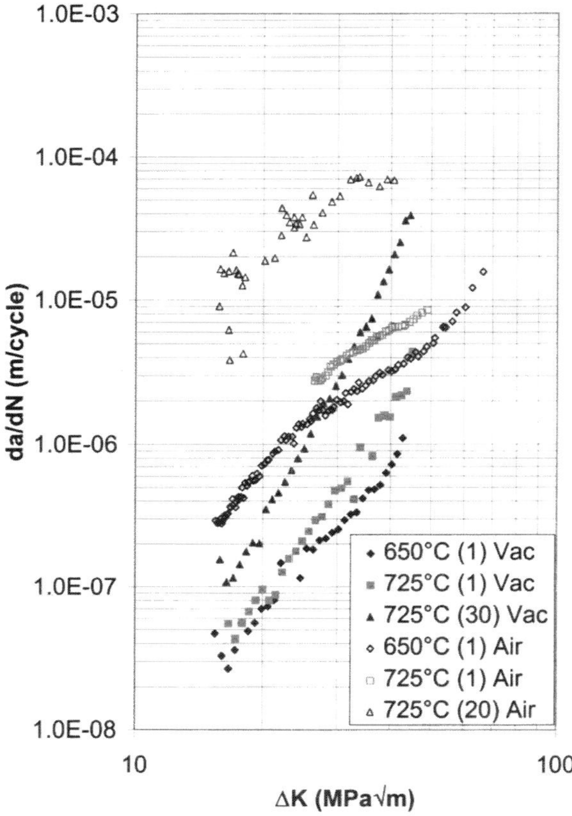

Fig. 6   Comparison of f.c.p rates in air and vacuum for U720Li.

processes can occur in an air environment as the oxidation assisted crack growth occurs at such a rate that there is insufficient time for creep processes to occur or that they are insignificant with respect to the effect of oxidation.

FATIGUE TESTING OF THE LARGE GRAIN AND PRECIPITATE
MICROSTRUCTURAL VARIANTS

The microstructural analysis work indicates that increasing grain size appears to have far less of an effect on flow stress than increasing γ' precipitate size. It is noteworthy that the U720Li material, despite having a slightly smaller grain size and a lower Cr content, appears to have improved creep-fatigue and oxidation-fatigue resistance than the U720. A limited number of tests have been carried out on a larger grain size variant of the U720Li. In Fig. 7 a comparison of the effect of an increase in grain size (from 5.9μm to 17μm) on the creep-fatigue behaviour of U720Li can be seen.

At 650°C there appears to be only a slight advantage observed in terms of f.c.p. resistance of the large grain size variant over the base-line material. The effect of grain size

104

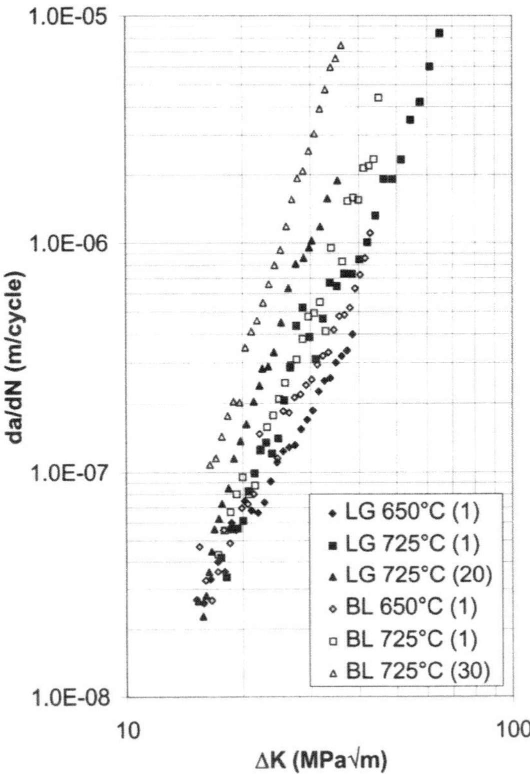

Fig. 7   Comparison of f.c.p rates in coarse and fine grain U720Li under vacuum conditions.

is also very slight for an increase in temperature to 725°C. It seems that increasing the grain size has a beneficial effect in the creep-fatigue regime (725°C, 20 second dwell), but little discernible effect in the cycle dependent fatigue regime when oxidation effects are eliminated.

The effect of increased precipitate size on f.c.p. characteristics in vacuum is shown in Fig. 8. At 650°C the LP variant shows little effect in terms of f.c.p. rates, whilst at 725°C with a 1 second dwell the base-line U720Li has an increased resistance to crack growth. This indicates that increasing precipitate size is of little benefit in terms of increasing fatigue/cyclic crack growth resistance. However, at 725°C (20 second dwell) it is apparent that the LP material does have a lower f.c.p. rate compared with the base-line material. Increasing precipitate size has been found to significantly reduce the flow stress of U720Li and may account for the results at 650°C and 725°C when a 1 second dwell is imposed. At 725°C with a 20 second dwell intergranular cracking modes appeared to predominate (indicating time-dependent behaviour) and it is this predominating behaviour that is thought to favour the LP microstructure in terms of increasing resistance to crack growth. It should also be noted that the grain size and volume fraction of primary γ' had been increased slightly over that of the base-line U720Li which may also contribute to increased creep-fatigue crack growth resistance as these features can help to restrict grain boundary sliding.

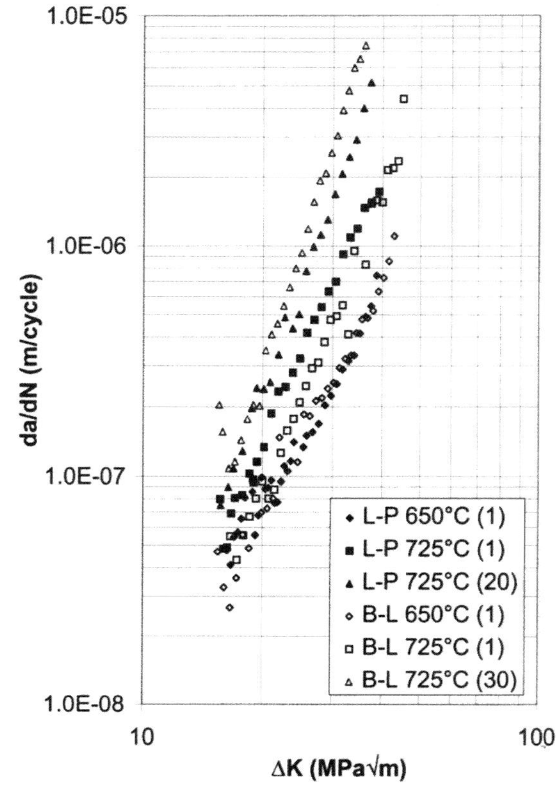

Fig. 8 Comparison of f.c.p rates in coarse and fine precipitate U720Li under vacuum conditions.

SUMMARY AND CONCLUSIONS

U720Li appears to exhibit improved high temperature fatigue and creep-fatigue resistance over that of the higher Cr U720 alloy in both air and vacuum. In this instance the higher Cr level does not appear to offer any improvement in crack propagation resistance. The time-dependent crack growth regime in air is primarily due to oxidation assisted crack growth, since any underlying 'true' creep-fatigue behaviour, such as that witnessed at 725°C in vacuum is swamped by rapid oxidation effects.

The effect of varying grain size (whilst maintaining similar γ' sizes) on f.c.p. rates in vacuum has been evaluated. A larger grain size appears beneficial in resisting creep-fatigue, but only offers a slight benefit in the cycle-dependent fatigue regime. This indicates that there is little reduced intrinsic driving force for fatigue crack growth for an increased grain size. Our results do not clearly vindicate the argument for less damage accumulation[4] due to increased slip reversibility to higher crack tip stress levels as crack tip plasticity is contained within a larger grain. The beneficial effect of larger grain size is seen in the 'true' creep-fatigue regime in vacuum, where the decreased number of grain boundaries offers more resistance to intergranular crack growth mechanisms. The larger grain sized material is expected to show improved oxidation fatigue resistance for the same reason. Altering grain size was found to have little effect on the flow stress of the material, whereas altering the coherent γ' size gave far greater effects.

In terms of f.c.p. behaviour, larger coherent precipitates only appear beneficial when static creep failure modes appear to dominate. This may be a result of increased creep-rupture strength through increasing the coherent precipitate size, although the slightly larger grain size and volume fraction of primary γ' may also be a factor within this regime where intergranular features are important. No beneficial effect of increased coherent γ' size on crack growth was found at test temperatures of 650°C and 725°C when a 1 second dwell was imposed and pure cyclic fatigue processes were a significant feature. This may be attributed to a decrease in the yield stress of this material. The next stage of the research will investigate the effect of environment on the large grain and precipitate variants with respect to creep-fatigue and high temperature fatigue behaviour.

LIST OF SYMBOLS

γ	= Matrix phase in Udimet alloys.
γ'	= Nickel rich precipitates, $Ni_3(Ti,Al)$.
Rratio	= Ratio of minimum to maximum loading stress.
ΔK	= Stress intensity factor range (MPa$\surd$).
V_f	= Volume fraction (%).

ACKNOWLEDGEMENTS

The work in this paper has been supported by the Mechanical Sciences Sector, DERA Farnborough under DTI-CARAD MOD Applied Research Package funding whose support is gratefully acknowledged. Rolls Royce plc are thanked for the supply of materials, use of equipment and technical discussion. The assistance of T.Powell and A.Tucker is also gratefully acknowledged for the air testing, conducted at DERA Farnborough.

REFERENCES

1 M.F. Henry, United States Patent No 5,124,123, 1992.
2 R. Reed, M. Jackson and Y. Na, *Met. Trans. A.*, 1999, **30A**, 521.
3 H. Ghonem, T. Nicholas and A. Pineau, *Fatigue Fract. Engng. Mater. Struct.* 1993, **16**, 577.
4 D. Krueger, S. Antolovich and R. Van Stone, *Met. Trans. A.*, 1987, **18A**, 1431.
5 H. Smith and D. Michel, *Met. Trans. A.*, 1993, **17A**, 370.
6 S. Floreen and R. Kane, *Fatigue Engng. Mater. Struct.*, 1980, **2**, 401.

Modelling of Creep Strain Behaviour of Udimet 720Li Wrought Superalloy

M. MALDINI and V. LUPINC

CNR-TEMPE, Milan, Italy

ABSTRACT

Constant load creep tests on isothermally forged Udimet 720 Li, an advanced superalloy for gas turbine disc application, have been run in the stress/temperature field 900–450MPa/650–700°C producing rupture times in the 20–5000 h range. The creep curves have shown a predominant tertiary creep stage, that has been described by the following equation:

$$\mathrm{d}\varepsilon/\mathrm{d}t = A° \exp(-Q/RT)(1 + C\varepsilon)$$

where $A°$ and C parameters are mainly functions of stress, Q is the activation energy, and RT has the usual meaning. A Norton type stress dependence has been utilised for $A°$ parameter. The damage parameter C dominates the tertiary creep stage at stresses lower than 750 MPa while its contribution to the creep strain acceleration is less important at higher stresses.

1   INTRODUCTION

The introduction of stronger and metallurgically more stable superalloys in advanced gas turbines has enabled increases to the turbine inlet temperature, enhancing thrust and operating efficiency levels. The knowledge of creep behaviour is becoming more and more useful for designing discs of advanced gas turbines as the applied temperatures and stresses are continually pushed to higher values.

This paper reports the creep behaviour and modelling of an isothermally forged Udimet 720 Li superalloy. This work was performed within a Brite EuRam II project.[1]

2   MATERIAL AND EXPERIMENTAL PROCEDURES

The alloy Udimet 720 Li is an advanced wrought superalloy for gas turbine disc application. The chemical composition of the examined alloy is given in Table 1.

Table 1   Udimet 720 Li nominal chemical composition (wt%).

Elem	Al	Co	Cr	Mo	Ti	W	Zr	Fe	Ni
wt%)	2.5	14.7	16	3.0	5.0	1.25	0.40	0.5	bal.

The material has been isothermally forged, heat treated and test piece blanks cut from disc forgings by Thyssen Umformtechnik. The heat treatment sequence was:

4 h / 1110°C / oil quench + 24 h / 650°C / air cool + 16h / 750°C / air cool

109

Constant load creep tests were performed on specimens having cylindrical symmetry with a gauge diameter of 4.0 mm and 20 mm gauge length. Three thermocouples were placed in the gauge length allowing control of the temperature gradients. Before the start of each test, cold (room temperature) and hot (testing temperature) loadings were carried out to check the Young's modulus values and bending of the specimen due to nonaxiality of the loading. The specimens were not allowed to spend more than three hours and less than an hour at the testing temperature before commencement of each test.

3   EXPERIMENTAL RESULTS AND DISCUSSION

Constant load and temperature creep tests have been performed at 650 and 700°C and the stresses were selected in the 450–900 MPa range to produce times to rupture of up to 5400 h as shown in Fig. 1. At both temperatures the majority of the single creep curves consist of an accelerating stage, tertiary creep, where the strain rate increases from the minimum creep rate obtained after a short and small primary creep stage (Fig. 2). Analysis of the experimental creep curves has shown a different behaviour for both the minimum creep

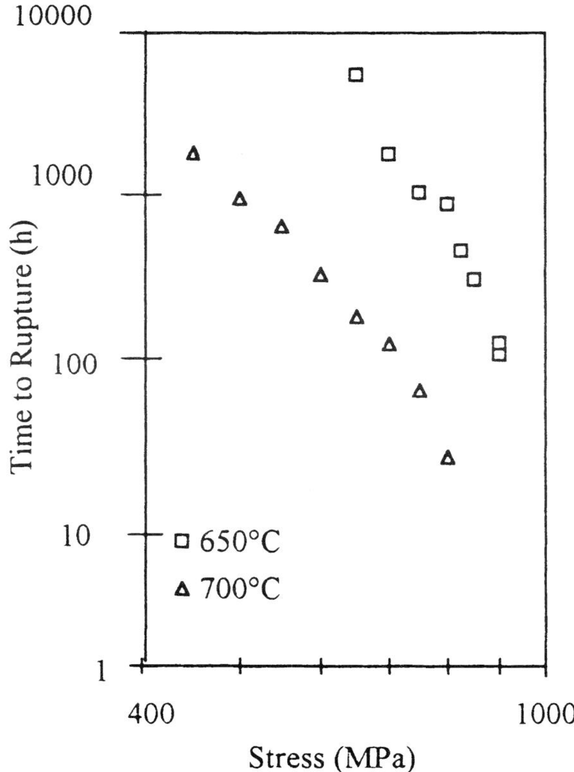

Fig. 1   Stress rupture experimental data.

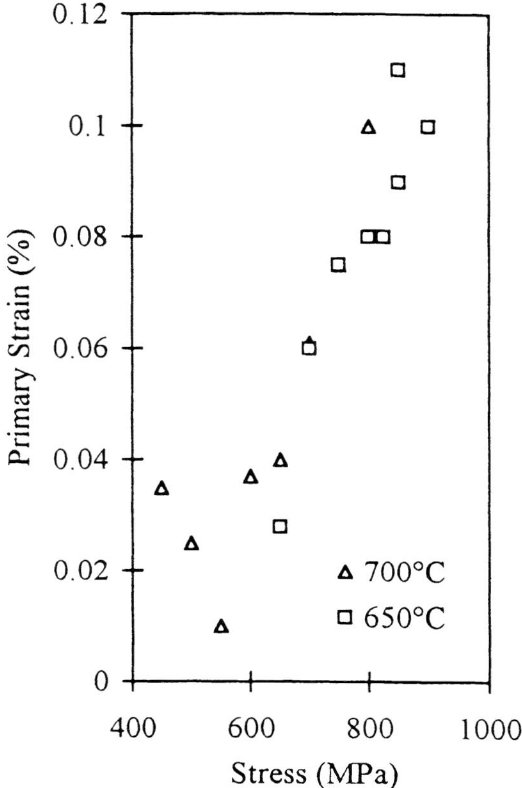

Fig. 2   Creep strain due to the primary stage.

rate and of the tertiary creep curve shape at high and low stresses at both the studied temperatures. They are analysed in the following two sections.

3.1   MINIMUM CREEP RATE

It is unclear, initially, from examination of the experimental results which relationship best describes the stress dependence of the minimum creep rate as the data do not lie on straight lines in the double logarithmic $(d\varepsilon/dt)_{min}$ vs σ graph (shown in Fig. 3) and excludes the possibility of simply adopting the Norton stress dependence. Furthermore, two different behaviours appear at low and high stress portions of the isotherms. To overcome these inconsistencies, the following relationship, that includes a stress and temperature dependent internal stress $\sigma^\circ(\sigma,T)$, has been adopted in this work:

$$(d\varepsilon/dt)_{min} = A(\sigma - \sigma^\circ)^n \exp(-Q/RT) \tag{1}$$

where A is a material constant, Q is the activation energy and R is the universal gas constant. The best fit of 700°C data occurs at $n = 4$ (Fig. 4). Since the data at 650°C are quite scattered, they have been fitted assuming a temperature independent value of parameter k defined below.

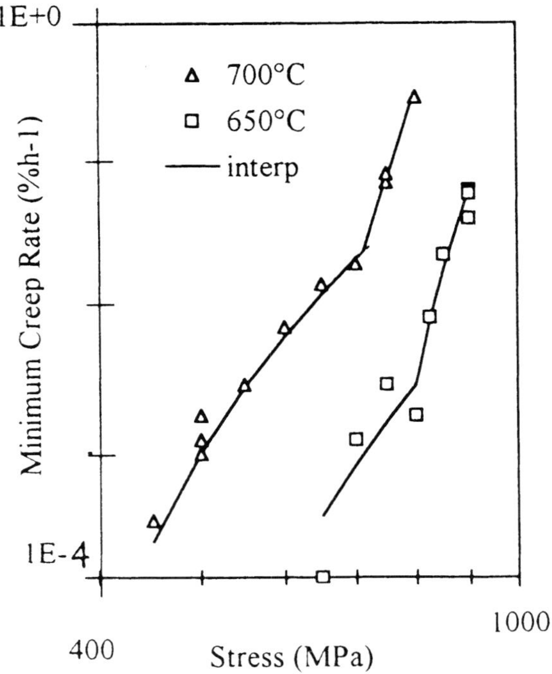

Fig. 3 Minimum creep rate, dependence on stress. The continuous curve interpolates experimental data using Eqns 1 and 2.

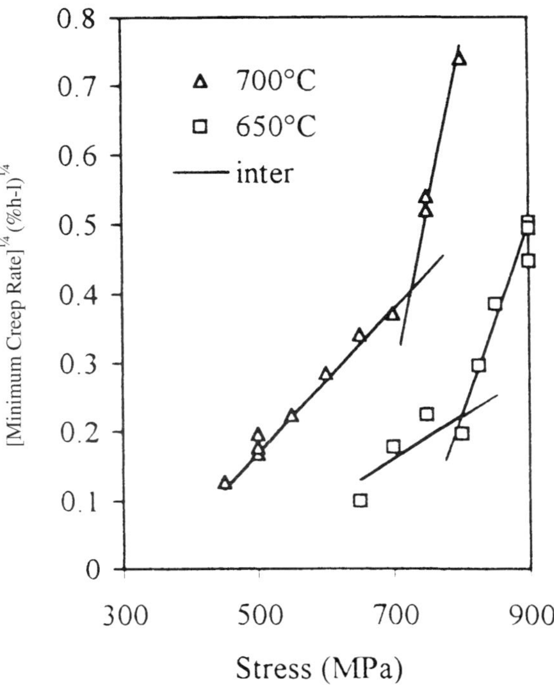

Fig. 4 Graphical representation determination of the parameters of Eqns 1 and 2.

The experimental results plotted in Fig. 4 are consistent with the following relationships between the internal stress and the applied stress:

$$\sigma^\circ = \begin{cases} \sigma_1(T) + k\sigma & \text{at low } \sigma, \text{ when the dislocation climb around particles is supposed to control the creep} \\[2em] \sigma_{th}(T) & \text{at high } \sigma, \text{ when Orowan dislocation bending or cutting of particles by dislocations is the supposed deformation mechanism.} \end{cases} \qquad (2)$$

After interpolating the values of these parameters using the experimental data, a double logarithmic plot of the $(d\varepsilon/dt)_{min}$ vs. $(\sigma-\sigma^\circ)$ can be produced (Fig. 5), showing that these equations are consistent with an activation energy value represented by the straight lines (250 kJ mol^{-1}) that is close to the self diffusion value. This interpolated curve is also represented by the continuous lines in the $(d\varepsilon/dt)_{min}$ vs σ graph of Fig. 3.

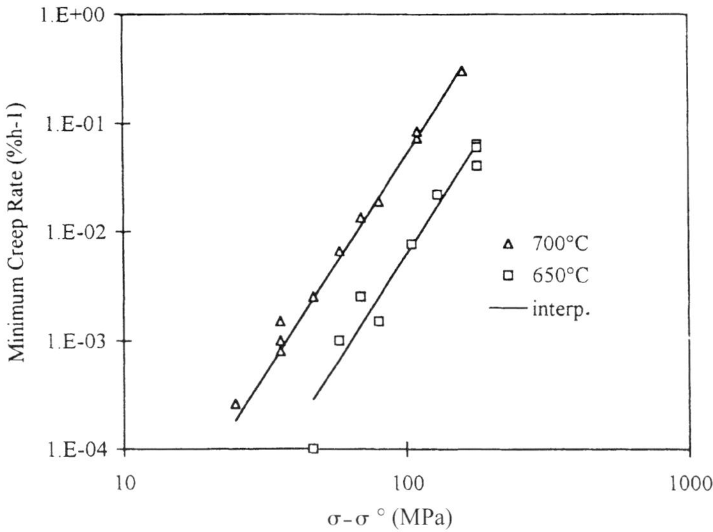

Fig. 5   Minimum creep rate vs. effective stress showing also the continuous line fit corresponding to Q = 250 kJ mol^{-1}.

3.2   MODELLING OF TERTIARY CREEP

In the stress-temperature field explored in this work, most of the creep strain is accumulated during tertiary creep, the stage that dominates the whole creep curve, as often observed in superalloys.[2]

It has already been shown that for this material the contribution of the primary creep stage to the total creep strain is small (Fig. 2) and of short duration.

Figure 6 shows the log($d\varepsilon/dt$) vs ε behaviour of some typical curves at 650°C. The great importance of the tertiary stage is apparent, although, for increasing applied stress levels, its contribution to the total creep strain tends to decrease. In fact at the highest applied stresses, the balance between the primary and tertiary stage seems to produce a quite large steady-like-state. As the creep curve shape is stress dependent, then the log strain rate vs. strain curves are generally not congruent, i.e. it is impossible to overlap them by translation.

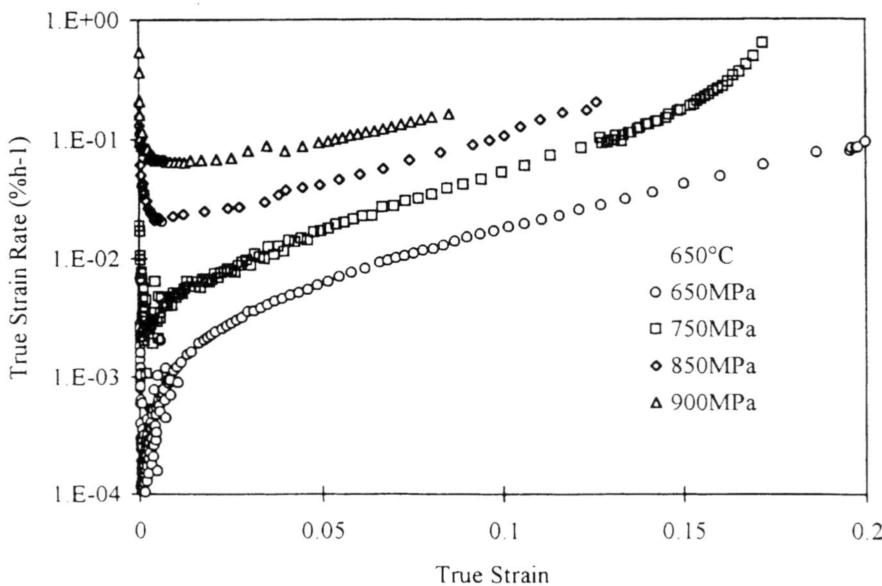

Fig. 6   Some examples of true strain rate versus true strain for the tests at 650°C.

In a constant load creep test the applied stress, σ, increases with the strain as $\sigma = \sigma_a e^\varepsilon$. where σ_a is the engineering stress, provided the reduction of the cross section area is homogeneous along the gauge length and the volume fraction of the specimen during the test is constant. Analysis of the experimental creep curves, together with relevant theoretical considerations,[3] shows that both the conditions are satisfied for most of the creep life. Only in the last few per cent of the specimen life does necking begin to influence the creep strain rate. The increment of the applied stress with the strain in a constant load creep test, produces an acceleration of the creep strain accumulation, not associated to a material damage mechanism. The contribution of the increasing applied stress with strain on the tertiary creep stage can be evaluated in a plot of the experimental creep strain rate as a function of the true stress σ as shown in Fig. 7.

The individual creep curves show similar behaviour at the two temperatures: at low stress as a small increment of the true stress causes a large increment of the strain rate, while such dependence decreases for tests conducted at higher stresses. Figure 7 shows that the increments of the strain rate during a creep test at high applied stress, almost overlap the interpolation of the minimum creep rate, obtained using Eqs 1 and 2. At high stresses, the long accelerating creep is mostly due to the increment of the stress with the strain, and the strain rate depends only on the applied true stress. At low applied stress levels, the

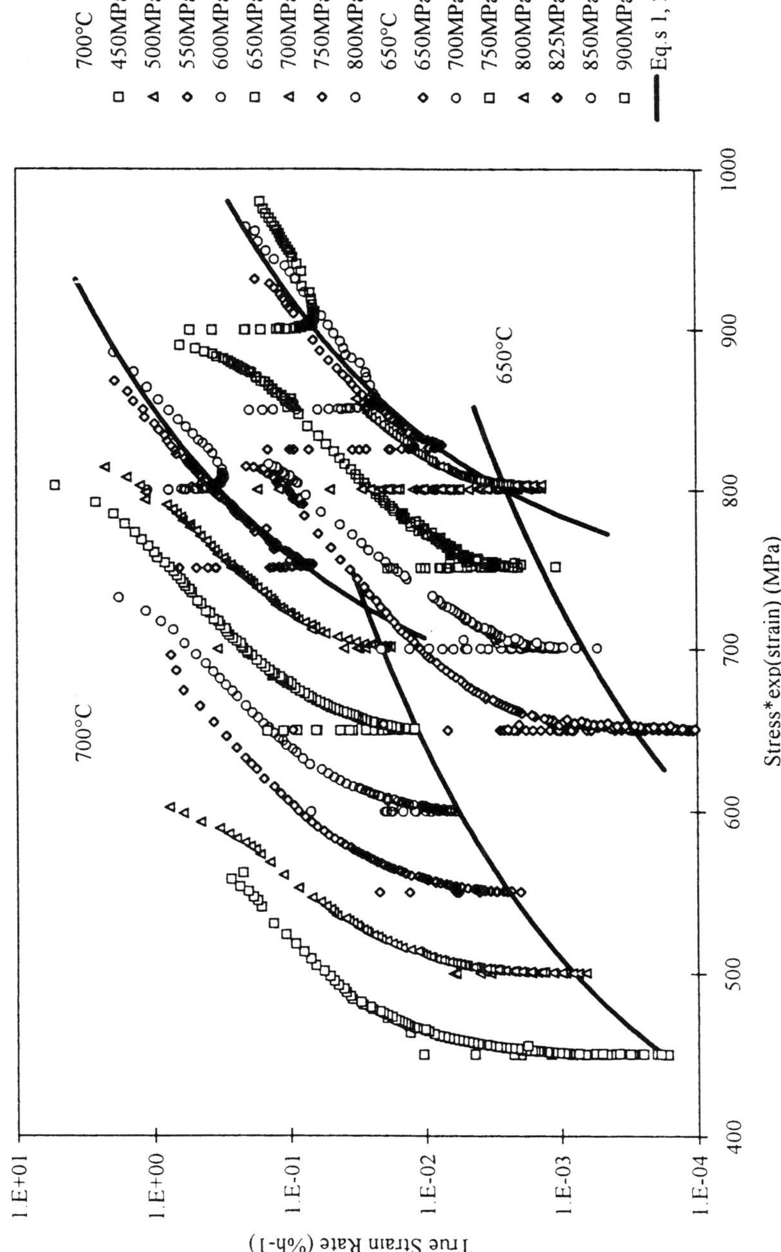

Fig. 7 True strain rate as function of the true stress and interpolation of the minimum creep rate values using eqns 1 and 2.

contribution of the applied stress increment with strain, represented by the minimum creep rate fit, is limited, and indicates that a material damage mechanism produces most of the strain rate increment.

Recently, various papers have pointed out that the long tertiary creep in nickel base superalloys is not associated to nucleation of fracture mechanisms,[2, 4–6] but to an increase of the flux of mobile dislocations with the strain due to an increase either of the mobile dislocations or the recovery rate, rather than the usual assumption of particle coarsening.[*] To describe the tertiary creep induced by this kind of damage in a constant stress test, the following set of differential equations has been proposed:[2, 4, 6]

$$d\varepsilon/dt = (d\varepsilon/dt)_{min} (\sigma, T)[1 + W]$$
$$dW/dt = C\,d\varepsilon/dt \tag{3}$$

where $d\varepsilon/dt$ is the instantaneous strain rate, $(d\varepsilon/dt)_{min}$ the minimum creep rate, ε is the tertiary creep strain, and W is a damage variable proportional to the tertiary creep strain and C is a stress dependent parameter that tends to vanish at high stress values.

Supposing the damage $W = 0$ at the beginning of the test, when $\varepsilon = 0$, then Eqn 3 can be written as follows for constant stress tests:

$$d\varepsilon/dt = (d\varepsilon/dt)_{min} (\sigma, T)[1 + C\varepsilon] \tag{4}$$

and they have the following analytical solution:

$$\varepsilon = \{\exp[C(d\varepsilon/dt)_{min}\, t] - 1\}/C \tag{5}$$

The differential formalism of the Eqn 3 allows the creep behaviour to be described under variable stress and temperature conditions, and for more complicated load/temperature histories as, for example, constant strain rate and stress relaxation tests.

Combining Eqns 1 and 4 and expressing true stress σ using engineering applied stress σ_a, a relation can be obtained that describes isothermal constant load creep. The stress sensitivities of the parameters $(d\varepsilon/dt)_{min}$ and C have then been defined interpolating the experimental creep curves.

These interpolated functions of stress, $(d\varepsilon/dt)_{min}(\sigma)$ and $C(\sigma)$, have been then utilised to calculate creep behaviour. In Fig. 8, two examples of the relation $\log(d\varepsilon/dt)$ vs ε are shown where interpolated curves are successfully compared with the experimental values.

4   CONCLUSIONS

An analysis of the creep behaviour of Udimet720Li alloy in the 650–700°C/900–450 MPa range shows the following:

---

*Among the possible damage mechanisms, the ripening of the reinforcing phase is not taken into account. Indeed, time softening due to particle instability cannot be excluded in early tertiary when creep time and temperature are, respectively, long and high enough to make particles grow, and stress is still high enough to activate particle size sensitive deformation mechanisms.

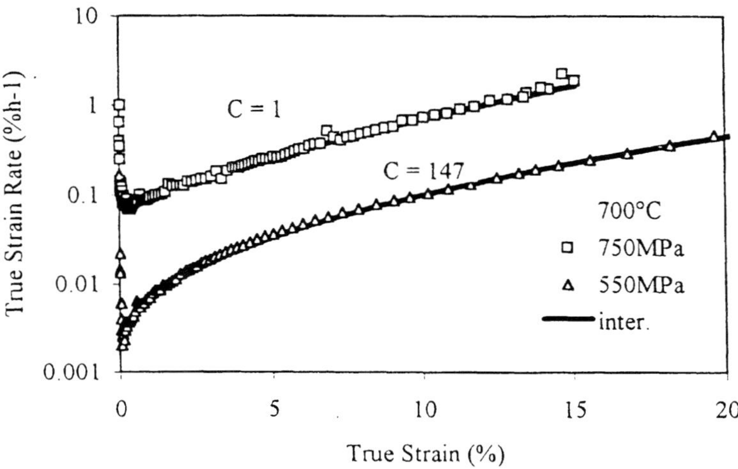

Fig. 8 Two examples of the true strain rate vs. true strain relation for experimental data and the continuous lines of the model calculation using Eqn 4.

- The primary stage of the creep curves is small and short and most of the creep strain is caused by tertiary creep.
- A change in the dominant creep mechanism seems to occur in the 700–800 MPa stress range. This change produces a discontinuity in the minimum creep rate stress sensitivity and a change in the creep curve shape.
- The accelerating creep, during tertiary stage, is due to: (1) a strain softening damage and (2) the increase of the applied stress with the strain in a constant load test.
- The strain softening mechanism is the most important damage mechanism at low stress and becomes less important with increasing stress where the effect of the reduction of the cross-section area of the specimen tends to prevail.
- The introduction of an internal stress parameter that depends linearly on the applied stress seems to rationalise the minimum strain rate stress sensitivity at low stresses. At high stress, a constant internal stress varying only with temperature seems to be more appropriate.

ACKNOWLEDGEMENTS

The main part of this work was performed within the Brite EuRam II project BREU 6021 (1993–1997). The partners were: Motoren und Turbinen Union, Munich D (coordinator), Fiat (formerly Alfa Romeo) Avio, Pomigliano d'Arco I, BMW/Rolls–Royce, Dahlewitz D, CNR-TEMPE (formerly ITM), Milan I, DERA, Farnborough UK, Rolls–Royce, Derby, UK, SENER, Las Arenas E, Thyssen Umformtechnik, Remscheid D, Turbomeca, Bordes F, Volvo Aero Corporation, Trollhättan S. The authors would like to thank the CEC for financial support.

REFERENCES

1. G.W. König, *Proc. Low Cycle Fatigue and Elasto-Plastic Behaviour of Materials*, K.-T. Rie and P.D. Portella eds, Elsevier Sci., p. 807, 1998.
2. M. Maldini and V. Lupinc, *Scripta Metall.*, 1988, **22**, 1737.
3. M. Maldini and V. Lupinc, *Scripta Metall. et Materialia*, 1995, **32**, 337.
4. F. Dyson and M. McLean, *Acta Metall.*, 1983, **31**, 17.
5. F. Dyson and T.B. Gibbons, *Acta Metall.*, 1987, **35**, 2355.
6. A. Barbosa, N.G. Taylor, M.F. Ashby, B.F. Dyson and M. McLean, in *Proc. 6th Int. Symp. on Superalloys*, D.Duhl et al. ed., Metall. Soc. A.I.M.E., 1988, 683.

Characterisation of Damage Accumulation During the Creep Deformation of CMSX-4 at 1150 °C

D.C. COX, C.M.F. RAE and R.C. REED

University of Cambridge/Rolls-Royce University Technology Centre, Department of Materials Science and Metallurgy, Pembroke Street, Cambridge CB2 3QZ, UK

ABSTRACT

The microstructure of <001>-oriented CMSX-4 superalloy single crystals crept under conditions of 1150° C and 100 MPa has been characterised using optical, scanning and transmission electron microscopies. Under these conditions, rafting of the γ structure is complete within the first 10 hours. At this stage and for a considerable time thereafter the creep strain rate decreases with increasing strain, implying a creep hardening effect. This seems to be associated with a reduction in the number of vertical γ' channels, and the formation of cellular dislocation networks in the remaining γ. Ultimate failure is associated with creep cavitation which occurs predominantly around casting porosity, although occasionally this is observed to take place at topologically close-packed phases. It is emphasised however that such creep cavitation is highly localised, being present only in the region of the specimens where necking occurs. Under the conditions examined, this instability is associated when a critical strain ε^* of (0.7±0.3)\% is reached; failure occurs within a few 10's of hours thereafter.

1 INTRODUCTION

Since the pioneering work of Versnyder in the 1970s, it has become commonplace to manufacture the turbine blading of modern aero-engines from nickel-base superalloys in single crystal form, often with intricate channels included so that cooler air can be forced to flow within and along the blades during operation. This situation represents a major challenge to the materials engineer, since there is a need for accurate and reliable models for the variation of accumulated creep strain with temperature, time and applied stress. Fortunately, much progress has been made in this regard and the behaviour of single crystals at temperatures below 1000°C is now reasonably well understood. At temperatures between 850°C and 1000°C, the loading of all known <001>-oriented superalloy single crystals yields a creep strain rate which increases monotonically with creep strain, there existing little evidence of a steady-state regime.[1,2,3] This strain softening behaviour has been termed tertiary creep,[4,5] and is associated with a proportionality of the creep strain rate and the accumulated creep strain;[6,7] the mode of deformation is octahedral slip with {111}<1–10> creep dislocations gliding and climbing around the γ' particles, which remain largely intact.[8] At lower temperatures, particularly in the vicinity of 750°C and provided that the threshold stress for γ' particle cutting is exceeded, a considerable amount of primary creep can occur *e.g.*[9,10] This is commonly associated with a macroscopic deformation of the form {111}<1–12>.[11]

In contrast, our understanding of the creep deformation behaviour of single crystal superalloys at temperatures in excess of 1000°C is incomplete, and in this high temperature regime some important questions remain unanswered. For example, what are the factors which give rise to the morphological instability of the γ' microstructure which has become known as the rafting effect?[12,13] To what extent does rafting affect the creep strain evolution, if at all? And what are the damage mechanisms occurring at the very late stages of creep which lead to ultimate failure of the material? The purpose of this paper is to address some of these outstanding isssues by considering the creep deformation behaviour of <001>-oriented CMSX-4 single crystals at very extreme conditions of temperature and stress: 1150°C and 100 MPa. The major aim is to further our understanding of the microstructural degradation mechanisms occuring at various points along the creep curve. The information reported in this paper is of use to those interested in designing models for the creep deformation of single crystal superalloys which are based on the general principles of damage mechanics *e.g.*;[14] there is a need to ensure that the expressions are used correctly to reflect the forms of creep damage which are occurring.

2   EXPERIMENTAL DETAILS

2.1   MATERIAL AND SPECIMEN PREPARATION

Creep strain testpieces of diameter 5.5 mm and gauge length 28 mm were machined from CMSX-4 single crystal material provided by Rolls–Royce plc, in the form of 1 cm diameter cast rods in the fully heat-treated condition. The chemical composition of the CMSX-4 material is given in Table 1. The axes of all rods were shown to lie within 10° of <001> by the indexing of back-reflection Laue patterns using the SCORPIO method.[15]

Table 1   Alloy composition, in wt% for CMSX-4.

Ni	Cr	Al	Ti	Mo	W	Ta	Co	Re	Hf
bal.	6.4	5.6	1.0	0.6	6.4	6.5	9.7	3.0	0.1

Creep strain testing was carried out using 20 kN constant load creep testing machines; such testing was compliant with the British Standard UDC 629.7,[16] to which the interested reader is referred. The specimens were tested at 1150°C and 100 MPa and some were interrupted at various amounts of accumulated creep strain. Since the tests were carried out at temperatures beyond 1000°C, some special precautions were taken. Platinum wound furnaces were used. Both pull rods and extensiometers were machined from creep resistant MA956 oxide dispersioned strengthened mechanically alloyed material; the adaptors were machined from MA754 and MA758.

2.2   ANALYSIS BY SCANNING AND TRANSMISSION ELECTRON MICROSCOPIES

Specimens were examined by scanning electron & transmission electron microscopies (SEM & TEM). For SEM analysis the creep specimens were sectioned on a {001} plane which contained the tensile axis. The surfaces were then polished according to standard metallurgical practice to a 1 μm finish, and then for a further twenty minutes with colloidal silica and etched electrolytically (6V dc) using a mixture of 12 ml perchloric, 47 ml sulphuric and 41 ml nitric acids. Microstructural examination was then carried out using a JEOL 6340 field-emmission gun scanning electron microscope (FEGSEM) using both secondary and backscattered electron signals.

For TEM analysis discs were cut from the specimens at low index orientations. After thinning by mechanical abrasion, these were electropolished at 50 volts in a solution of 10% perchloric acid in acetic acid. The resulting foils were examined in a JEOL 2000CX transmission electron microscope operating at 200 kV.

3 RESULTS FROM CREEP TESTING

Figure 1 illustrates the creep curves generated at 1150°C at an applied stress level of 100 MPa. The identities of the various samples, some of which were not taken to failure, are also given. The following observations can be made: (i) initially a small amount of strain is accumulated but there exists a plateau in which the creep strain does not vary strongly with time; (ii) at later times, the creep strain increases rapidly, with rupture eventually occurring and (iii) the time at which the creep rate starts to accelerate, i.e. the time to rupture is not well defined and varies by 100 hours for the samples tested. The behaviour is best examined via a plot of strain rate versus strain as shown in Fig. 2. Initially, there exists a 'decreasing strain rate' regime with the strain rate eventually reaching a minimum which occurs within the creep plateau seen in Fig. 1. Thereafter the creep strain rate increases dramatically in what will be referred to as the 'increasing creep rate' regime. As can be seen, the creep experiments have indicated that the maximum creep rate is reached consistently at a critical creep strain ε^* of (0.7±0.3)%. In fact, examination of our results from a wider range of testing conditions has indicated that this figure is quite insensitive to the magnitude of the applied stress and temperature.[17]

To understand the reasons for this behaviour, microstructural analysis is required so that the underlying mechanisms can be understood and rationalised. This is presented in the following section.

4 CHARACTERISATION OF CREPT MICROSTRUCTURES

Figure 3 shows the microstructure of sample A as seen via scanning electron microscopy (SEM). Such analysis confirmed that under these testing conditions the evolution of the γ/γ′ microstructure from cuboidal to plate-like structure takes place at an early stage of creep deformation, as observed in;[18] in the present case well-developed rafts have already formed after 10 hours. Further SEM microscopy on samples B and C confirmed that substantial change in the rafted structure occurs only in the very late stages of deformation and very close to the final fracture. This is illustrated in Fig. 4, which shows the variation in microstructure of sample C at several distances from the fracture surface. It can be seen

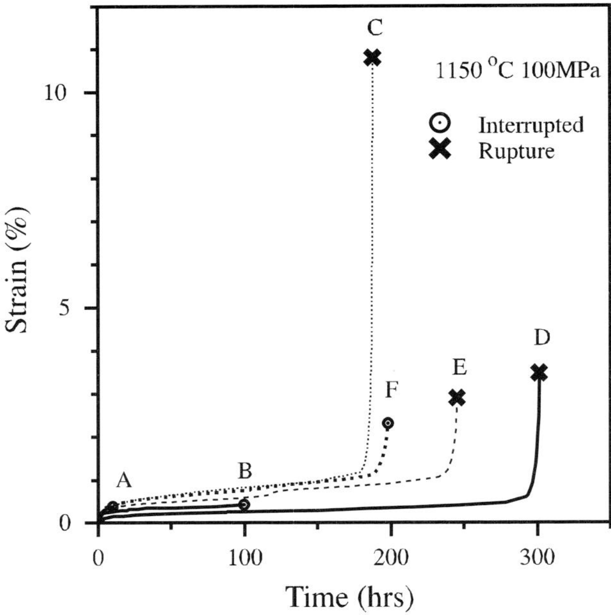

Fig. 1 Experimental creep curves for CMSX-4 tested at 1150°C and 100 MPa.

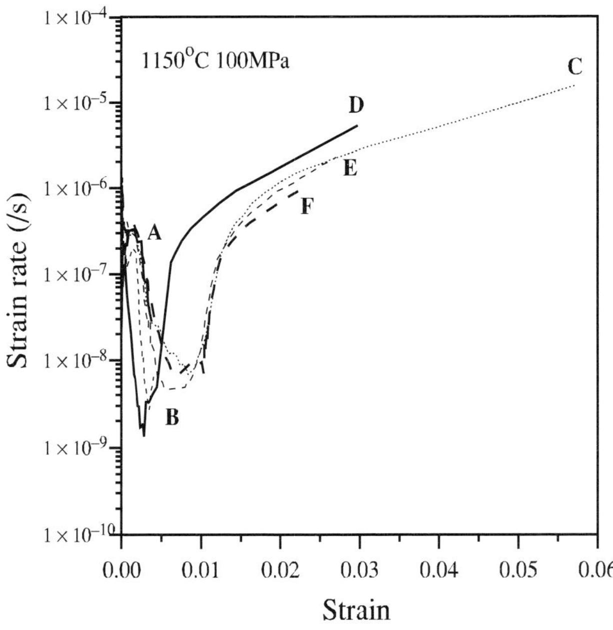

Fig. 2 Plot of creep rate vs creep strain for CMSX-4 tested at 1150°C and 100 MPa.

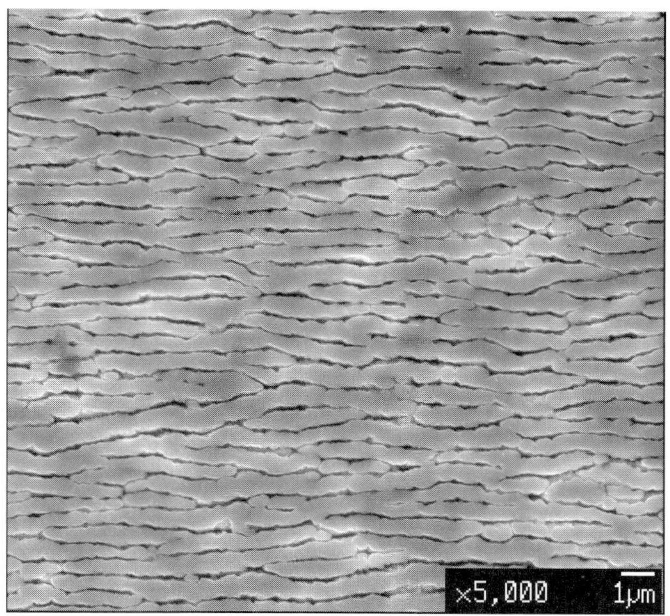

Fig. 3   Secondary electron image of sample A tested at 1150°C and 100 MPa, interrupted after 10 hours.

from this figure that the γ' near to the fracture has been broken up and the precipitates have rotated to approximately 45° to the tensile axis, whereas by 4 mm from the fracture the microstructure is essentially unchanged and the γ' rafts are still complete.

Sample C also reveals wide γ channels filled with significant amounts of coarse secondary γ' that is not observed in samples A and B. This is shown in detail in Fig. 5 where particles of secondary γ' of approximately 50 nm can be seen. On cooling from 1150°C the volume fraction of γ' increases substantially: Figure 6 is a plot of the calculated volume fraction of γ' as a function of temperature obtained using the commercial software package Thermo-Calc.[19] It can be seen that at the test temperature of 1150°C the γ' volume fraction has reduced to approximately 45% from a usual volume fraction of around 70% found in this alloy at room temperature. It is believed that only sample C exhibits this secondary γ' because it cooled more quickly from the test temperature following fracture during the creep test. By comparison samples A and B were interrupted prior to failure and cooled more slowly, allowing most of the excess γ' in solution to re-precipitate onto the existing γ' rafts.

It was also observed that both samples B and C exhibited the formation of topologically close-packed (TCP) phase particles. Figures 7a, and b show typical TCP particles observed in samples B, and C respectively in the SEM, using back-scattered signals to obtain atomic number contrast. As can be seen the TCP phase particles do not grow substantially in size after 100 hours, but act as sites for the nucleation of pores in addition to the casting pores present in the initial structure. Although the casting pores outnumber the TCP pores by

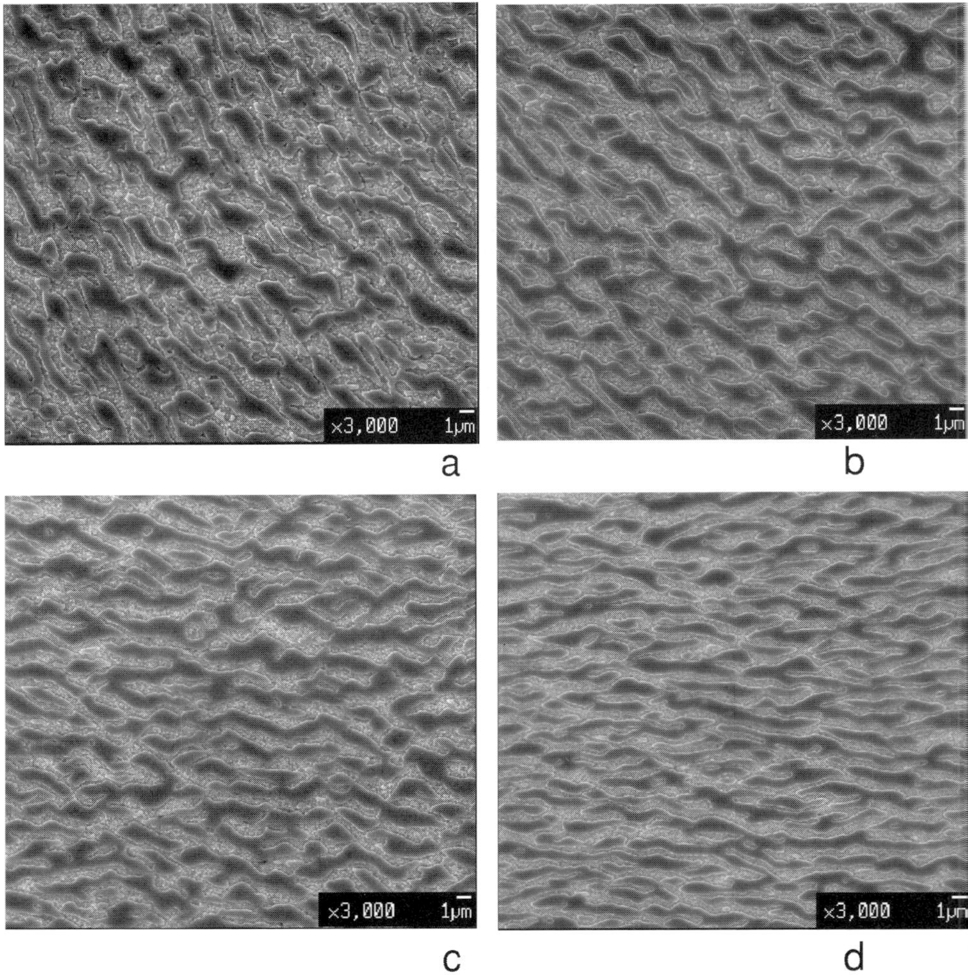

Fig. 4 Secondary electron micrographs from sample C showing the break-up of the γ' rafts near to the fracture surface, at 200 μm (a), 0.5 mm (b) 2 mm (c) and 4 mm (d).}

25:1, they form predominantly in the tungsten and rhenium rich inter-dendritic regions. Initially in sample A, the casting pores are spherical, but as they grow many appear to facet as shown in Fig. 8a. Others develop conical 'tunnels' extending along the <100> directions perpendicular to the tensile axis; Fig. 8b. This has the effect of increasing the cross-sectional area of the pores.

Examination of Sample A by transmission electron microscopy (TEM), see Fig. 9a, revealed that well-developed networks of dislocations had already formed after 10 hours. Further analysis of samples B & C confirmed that with increasing deformation the dislocation networks become more regular, but that the cell spacing remains largely unchanged at ~50 nm, see Figs 9b & c. There is a gradual transition in the network structure from the somewhat irregular configuration resulting from dislocations in the γ becoming trapped at

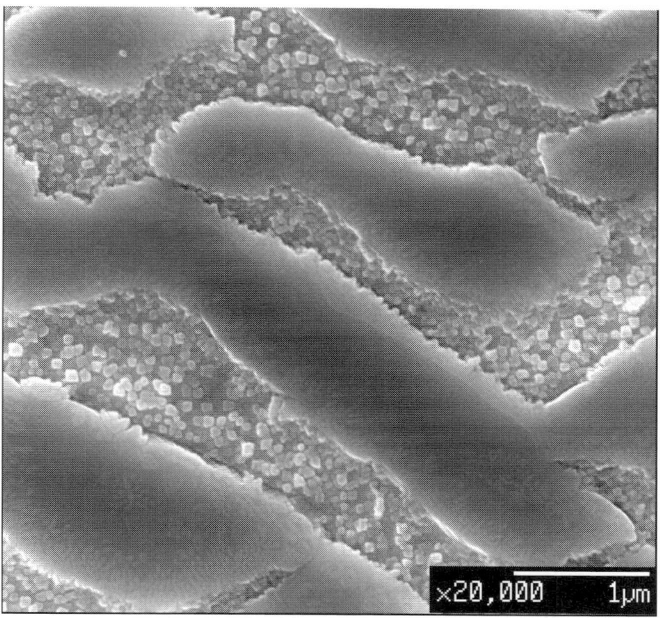

Fig. 5 Secondary electron micrographs from sample C showing the precipitation of secondary γ′ on cooling.

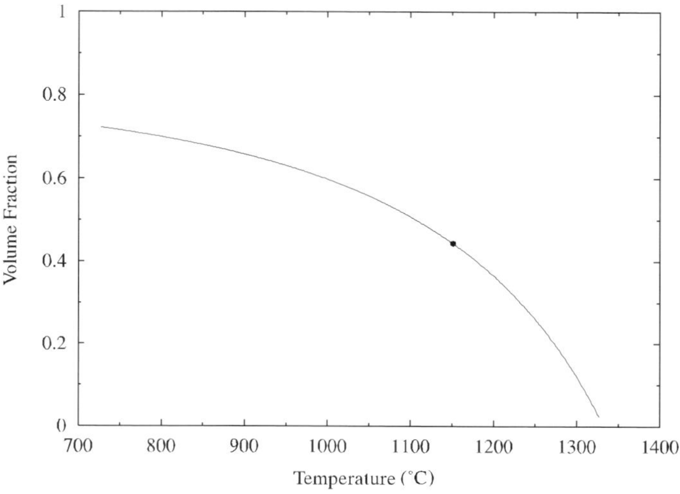

Fig. 6 Plot of volume fraction of γ′ as a function of temperature. The filled circle denotes the temperature of the creep tests undertaken in the present study.

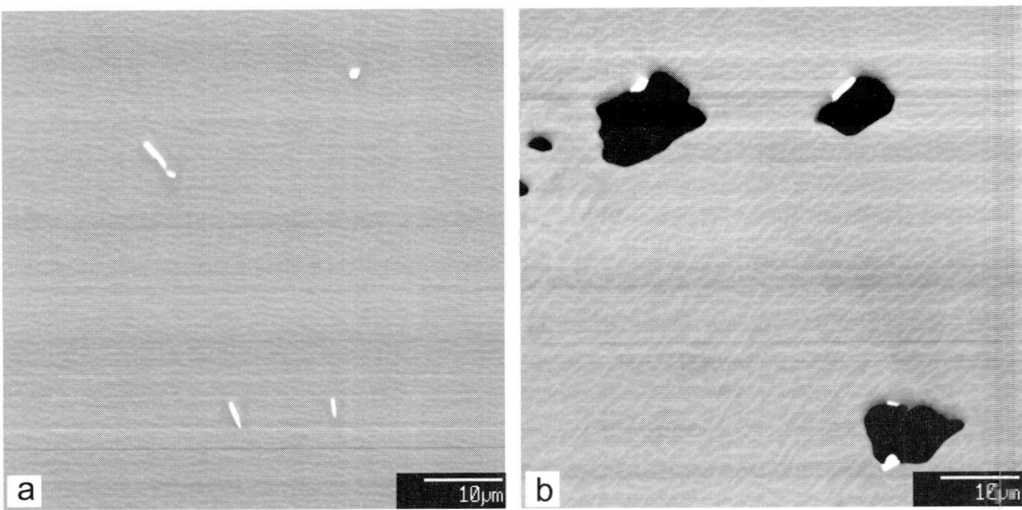

Fig. 7 Backscattered electron images showing TCP phase formation and porosity in sample B (100 hours) and Sample C (170 hours), (a and b respectively).

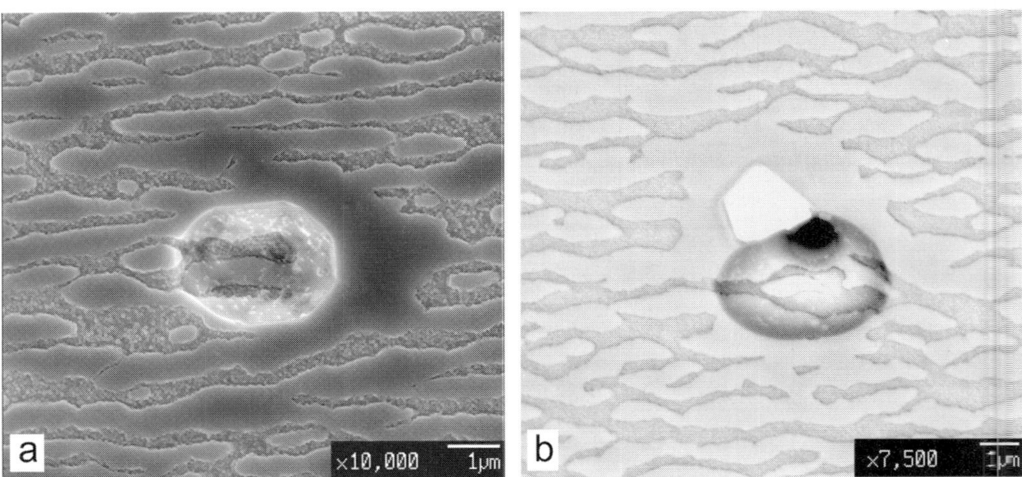

Fig. 8 Secondary electron image of casting pore (a), and backscattered electron image of TCP pore sample (b) showing growth of both types observed in sample C (170 hours).

the γ/γ' interface to the 'equilibrium mismatch configuration' of octahedra and diamonds which is described in.[20,21] The nature of the network varies with the orientation of the γ/γ' interface but a good example of the equilibrium network can be seen in the bottom left of Fig. 9c. The contrast from the six different dislocations which make up the network in the imaging conditions of Fig. 9 are shown in bold in the schematic illustration in Fig. 10. Activity in the γ phase decreases as the deformation proceeds; for example at 10 hours this was found in all areas but in sample C which was taken to fracture, dislocations in either phase outside the γ/γ' interface networks are rare. The defects which were identified included single a/2<110> dislocations, pairs of similar dislocations separated by anti-phase boundaries (APBs), dislocations of Burger's vector a<100>[22] as well as complex combinations of these. It is clear from the sequence of specimens A, B and C that outside of the fracture zone dislocation activity decreases and networks become more stable with increasing strain.

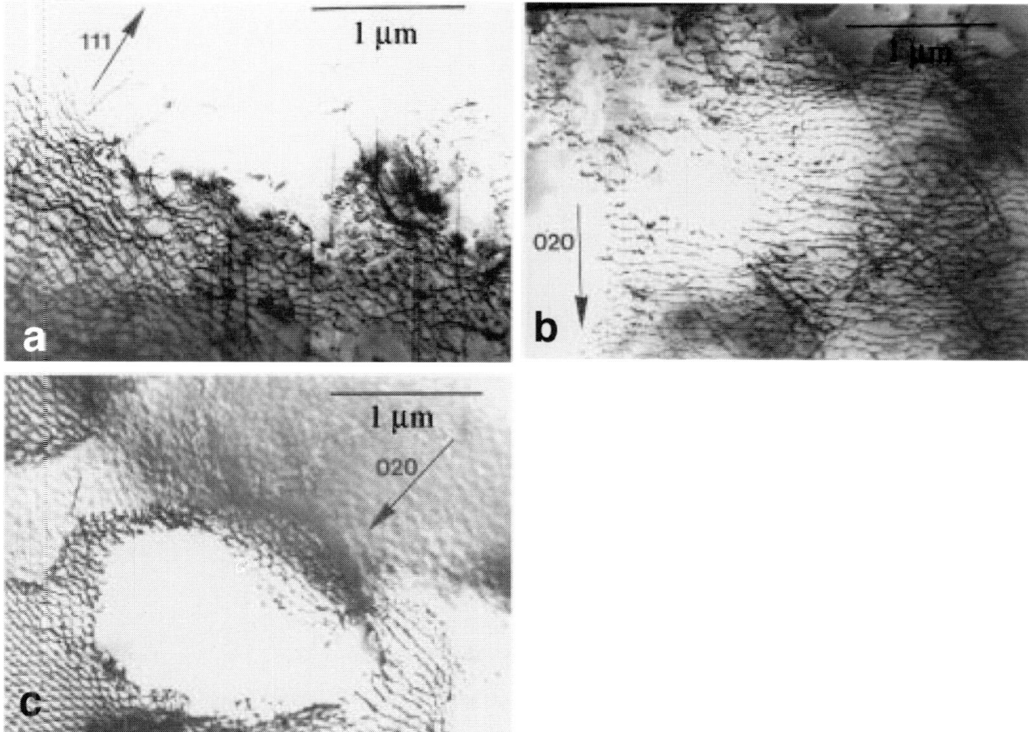

Fig. 9   TEM micrographs of the CMSX-4 material crept at 1150°C and 100 MPa showing the evolution of stable dislocation networks in samples A (10 hours), B (100 hours) and C (170 hours), a, b and c respectively.

What are the mechanisms responsible for the rapid increase in the creep strain rate and the onset of fracture? In the vicinity of the fracture surface the deformation is severe, but a build of dislocation activity does not appear to be the trigger for the increase in strain rate immediately prior to failure. Under the conditions examined, we have noted that the rapid increase in the creep strain in the 'increasing creep strain rate' regime is associated with highly localised deformation in a region extending 4 mm either side of the fracture surface, Fig. 11. Within this highly strained zone the deformation is associated with cavitation of both the casting porosity and porosity nucleated from the TCP phases, resulting in the tearing of pores perpendicular to the tensile axis, see Fig. 12. In addition the rafts undergo a change in morphology from lying perpendicular to the tensile axis to lying about 45° to

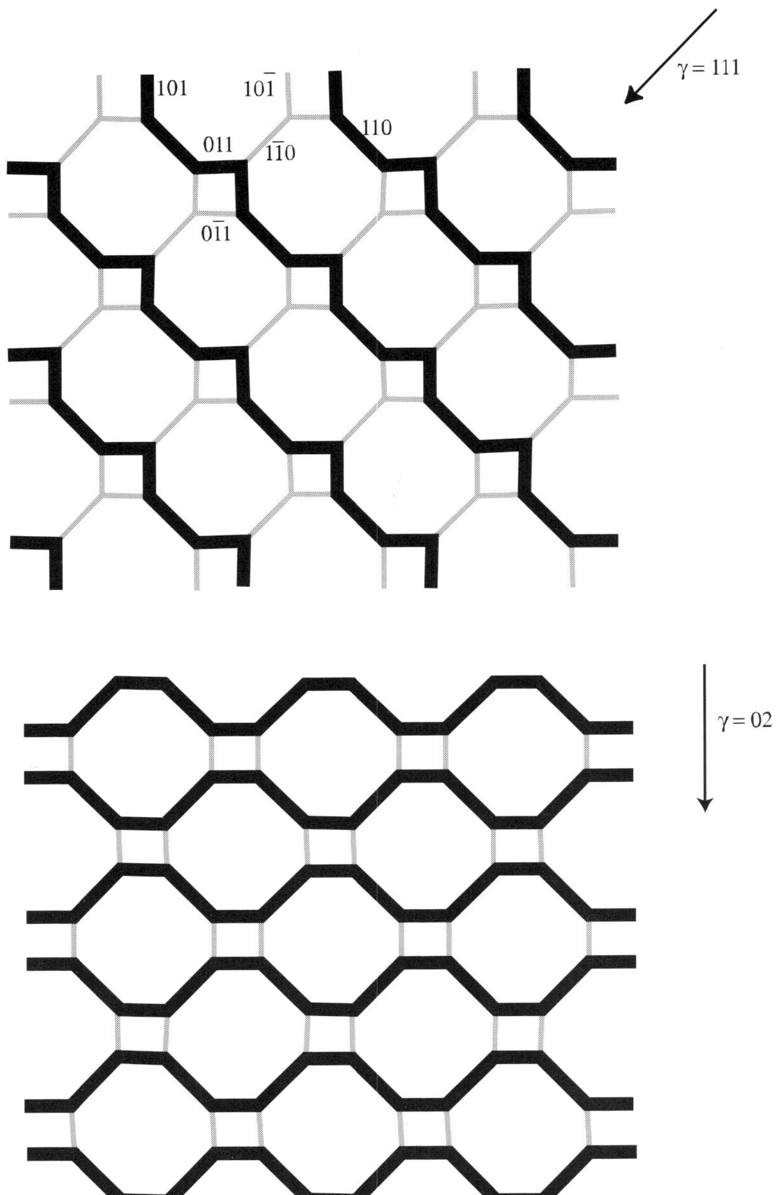

Fig. 10   Schematic illustration of the imaging conditions which give contrast from the six different dislocations which make up the network configuration.

the axis. This occurs without the rotation of the matrix, the orientation of which remains aligned approximately with the [001] axis as shown by the orientation of the secondary precipitates in Fig. 5. TEM shows that this is achieved by intense dislocation activity in the γ/γ′ during the final localised deformation but appears to be triggered by the increased stress resulting from the growth of porosity.

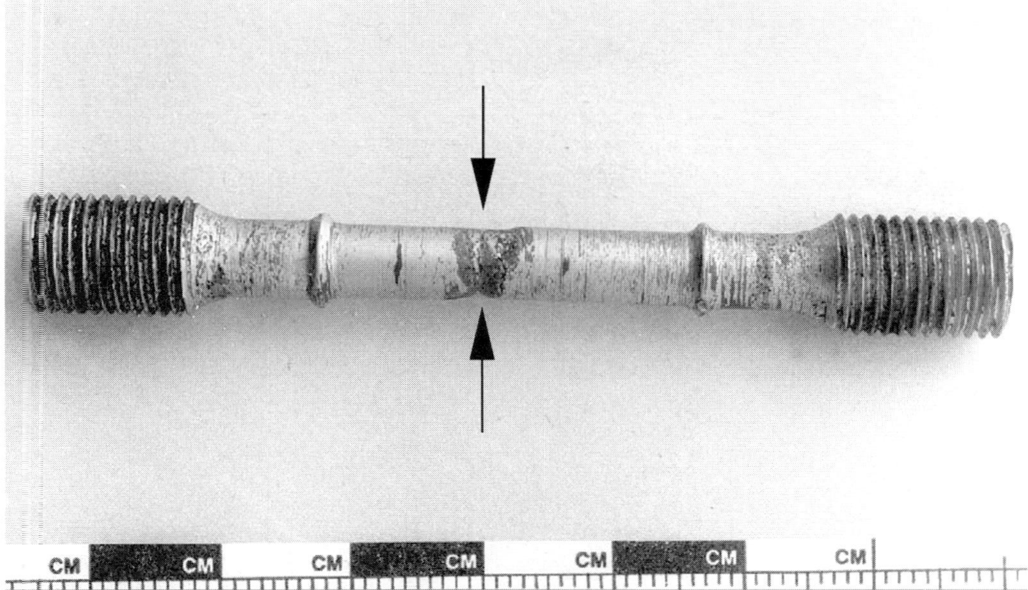

Fig 11 Photograph of specimen F crept at 1150°C, and 100 MPa interrupted just before rupture at 2.32% strain after 198 hours.

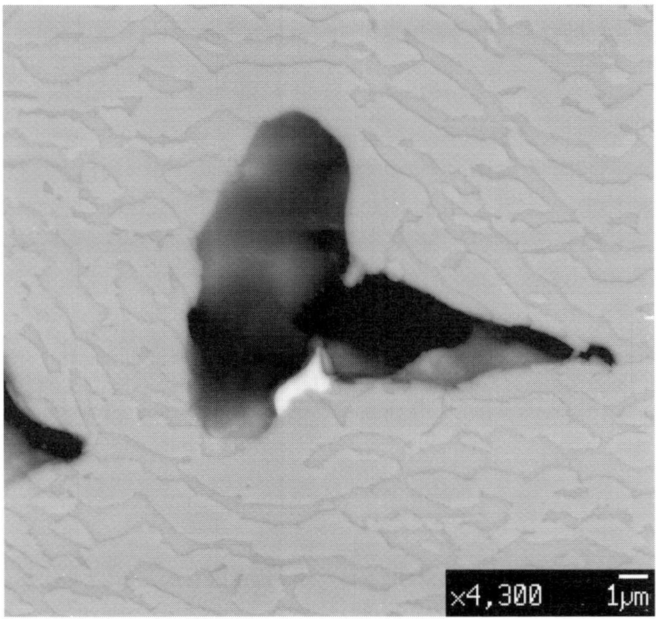

Fig. 12 Backscattered electron image showing creep cavitation of porosity in sample C.

Examination of the specimen using optical microscopy revealed large variation in the number of pores along the gauge length of sample F. Fig. 13 shows the total number of pores counted across each 2 mm section of the gauge length of sample F plotted together with the number of pores observed to be torn. This shows that the total density of pores can vary by a factor in excess of four along the gauge length of the sample, but the region where pores were found to be tearing coincides with the highest local density of porosity. Fig. 14 plots the fraction of all pores creep-cavitated, as a function of the gauge length of sample F. This figure clearly demonstrates that the damage is restricted to a narrow region around the fracture site. Away from the fracture zone but within the central section of the

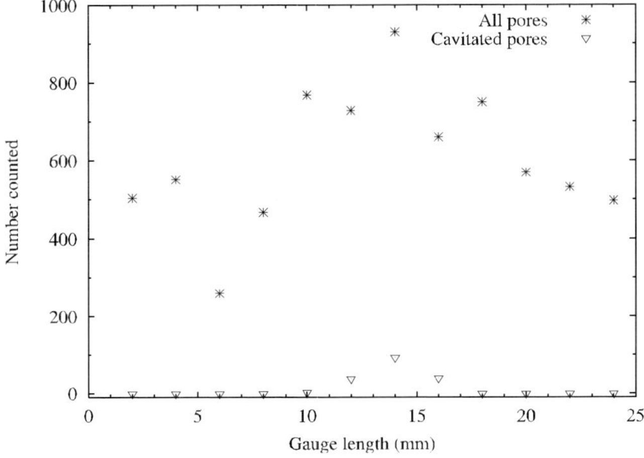

Fig. 13   Plot of total and cavitated number of pores counted as function of sample gauge length in sample F.

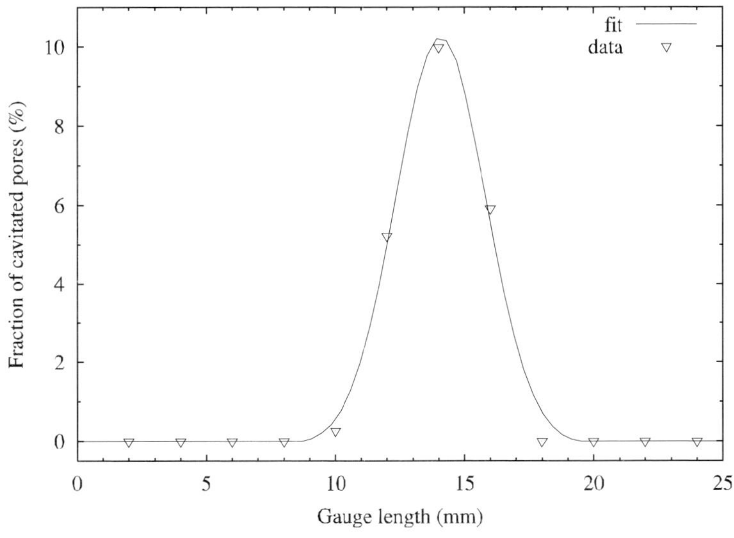

Fig. 14   Plot of fraction of pores cavitated as function of sample gauge length from sample F.

gauge length the microstructure shows a regular rafted structure with little or no dislocation activity and pores which show no sign of tearing, Fig. 8a & b and Fig. 9c.

5 DISCUSSION

These observations support the view that the reduction of the creep rate in the 'decreasing creep rate' regime arises as a consequence of the rafting effect. It appears that rafting of the γ/γ' structure prevents the glide/climb of $\{111\}<1-10>$ creep dislocations in the γ past the γ' particles. As creep dislocations of a given $\{111\}<1-10>$ form are introduced into the structure, the stress field from the applied load and the misfit stresses interact with the internal stress fields of the dislocations themselves, such that a build-up of back stresses occurs. At this point other $\{111\}<1-10>$ forms become favored with the cellular dislocation structures eventually being formed.[20,21] The conclusion that the 'decreasing creep rate' regime arises because of the formation of the rafted structure is supported further by the quantitative analysis presented in.[17] The observation that the γ' rafts contain little or no dislocations and remain as continuous unbroken rafts except in the region very close to fracture in the necked region, where the local strain will be very high compared to the overall recorded strain, shows that the γ' precipitates continue to act as barriers to the movement of dislocations right up to the onset of fracture.

The observation that creep cavitation at casting porosity in the region of the fracture surface is consistent with the variation in the time to rupture of the samples C, D, E & F, see Fig. 1, since it is to be expected that each test-piece will have different statistical distributions of pores. The implication of these observations is that a reduction of the porosity would have a beneficial effect on the time to rupture. In the case of the casting porosity this could be achieved by modification of the conditions pertinent to the casting process or else by subsequent hot iso-static pressing of the as-cast structure. However it should also be emphasised that the formation of TCP phases and the additional porosity associated with them occurs as a result of the local alloy chemistry. Either changes to alloy composition or increased solution heat treatment duration to re-distribute the TCP forming elements more homogeneously would be required to reduce this type of porosity. If one assumes that the cavitation of porosity is indeed responsible for the ultimate creep rupture it is suggested that experiments be carried out to test whether the removal or reduction of porosity can extend ultimate creep life.

The observations described give an insight into the microstructural mechanisms in operation throughout the creep of CMSX-4 at 1150°C and 100 MPa. The evidence suggests the following scenario: (i) the γ' particles coalesce within the first ten hours to form rafts which in turn lead to the 'decreasing creep rate regime' via the blocking of dislocation movement; (ii) during the period of the long stable creep rate, equilibrium networks of dislocations form and overall dislocation activity falls; (iii) also during this period TCP phase particles form in the sample in the rhenium and tungsten rich dendrite cores; (iv) formation and growth of TCP porosity and growth of casting porosity leads to regions that contain high local pore density experiencing local stresses higher than the overall stress the sample is experiencing; (v) the growth of the porosity, and subsequent reduction in cross sectional area, increases the local stress between the pores above the threshold required for the movement of dislocations through the γ' rafts, allowing the onset of the 'increasing creep rate regime'; (vi) finally with dislocations passing through the γ' and the accumulated local strain accelerating, the test bar fails within a few hours as the γ' rafts break down and pores tear and coalesce increasing local strain further.

6 SUMMARY AND CONCLUSIONS

The following conclusions can be drawn from this work:

1. At 1150°C and 100 MPa, rafting occurs rapidly and is completed within 10 hours; whilst this is occuring and for a considerable period thereafter the creep rate *decreases* with increasing strain. Thus a creep hardening effect is operative.
2. This creep hardening is attributed to the reduction in the number of vertical γ channels, which is caused by the rafting effect. The glide/climb of a/2<1–10> dislocations is hindered and stable networks of dislocations are formed.
3. Once a critical strain ε^* of $(0.7\pm0.3)''$ is reached, the creep strain increases dramatically and failure occurs within a few 10's of hours.
4. Ultimate failure is highly localised. The damage mechanism which is prevalent at this stage is creep cavitation at casting porosity.
5. There is a need to clarify the relationship between the various contributions to the macroscopic creep rate occurring in the increasing creep rate regime, and the factors which determine its initiation.

7 ACKNOWLEDGEMENTS

The authors acknowledge financial support from the Engineering & Physical Sciences Research Council (EPSRC), Rolls–Royce plc and the Defence Evaluation Research Agency (DERA). Helpful discussions with Bob Broomfield & Neil Jones of Rolls-Royce plc and Mike Winstone, Mike Henderson & George Harrison of DERA have taken place over the course of this project.

REFERENCES

1. F.R.N. Nabarro and H.L. de Villiers, *The Physics of Creep*. Taylor and Francis, 1995.
2. *Superalloys 1996*, edited by R.D. Kissinger, D.J. Deye, D.L. Anton, A.D. Cetel, M.V. Nathal, T.M. Pollock and D.A. Woodford eds, 1996.
3. P. Caron and T. Khan, 'Improvement in creep strength in a nickel-base single-crystal superalloy by heat treatment', *Mater. Sci. Eng.*, 1983, **61**, 173–184.
4. B.F. Dyson and M.McLean, 'Particle coarsening, σ_o and tertiary creep', *Acta metall.*, 1983, **31**, 17–27.
5. P.J. Henderson and M. McLean, 'Microstructural contributions to friction stress and recovery kinetics during creep of the nickel-base superalloy in 738lc', *Acta metall.*, 1983, **31**, 1203–1219.

6. R.N. Ghosh, R.V. Curtis and M. McLean, 'Creep deformation of single crystal superalloys – modelling the crystallographic anisotropy', *Acta metall. mater.*, 1990, **38**, 1977–1992.
7. L.M. Pan, I. Scheibli, M.B. Henderson, B.A. Shollock and M. McLean, 'Asymmetric creep deformation of a single crystal superalloy', *Acta metall.*, 1995, **43**, 1375–1384.
8. T.M. Pollock and A.S. Argon, 'Creep resistance of CMSX-3 nickel-base superalloy single crystals', *Acta metall.*, 1992, **40**, 1–30.
9. G.R. Leverant, H.B. Kear and J.M. Oblak. Creep of precipitation-hardened nickel-base alloy single crystals at high temperature. *Metall. Trans.*, 1973, **4**, 355–362.
10. V. Sass, U. Glatzel and M. Feller-Kniepmeier, 'Creep anistropy in the monocrystalline nickel-base superalloy CMSX-4', in *Superalloys 1992*, R.D. Kissinger, D.J. Deye, D.L. Anton, A.D. Cetel, M.V. Nathal, T.M. Pollock, and D.A. Woodford, eds, pages 283–290, TMS, 1996.
11. N. Matan, D.C. Cox, P. Carter, M.A. Rist, C.M.F. Rae and R.C. Reed, 'Creep of CMSX-4 superalloy single crystals: Effects of misorientation and temperature', *Acta metall.*, 1999, **47**, 1549–1563.
12. F.R.N. Nabarro, 'Rafting in superalloys', *Metall. Trans. A*, 1996, **27A**, 513–530.
13. N. Matan, D.C. Cox, C.M.F. Rae and R.C. Reed, 'On the kinetics of rafting in CMSX-4 superalloy single crystals', *Acta metall.*, 1999, 47 2031–2045.
14. R.W. Evans and B. Wilshire, in *Structural Materials: Engineering application through scientific insight, the Donald McLean symposium*, The Institute of Materials, London, 1996.
15. M.J. Goulette, P.D. Spilling and R.P. Arthey, 'Cost effective single crystals', in *Superalloys 1984*, M. Gell and C.S. Kortovich et al., editors, TMS, 1984, 167–176.
16. British Standard Udc 629.7, The British Standards Institution, 1965.
17. R.C. Reed, N. Matan, D.C. Cox, M.A. Rist, and C.M.F. Rae, 'Creep of CMSX-4 superalloy single crystals: effects of rafting at high temperature', *Acta metall.*, in press.
18. A. Royer, P. Bastie and M. Veron, 'In-situ determination of gamma prime phase fraction and of relations between lattice parameters and precipitate in ni-based single crystal superalloys', *Acta metall.*, 1998, 46 5357–5368.
19. B. Janssen B. Sundman and J.O. Andersson, 'The thermo-calc databank system', *Calphad*, 1985, 9(2), 153–190.
20. D.F. Lahrman, R.D. Field, R. Darolia and H.L. Fraser, 'Investigation of techniques for measuring lattice mismatch in a rhenium containing nickel-base superalloy', *Acta metall.*, 1988, **36**, 1309–1320.
21. R.D. Field, T.M. Pollock and W.H. Murphy, in *Superalloys 1992*, S.D. Antolovich, R.W. Stusrud, R.A. Mackay, D.L. Anton, T. Khan, R.D. Kissinger, and D.L. Klarstrom, eds, TMS, 1992, 557–566.
22. G. Eggeler and A Dlouhy, 'On the formation of <010>-dislocations in the gamma prime phase of superalloy during high temperature low stress creep', *J. of Metals*, 1997, **45**, 4251–4262.

Coating Life Assessment

J.R. NICHOLLS

Cranfield University, Cranfield, Bedford MK43 0AL, UK

ABSTRACT

Coating systems to protect high temperature components are now an integral part of gas turbine engine design in many current generation aero, industrial and marine gas turbines. Future engines will become increasing reliant on such protection systems as higher firing temperatures are used to improve both engine performance and efficiency. Thus accurate models will be required of the coating degradation processes to allow the life of such coated components to be predicted.

Coating life at high temperatures is determined by a complex interplay of many degradation mechanisms; including primarily oxidation, spallation, hot corrosion, erosion and foreign object damage, interdiffusion and thermomechanical fatigue damage.

Models for the life prediction of coatings, based on high temperature oxidation and spallation, are largely empirical or semi-empirical, and are based on laboratory furnace tests, burner rig tests etc, which are then related through 'experience' factors to the life of the engine. Models for the hot corrosion of coatings have been developed, but these need a knowledge of the rates of salt deposition within the engine if accurate corrosion rate predictions are to be made.

Similarly, models for the erosion behaviour of coatings are under development, but now the complex interaction between oxidation/corrosion rate and the erosive particle flux must be addressed and this will determine whether the coating erosion behaviour is dominated by the substrate mechanical properties or the properties of the oxide scale.

The above mechanisms have addressed degradation at the coating/gas interface. However, as the temperature rises in the engine diffusion across the coating/substrate interface can play an increasingly important role. These diffusion processes can lead to a reduction in scale forming elements, 'coating exhaustion', and thus lead to early coating failure. Compositional changes at this interface can also compromise the structural integrity of the substrate.

This paper will review the current status of coating life assessment models in each of these areas, identify current deficiencies and recommend where future modelling work should be directed.

1   INTRODUCTION

Coating systems to protect high temperature components are now an integral part of modern gas turbine engine design. This applies whether the engine is an aero-marine or industrial power unit and this trend is expected to continue in future engine design, as higher firing temperatures are used to improve both engine performance and efficiency. Thus coating technologists will have to be increasingly innovative in the design of new protection systems, often deviating from or extending traditional design concepts such that the coating system becomes an integral part of modern component design, 'the designed in coating', rather than a functional add-on aimed at extending component life.

Such design concepts require that the failure modes of coating systems are known and that models for coating life exist, such that future components can be designed with the best surface protection systems for the application. This requires a knowledge of surface related coating performance, i.e. oxidation, corrosion and erosion by solid particulates, together with substrate interdependent properties including metallurgical stability, interdiffusion and inter-related thermomechanical properties.

This paper will review the most important of the coating failure modes and the current status of life assessment models in each of these areas, with the aim that at some future time all of these models may be integrated as part of a suite of tools for advanced coating design.

2 HIGH TEMPERATURE SURFACE ENGINEERING – THE DESIGN OF OXIDATION, CORROSION AND THERMAL PROTECTION SYSTEMS

The recognition that future engine designs will be increasingly reliant on coating systems as an integral part of component design, has prompted much research in the design of oxidation, corrosion and thermal protection coatings over the last three decades. Much of this information is well documented in reviews and critiques in this area.[1–8]

The many types of high temperature coatings that are currently in use, broadly fall into one of three classes.

2.1 DIFFUSION COATINGS/SURFACE TREATMENTS

These are treatments in which the surface properties of a component are modified by surface enrichment with elements that are desirable for the production of a slow growing, stable protective oxide when in service. To this end aluminising, chromising and siliconising have been widely used. If intermetallic compounds result from such diffusion treatments, then aluminide, chromide and silicide based coating are produced.

Much interest over recent years has focused on modified aluminide coatings,[6–8] particularly the precious metal aluminides, based on the platinum group metals, for high temperature oxidation and bond coat applications and the silicon-aluminides and chromium-aluminides for type II hot corrosion resistance.

2.2 OVERLAY COATINGS – THE MCrAlY SERIES

The deposition of custom designed corrosion resistant coatings onto a component are known as overlay coatings. This concept was first reported by Talboom,[9] who used electron beam physical vapour deposition (EB-PVD) to deposit oxidation/corrosion resistant CoCrAlY coatings onto turbine hardware. Since this milestone paper, many MCrAlX type coatings have been produced and are available commercially.

In the MCrAlX series, M is the base metal, usually Ni, Co or a combination of these, chromium contents lie in the range 18–40 wt%Cr; aluminium contents lie in the range 3–20 wt%Al; and X is one, or a combination, of reactive elements added to stabilise oxide scale growth, slow short circuit diffusion, and/or getter tramp elements that are present in the coating or diffuse from the substrate both during manufacture and service. Yttrium (Y) is the most widely used active element addition and was included within the original CoCrAlY coating proposed by Talboom; this had the composition Co25Cr14Al0.5Y.[9]

To date, a wide variety of processing routes have been used to apply MCrAlY coatings, these include electron beam physical vapour deposition, plasma spraying (air; APS; vacuum, VPS, or low pressure plasma spraying, LPPS), thermal spraying (high velocity oxyfuel, HVOF, or high velocity air fuel, HVAF) or electro-plating. Overlay coatings are generally more expensive and thicker than diffusion coatings but offer a greater flexibility in alloy design.

2.3   THERMAL BARRIER COATINGS (TBCs)

Thermal barrier coatings are a multilayered coating system, designed to lower the metal surface temperature in conjunction with advanced cooling technologies.[1, 2, 4, 5] The coating consists of an outer ceramic top coat (usually ZrO_2–8wt%Y_2O_3) overlayed onto an oxidation resistant bond coat (MCrAlY alloys or platinum aluminide diffusion coatings are widely used). The ceramic top coat can be deposited by plasma spraying or EB-PVD coating technologies with thicknesses between 200–400μm generally, although thicker coatings have been suggested for combustor and diesel engine applications.

This class of coating is capable of lowering metal surface temperatures by 40–120°C[5] and thus improve component durability.

The performance of TBC systems is critically dependent on the quality and state of the thermally grown oxide (TGO) at the bond coat ceramic interface. Failure of this oxide during thermal cycling will ultimately lead to loss of the TBC, the development of a local hot spot and then rapid degradation of the coating system locally as a result of oxidation, corrosion and interdiffusion of the coating with the substrate.

3   FACTORS THAT INFLUENCE COATING LIFE

Coating life at high temperatures is determined by a complex interplay of many degradation mechanisms. The degradation modes are a function of the engines operating conditions, the material from which the component is manufactured and the chosen coating system. Degradation modes common to superalloy hot section components include, to varying degrees:- oxidation, spallation, hot corrosion, erosion, foreign object damage (FOD), interdiffusion, creep, low-cycle thermo-mechanical fatigue (TMF), and high cycle fatigue (HCF).

During service, coatings degrade at two fronts – at the coating/gas path surface and at the coating/substrate interface. Degradation of the coating surface – at the coating/gas path interface – is a consequence of environmental attack and this depends primarily on the engines operating condition, particularly the metal surface temperature, the gas composition, whether deposition occurs (hot corrosion) and the level of particulates generated during engine operation. The duty cycle of the engine is also a prime factor as rapid thermal cycles can stress the surface oxides/coating systems leading to spallation. For thermal barrier systems this would lead to the loss of the TBC, while for environmental protection coatings this may lead to a more rapid consumption of scale forming elements or trigger the onset of hot corrosion reactions through scale fracture and the deposition of molten sulphate and chloride rich deposits.

Solid state diffusion, across the coating substrate interface, occurs at high temperatures and has become increasingly important in the modern generation of high performance engine. The higher operating temperatures result in enhanced diffusion and thus cause compositional changes at this interface, which may compromise the structural integrity of the component.

Furthermore, such diffusion processes can deplete the coating of critical scale forming elements resulting in early scale failure or can lead to the diffusion of substrate elements into the coating which may result in less protective oxides, poor oxide-interface adhesion and hence early scale/TBC spallation.

Mechanical compatibility is also an important issue. The role of oxide spallation, in both enhanced oxidation and TBC loss, has been eluded too. Maintenance of a protective oxide scale requires that the oxide remains adherent and is free of through thickness cracks. As all gas turbines will see severe themal cycles both during start up and shutdown, then the degree of match between the coefficient of thermal expansion (CTE) of the coating, the substrate, the surface scale and any thermal barrier – if used – becomes an important consideration.

Mismatch in CTE, between the coating and substrate can lead to coating cracking, crack propagation into the substrate and ultimate component failure. Mismatch in CTE between the coating and the thermally grown oxide can lead to oxide spallation or through thickness cracking with the loss of environmental protection and the likely loss of any thermal barrier coating if used. Thus strain accommodation mechanisms within the coating, coating cohesion, coating and oxide scale adhesion become important life cycle issues. The CTE mismatch, coupled with the oxide and coating mechanical properties – particularly creep and fracture resistance – are important factors determining the onset of mechanical failure within these coating systems.

4.0   COATING FAILURE MODES AND LIFE ASSESSMENT MODELS

4.1   ISOTHERMAL OXIDATION, CYCLIC OXIDATION AND OXIDE SPALLATION

Alloys/coatings which contain concentrations of aluminium or chromium in excess of 10 and 16 wt% respectively*, will in an oxidising atmosphere eventually form a scale in which alumina and/or chromia predominates. Both of which are protective, providing a barrier between the environment and the underlying alloy. Alumina is the slowest growing, is stable at temperatures above 1000°C, and is the oxide scale formed by most high temperature coating systems, for example the MCrAlY alloys and many diffusion aluminides*. Thus alumina forming alloys are attractive as high temperature oxidation resistant coatings. However, under hot corrosion conditions (see Section 4.2) alumina does not provide sufficient protection against the attack by alkali metal sulphates[7] and then coatings that are chromia or silica formers find increased usage.

When aluminium and chromium are present in combination within a corrosion resist-ant alloy then generally much lower aluminium contents result in the formation of a stable

---

*Intermetallic compounds will in general require high levels of aluminium to form stable alumina scales.

alumina scale; 4–5wt% Al generally being sufficient if the chromium level is above 15 wt%.[10–12] The stability of high temperature alumina scales as a function of alloy composition is illustrated in Fig. 1.[12, 13]

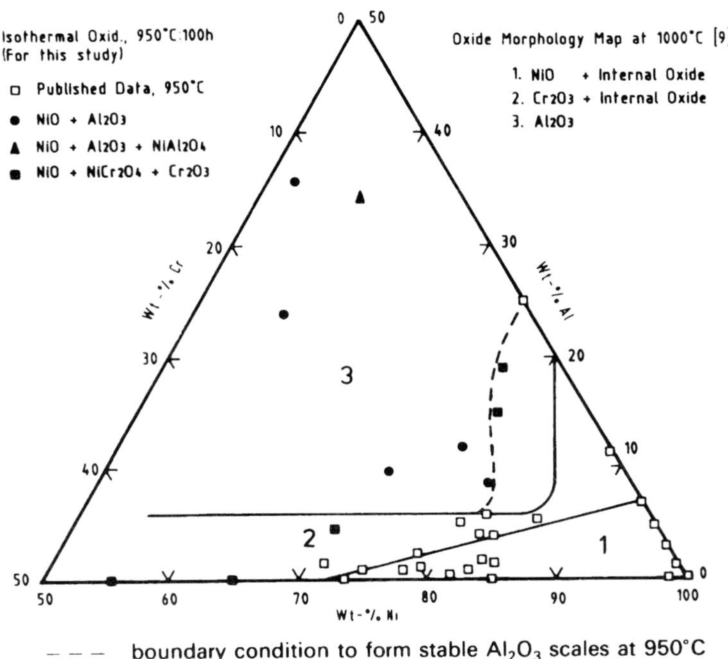

Fig. 1 Ternary map of stable protective oxides at 950°C and 1000°C as a function of alloy composition.[12, 13]

An implication of these studies is that the lifetime of high temperature coatings, under oxidising conditions, is governed by alloying element exhaustion. For example, as deposited a coating may have 25%Cr and 14%Al (the original Talboom CoCrAlY coating optimum composition). During service this grows an alumina scale, which may thicken, spall and repair throughout the components life. Ultimately the aluminium level will drop below 4–5wt% and then less protective oxides will form. The coating will then be rapidly consumed and the substrate attacked.

This introduces the concept of a 'critical aluminium reservoir' – 9–10wt% in the above example – which when consumed defines the life of the coating. One must further bear in mind that at elevated temperatures aluminium consumption can occur at both the coating/oxide interface, due to oxidation and corrosion, and at the coating/substrate interface, due to interdiffusion. Figure 2 presents a micrograph of a partly degraded MCrAlY coating, due to high temperature oxidation and interdiffusion with the substrate. The sample had been exposed for 1h at 1150°C, giving a β depletion zone of some 20μm. The degree of aluminium loss can be simply monitored by measuring the extent of β depletion in these MCrAlY coatings.

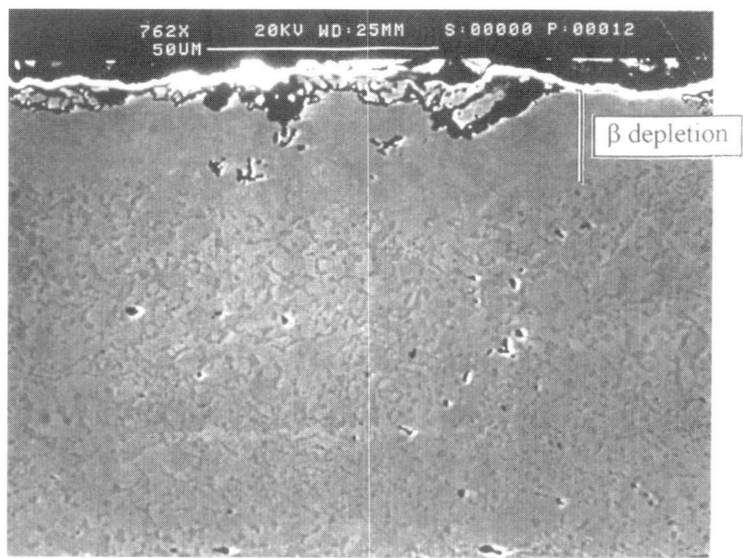

Fig. 2 β-phase depletion from an MCrAlY coating following 1 h at 1150°C.

A number of models, based on this alloy element depletion concept, are available in the literature.[14–16] All are largely empirical or semi-empirical and are based on curve fits (or computer fits) to laboratory furnace test data, burner rig tests etc., which are then related through 'experience' factors to the life of an engine or high temperature plant. Many of these models have been derived to predict alloy oxidation, but are equally applicable to coating systems. One such model is that derived by Quadakkers and coworkers[14, 17] to predict the time to breakaway corrosion of alumina forming FeCrAl based alloys.

The model is based on the assumption that the time to breakaway is controlled by aluminium consumption. A hypothesis that is well supported by experimental evidence in both laboratory and pilot plant tests.[17] Parabolic, or sub-parabolic, oxidation kinetics are assumed, with breakaway occurring when the aluminium level falls below some critical value C_B. Metal thickness (h), original aluminium content (C_o), plus oxidation rate constant and exponent (k and n respectively) are important model parameters as they define both the available aluminium and the rate of consumption. The model results in an equation of the form:

$$t_B = 4.4 \times 10^{-3}. (C_o - C_B). \rho \; h \; k^{-(1/n)} \; (\Delta m^*)^{1/n - 1} \tag{1}$$

where t_B is the time to breakaway and Δm^* is the mass gain at which spallation starts. The other parameters have been previously defined. Figure 3 illustrates a plot of the goodness of fit of this model to experimental data measured on a range of FeCrAl based alloys. The plot presents data for the $t_B \times K$ product as a function of the available aluminium reservoir ($h \times (C_o - C_B)$ over a range of temperatures from 1100–1400°C.

This model could equally be applied to coating systems. In this case, C_o would be the mean available aluminium level and h would be the coating thickness. A correction would

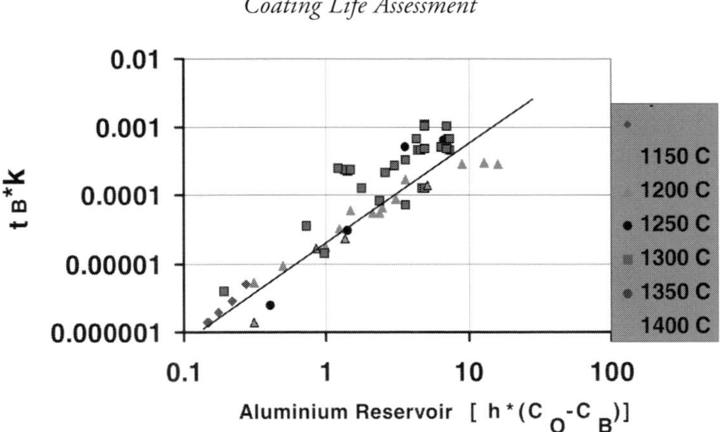

Fig. 3 Oxidation life prediction model on an aluminium exhaustion model.

have to be made for the loss of aluminium by interdiffusion, possibly by reducing the apparent coating thickness or increasing the rate of aluminium consumption by applying a correction to k, the oxidation rate constant.

The approach developed at NASA Lewis, particularly addresses cyclic oxidation.[15, 18] This uses a computer based programme, COSP (cyclic oxidation spalling programme), to model the growth, local spallation and regrowth of the scale. The model allows the input of experimental data, then determines the best fit to the experimental curve yielding values for k_p, the parabolic rate constant, and Q_o, the spall coefficient. With these parameters known (from least square fits to experimental data), the model allows the rate of metal consumption to be calculated and therefore the component/coating life. Figure 4 illustrates the

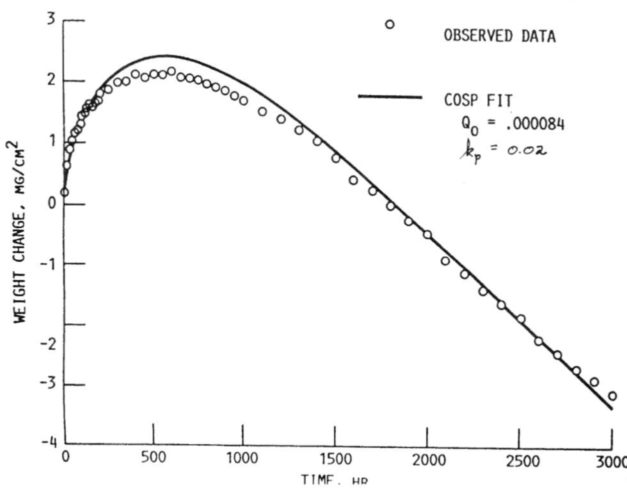

Fig. 4 Modelling cyclic oxidation using the COSP software: cyclic oxidation of β-NiAl (Ni48Al-0.1Zr) oxidised at 1200°C for 3000 1 h cycles and the COSP Model fit for $k_p = 0.02$ and $Q_o = 8.4 \times 10^{-5}$.

degree of fit between the COSP model and experimental data for β-NiAl (Ni48Al 0.1Zr) oxidised at 1200°C for 3000 × 1h cycles.[19]

This NASA cyclic oxidation model has recently between modified by Chan[16] to include effects arising from an increase in stress in the scale on the scale properties and fracture damage. Thus the spallation constant, Q_o, may incorporate a number of scale material parameters including the fracture toughness, elastic modulus, difference in thermal expansion coefficient and microcrack density within the oxide scale.

Recent interest in stress effects has been heightened by the need to develop models capable of predicting the life of thermal barrier coatings. The HOST models for TBCs all include terms that account for the strain imposed on the ceramic layer during thermal cycling.[20–22] These models are derived from a Manson Coffin relationship, with one of the more sophisticated[21] fitting equations of the form:

$$N_f = A \left(\frac{\Delta\varepsilon_p}{\varepsilon_f} \right) \tag{2}$$

$$\varepsilon_f = \varepsilon_c \left(-\frac{M}{M_c} \right) + \Delta\varepsilon_p \left(\frac{M}{M_c} \right) \tag{3}$$

where N_f is the number of cycles to failure, $\Delta\varepsilon_p$ is the in elastic strain range, ε_f is the expected strain to failure in one cycle for a mass gain (M) and this value changes with the degree of oxidation. ε_c is the value of ε_f for zero oxidation mass gain and M_c is the critical mass gain that causes failure in one oxidation cycle.

4.2   Hot Corrosion – Modelling Extreme Corrosion Damage

Hot corrosion involves attack by molten salts, which form from contaminants that may enter the turbine hot section with the air or fuel. These molten salts, mainly sodium and potassium sulphate, condense onto hot gas path components and can lead to rapid material loss.[7, 23–25] Two forms of hot corrosion have been identified: type I, high temperature hot corrosion, which typically occurs at temperatures between 800–950°C; and type II, low temperature hot corrosion, which occurs at temperatures in the range 600–800°C. The basic mechanisms of hot corrosion are relatively well understood,[23] although the role of some minor contaminants in the combustion environment are still being researched (for example Pb and Zn).[24]

Successful protection against hot corrosion, requires that the protective scale remains intact and the hot corrosion process remains in its incubation phase.[23] Rapid loss of any coating, and the underlying substrate, will occur if the scale is lost or locally penetrated. Thus to model hot corrosion attack, one must be able to predict the rate of attack at local failures in the oxide scale, where corrosion rates may be extreme. Few quantitative models for hot corrosion exist in the open literature. Those that do exist follow one of two approaches. The first proposed by Nicholls and Hancock in 1983[26] looks at modelling the regions of maximum attack (areas where the coating is expected to penetrate first) using Extreme Value Statistics. The second, builds on oxidation rate models but includes factors to account for salt deposition, salt corrosivity and engine operating factors.[27]

The extreme value statistics approach has proved particularly successful in predicting type II hot corrosion under conditions simulating marine service[25] and in predicting vanadic corrosion within diesel engines operating on residual fuel oils.[28, 29] Both of these examples modelled deep local pitting attack, using laboratory conditions aimed at simulating service behaviour.

The problem with laboratory simulations is that furnace tests are unable to simulate the high velocity gases, high heat fluxes and contaminant fluxes seen in service. While burner rig test, whilst producing deposition, do not allow these deposition fluxes to be accurately controlled. As a result there may be considerable scatter in experimental tests from one rig run to another.

To overcome these problems, recent modelling work[24, 30–32] has looked at combining the approaches of Nicholls and Hancock[26] with that of Strangman.[27] Reference 24 in this proceedings volume is a good example. The approach is to set up a matrix of laboratory furnace experiments where the salt deposition rates, salt compositions, gas chemistry and temperature are accurately controlled. This gives samples with a broad spectrum of corrosion damage. Each sample is accurately measured and the data set that results is then analysed using Extreme Value Statistics. Figure 5 shows micrographs of type II corrosion of an RT22 platinum aluminide coating after 200h and 500h hot corrosion testing at 700°C, using an ash replacement rate of 0.3mg every 20h (i.e. 0.015mg h^{-1}). The corresponding metal loss data, plotted as a cumulative probability plot, are illustrated in Figure 6, here the median recession is 10.5µm after 200h (460µm per year), whilst the maximum recession

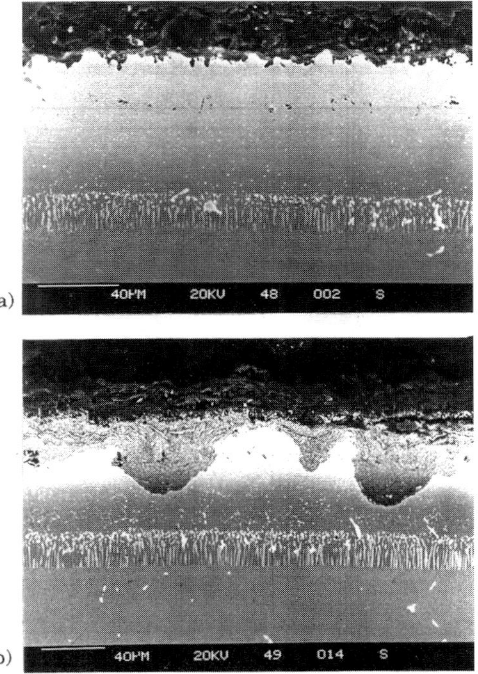

Fig. 5   Type II hot corrosion of an RT22 platinum aluminide coating, using an ash replacement test procedure (0.3 mg every 20 h) (a) after 200 h, (b) after 500 h.

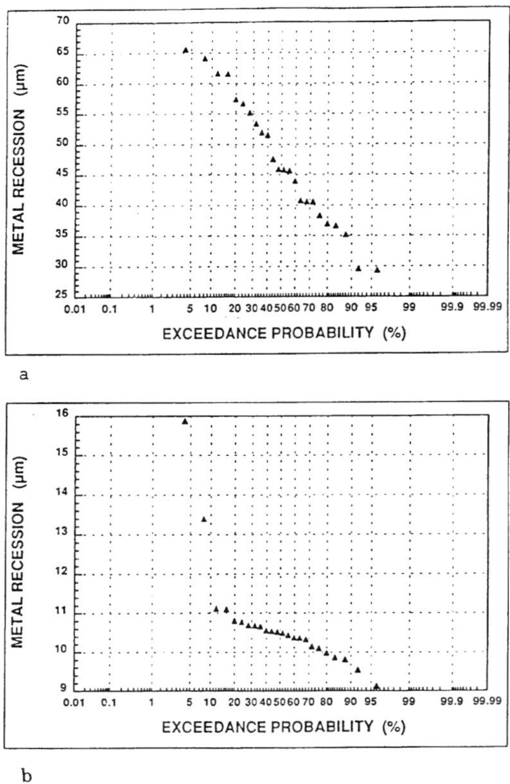

Fig. 6 Cumulative probability plots of coating loss for an RT22 coating on IN738 after type II hot corrosion testing at 700°C (a) after 200 h, (b) after 500 h.

was 15.9μm (697μm per year). Some 10% of the measured section loss data in this instance can be considered extreme. By 500h (Figure 6(b)) over 95% of the data corresponds to a pitting mode of attack, with the median recession rate of 46μm in 500h (800μm per year) and a maximum of 66μm (1150μm per year).

By fitting this extreme data to a Gumbal type I model of maxima[26] one can determine how the rate of growth and spread of these extreme regions of attack change with changes in hot corrosion parameters such as deposition rate, salt composition, gas chemistry and temperature.[24, 32]

The resulting model is of the form:

$$\frac{\mu_e}{t} = \{a + bx + cx^2\} + \{d + ex + fx^2\}.\ln\left([SO_2]\right) \tag{1}$$

where μ_e/t is the most likely extreme corrosion rate* (in μm h⁻¹)

x is the salt deposition rate (in mg cm⁻² h⁻¹)

$[SO_2]$ is the concentration of SO_2 in the test atmosphere (in vpm), and

$a, b, c, d, e,$ and f are experimentally determined model coefficient.

---

*See reference 26 for a definition of the statistical terms.

Using the terminology of Strangman, coefficient *a* is a measure of the corrosion rate, without slat deposition or gas phased contaminants; coefficients *b* and *c* determine the salt corrosivity; coefficient *d* the gas phase corrosivity; and coefficient *e* and *f* account for any synergistic effects between salt and gas chemistry.

Figure 7 illustrates the degree of fit between this model and laboratory data for tests undertaken at 650 and 700°C, with a range of salt deposition rates from 0.0015 to 0.05mg cm^{-2} h^{-1}, and SO$_2$ contents in the combustion gas of 44–2870 vpm. The HCl content was held constant at 270 vpm.[32] The precision of this modelling approach is better than ± 0.006μm h^{-1} (~50μm per year), even when extrapolated beyond the modelling range. Comparison of this model with burner rig and plant data is present elsewhere in this volume by Simms et al.[24]

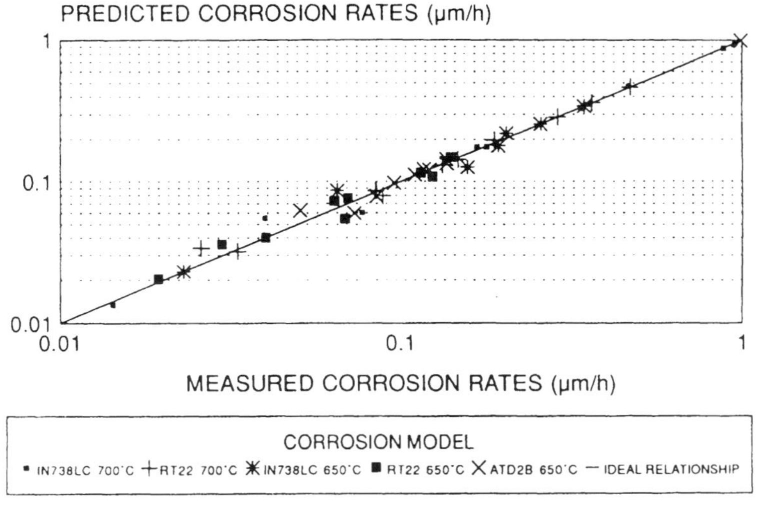

Fig. 7 Goodness of fit between the hot corrosion model and laboratory data for coated and uncoated materials that has been subject to type II hot corrosion tests at 650 and 700°C for 500 h.

4.3 MODELLING EROSION BEHAVIOUR

The degree of erosion that a gas turbine engine sees depends on many factors, the more important of which include the density of particles in the working gas, particle properties (size, angularity, hardness), impact angle and impact velocity. Thus from an engine operators perspective, particle loadings are expected to be higher for engines operating in dusty environments; thus aircraft engines inject debris on takeoff and landing, while land based turbines operating in deserts and dusty sites may see considerable particle injestion on a continuous basis. Particulates can also be generated within the engine. Coking as a result of poor combustion, loss of TBCs from the combustor and salt shedding from the compressor have all been reported as sources of particulates.

Erosion and erosion-corrosion of turbine materials under high temperature, high velocity impact conditions have been systematically studied at Cincinnati[33, 34] in the USA and Cranfield[30, 35, 36] in the UK. Figure 8 illustrates the influence of velocity, particle size, particle loading and impact angles on the erosion of IN738LC at 700°C.

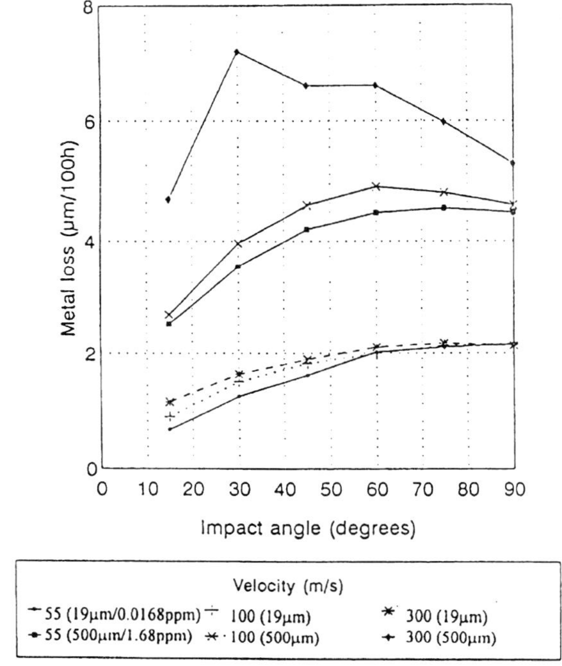

Fig. 8 Influence of velocity, impact angle and particle loading on the erosion-corrosion on IN738LC.

From this figure it can be seen that, when impacted by small particles (19µm) maximum wastage is observed at 90°. The maximum wastage rate was 21µm per 100h, for a particle loading of 0.0168ppm, and was found to be independent of particle velocity over the range 55–300ms^{-1}, suggesting that materials loss by impaction under these conditions may well be corrosion dominated. At more glancing angles (15–45°) the wastage rate with these small particles is observed to depend on impact velocity, increasing with increase in velocity.

Large particles are found to do much more damage. For 500µm particles travelling at 300ms^{-1} peak erosion is observed at 30°, with the wastage rate 4–5× that observed with the smaller particles at the same impact angle and velocity. This behaviour reflects the role a thin surface oxide plays in modifying the erosion behaviour at high temperatures. Thus at high particle loads, and high velocities, erosion is controlled by the substrate properties, with peak erosion observed at 30° (ductile erosion). Reducing the impact velocity, the particle size and/or particle loading results in a shift in mechanism to the scale-modified regime with peak wastage rates now at 90°, reflecting impact with a more brittle surface oxide.

Thus modelling high temperature erosion is complicated by the need to account for simultaneous oxidation along with the erosion process. In this light, it is understandable th_t although the modelling of pure erosion now stands at a relatively sophisticated level (se e.g. the review of solid particle erosion models by Majumdar et al.,[37] that of oxidation-er sion remains, by contrast, significantly less advanced. There are relatively few erosion-corrosion models available in the literature, and those that do exist are often over simplistic However, even a relatively simple erosion-corrosion model can be useful at least for describing gross features of the overall process, within a given erosion regime.

A large empirical oxidation-erosion model was presented by Birks and Pettit[38] and again by Kan et al.[39] in which parabolic oxidation kinetics were assumed together with a constant rate of erosion-induced removal of material. These models predicted a constant scale thickness and constant rate of metal removal at asympototically large time. It will be shown later that this is only one of the erosion regimes that can exist.

Wright and coworkers[40] extended the empirical modelling work of Birks and Pettit.[38] They noted that both partial oxide removal can occur and that substrate material could also be removed as part of the erosion event, but limited their modelling work to the former. This model also predicts a constant rate of erosion-induced removal of material and a limiting oxide thickness on the component surface, after some extended erosion test duration, with

$$M = -\rho_m k^{1/n} \left(\frac{b}{c}\right)^{\frac{n-1}{n}} \tag{5}$$

where M is the erosion rate; ρ_m is the metal density, k is the oxidation rate constant, n is the oxidation rate exponent (where $n = 1$ is linear and $n = 2$ is parabolic etc), b is the instantaneous erosion rate of the oxide layer, which Wright and coworkers suggest is proportional to the instantaneous oxide thickness and c is the Pilling-Bedworth ratio, which defines the volume change on converting metal to oxide.

This simplified model of Wright and coworkers[40] appears to predict the gross features of erosion-oxidation when the erosion processes is controlled by the rate of growth and removal of the surface oxide. In reality, the oxide layer is probably removed incrementally and piece meal, rather than continuously as assumed. Thus to further extend such modelling it would be necessary to introduce such stoichastic processes into the model. Wei and Deffengbough[41] in their model as well as Hancock, Nicholls and Stephenson[42] have highlighted the need for such statistical analysis in modelling erosion processes.

A Monte Carlo model has been used at Cranfield to model the interaction of erosion and oxidation.[36, 43, 44] Figure 9 illustrates the predicted erosion behaviour for IN738LC, under normal impact conditions, at 700°C and a particle flux of 1×10^{-4} g mm^{-2} h^{-1}. Thus, for small particles at low impact velocities, erosion rates are dominated by scale removal, whereas for larger particles at intermediate velocities or small particles a high velocities erosion rates are controlled by substrate deformation and ductile erosion mechanisms. Such Monte Carlo modelling methods show good agreement with experiment over a wide range of impact conditions and measured erosion rates, by allowing for the stoichastic nature of erosion-corrosion processes, i.e. variable size of particles, impact velocities, loading rates and impact condition. In the Monte Carlo modelling approach this variability is allowed for by ascribing distribution functions to each of the variables. To date such Monte Carlo methods have been used to model erosion under turbine and heat exchanger conditions.[30, 44] Figure 10, illustrates the good agreement between predicted and measured erosion rates for a range of metals, ceramics and coating systems of interest within ductworks, heat exchangers and turbines of a combined cycle power generation system.[30] The experimental data were determined using Cranfield's high velocity gas gun and centrifugal erosion rigs.[45]

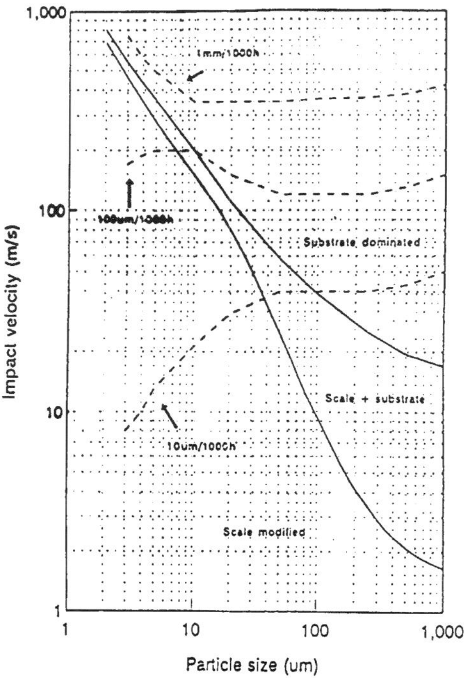

Fig. 9 High temperature erosion map for IN738 impacted by silica particles at 90° impingement angle and 700°C at a particle loading of 1×10^{-4} g mm^{-2} h^{-1} illustrating damage regimes and metal recession rates predicted by Monte Carlo modelling.

4.4 MODELLING COATINGS/SUBSTRATE INTERDIFFUSION

The previous three sections have addressed coating life modelling due to the loss of coating through surface environment interactions, namely oxidation, hot corrosion and erosion-corrosion processes. However, at sufficiently high temperatures, the loss of aluminium in the coating due to solid state diffusion of the coating with the base metal must also rank as a significant degradation mode. Such diffusion processes, between coating and the substrate, reduce the concentration of aluminium in the coating that is available to form a protective alumina scale. Thus, under cyclic oxidation conditions which may lead to spallation, a protective alumina scale may no longer reform once the aluminium concentration has fallen below a critical level. As a result, less protective oxides form and the coating

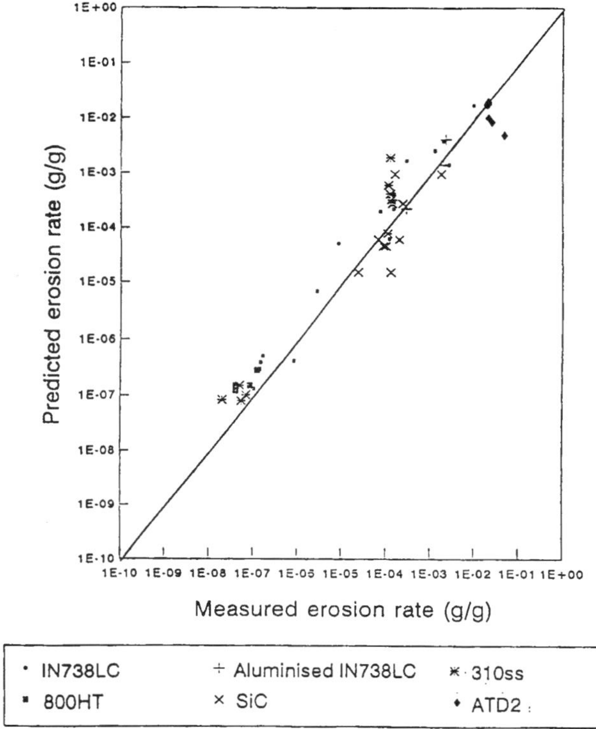

Fig. 10 Comparison of measured erosion rate with predicted values using the Monte Carlo erosion model for various materials.

is rapidly consumed by breakaway oxidation. Modelling of such breakaway phenomena has been discussed in Section 4.1, however, the role played by interdiffusion is to reduce available aluminium reserves triggering early breakaway.

The basic concepts of solid state diffusion are well understood for simple systems. Even for complexed systems, such as coatings on superalloys, the movement of multi-elements within a multi-phased structure can be formally described.[46, 47, 48] Two of the more comprehensive studies in this area, are that of Copper and Strang[47] for diffusion coatings and Mazaar et al.[48] for MCrAlY overlay systems.

However, apart from such descriptive models of the degradation process and the broad ranking of systems (coating plus substrate) performance few studies have been attempted to predict interdiffusion in such systems. The measurement of actual interdiffusion in such complex, alloy and coating, systems is a formidable task. However, for many systems, the measurement of phase depletion profiles – as can be seen in Fig. 1 for the β depletion after cyclic oxidation of a NiCoCrAlY coatings – may well suffice and allow aluminium fluxes to be calculated.

This approach, aimed at modelling depletion profiles, has been adopted in studies at NASA Lewis[46, 49, 50] to calculate the loss of scale forming elements – Al and Cr – due to interdiffusion with the substrate. A failure criterion must be derived which can be based on total surface recession,[46] depletion to a critical level of solute[46] or depletion to form a non protective surface layer.[50] For the latter two criteria, diffusion equations have to be set up and solved to predict the concentration profiles of aluminium and chromium into the substrate. Even when predicting the movement of two elements, finite-difference diffusion models have to be developed to predict the concentration profiles, during cyclic oxidation.[46, 51] This is because the metal/oxide interface provides a complex boundary condition with the element consumption and composition at this interface changing with time as the oxide scale grows and spalls. Figure 11 illustrates the measured and predicted aluminium and chromium profiles for the cyclic oxidation at 1150°C of a Ni–16Cr–25Al coating onto a Ni–22Cr model substrate.[15] The measured and predicted profiles agree well, but demonstrate that for this system considerable aluminium movement from the coating into the substrate occurs – over 200μm – with the coating near-surface composition dropping from 25wt% to 6wt% after 100h at 1100°C.

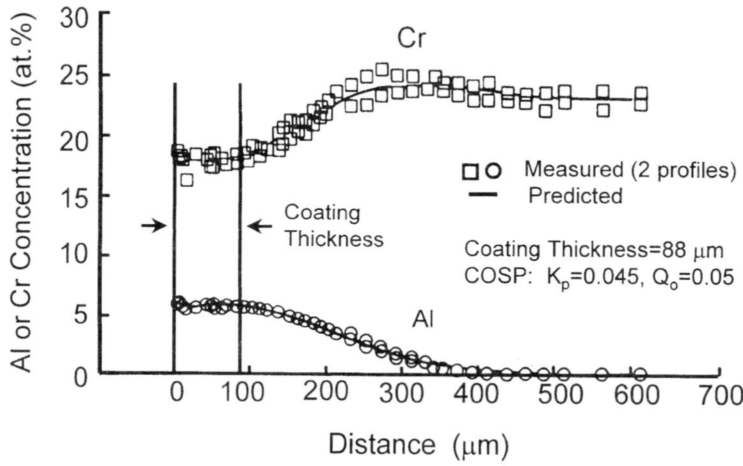

Fig. 11 Measured and predicted Al and Cr profiles after 100 1 hr cycles at 1150°C for a Ni–16Cr–25Al coating on a Ni–22Cr substrate.[15]

This loss of aluminium in the coating, caused by diffusion of the coating with the base alloy together with the loss due to oxidation, can result catastrophic failure of the coating – termed coating exhaustion – a result of the aluminium level falling below a critical level necessary to reform a protective scale. It is clear from Fig. 11 that any life modelling of coating performance, when operating at elevated temperatures – above 1050°C – must take into account the aluminium consumption due to interdiffusion with the substrate.

5 CONCLUDING REMARKS

In this paper the major degradation mechanisms of coating systems and the status of life prediction models for each of these mechanisms have been reviewed. Some are relatively advanced, for example the prediction of breakaway oxidation due to alloy/coating exhaustion, others are still in their infancy, for example the prediction of erosion/corrosion or the prediction of interdiffusion.

It is clear that for many applications, more than one mechanism will be depleting the reservoir of scale forming elements, therefore any mature model must be able to sum the damage from multiple degradation sources. Currently models are not at this stage, although modern computer systems should permit such model integration and allow damage accumulation through the operational life of the coating to be predicted.

For conditions, where mathematical analysis is unlikely to predict behaviour – as has been highlighted for multi-element diffusion in multi-phased structures – then parametric equations based on well planned experiments will have to be used. For the particular case of coating interdiffusion, the measurement of phase depletion profiles provides a promising solution.

This science based, parametric equation route has proved successful in predicting the rate of extreme corrosion attack under hot corrosion conditions.

In conclusion, accurate modelling of the service life of coatings is becoming increasingly important as the designers strive for increased power and efficiencies from gas turbine engines, making them increasingly reliant on coating systems to protect turbine components. Coating life models are in their infancy. More research is required to:

(1) Improve existing models; through the better understanding of failure mechanisms as a function of service conditions.
(2) to develop new quantitative models in areas where there is now only qualitative understanding, such that models exist for all active failure mechanisms, and
(3) To integrate the best of current modelling approaches into a single software tool that can be used to aid designers in coating selection/optimisation and provide accurate estimates of remaining coating life after taking into account all known degradation mechanisms.

At present we are someway from this ideal.

One final remark, no attempt has been made in this paper to assess damage that may result from thermomechanical interactions between the coating and the substrate, or to assess stress effects on the failure of the coating system. Much work aimed at mechanistic understanding the effects of thermal and mechanical load on the coating/substrate system is in progress. Life prediction models are needed in these areas to allow designers to avoid thermomechanical fatigue cracking of coated superalloy components. Their development is hampered by a lack of knowledge as to the mechanical properties of the as-deposited and aged coating systems.

REFERENCES

1. J.R. Nicholls. 'Design of Oxidation-Resistant Coatings', *JoM*, to be published.
2. *Coatings for High Temperature Structural Materials: Trends and Opportunities*, National Materials Advisory Board, National Academy Press, Washington DC, USA, 1996.
3. G.W. Goward and L.W. Cannon. ASME Paper, 87-GT-50 ASME, New York, 1988.

4. S.M. Manning-Meier and D. Gupta. ASME Paper, 92-GT-203, ASME, New York, 1992.

5. 'Thermal Barrier Coatings'. AGARD Report 823, NATO Neuilly-sur-Seine, France, 1998.

6. J.R. Nicholls and D.J. Stephenson. *Intermetallic Compounds*, Vol. 2, J.H. Westbrook and R.L. Fleischer eds., John Wiley & Sons, New York, 1994, Chapter 22.

7. J. Stringer and R. Viswanathan. In *Proc. of A.S.M. 1993, Materials Congress*, Pittsburg Pennsylvania. ASM International, Ohio, 1993.

8. D.F. Bettridge, R. Wing and S.R.J. Saunders. In *Materials for Advanced Power Engineering 1998*, part II, J. Lecomte-Beckers, F. Schubert and P. J. Ennis, eds. Forschungzentrum, Julich, Germany, 1998, 96–976.

9. F.T. Talboom, R.C. Elam and L.W. Wilson. NASA Report NASA-Cr-7813, 1970.

10. F.H. Stott and J.V. Wood. *Critical Al for alumina formation.*

11. G.S. Giggins and F.S. Pettit. *J. Electrochem. Soc.*, 1971, **118**, 1782.

12. G.R. Wallwork and A.Z. Hed. *Oxid. Met.*, 1971, **3**, 171.

13. J.R. Nicholls, P. Hancock and L.H. Al-Yasiri. *Mater. Sc. Technol.*, 1989, **5**, 799–805.

14. W.J. Quadakkers and K. Bongartz. *Mater. and Corr.*, 1994, **45**, 232–241.

15. J.L. Smialek, J.A. Nesbitt, C.A. Barrett and C.E. Lowell. In *Cyclic Oxidation of High Temperature Materials*, EFC Corrosion Publication 27, M. Schütze and W. J. Quadakkers, eds. IOM Communications, 1999, Chapter 9.

16. K.S. Chan. *Met. Mater. Trans. A.*, 1997, **28A**, 411–422.

17. W.J. Quadakkers, D. Clemens and M.J. Bennett. In *Microscopy of Oxidation* 3, S. B. Newcomb and J. A. Little, eds., Institute of Materials, 1997, 195–206.

18. C.E. Lowell, C.A. Barrett, R.W. Palmer, J.V. Auping and H.B. Probst. *Oxid. Met.*, 1991, **36**, (1/2), 81–112.

19. C.A. Barrett. In *Oxidation of High Temperature Intermetallics*, T. Großstein and J. Doychak, eds., TMS-AIME, Warrendale, PA, 1988, 67–83.

20. R.A. Miller. *Jour. Am. Ceram. Soc.*, 1984, **67**, 517.

21. J.T. DeMasi, K.D. Sheffler and M. Ortiz. NASA Report NASA-CR-182230, 1989.

22. S. Manning-Meier, D.M. Nissley, K.D. Sheffler and T.A. Cruse. ASME Paper 91-GT-40, American Society of Mechanical Engineers, 1991.

23. C.S. Giggins and F.S. Pettit. 'Hot Corrosion Degradation of Metals and Alloys – A Unified Theory', PWA report no FR11545, Pratt & Whitney Aircraft, Florida, 1979.

24. N.J. Simms, J.E. Oakey and J.R. Nicholls. 'Predicting the Remnant Life of Corrosion Resistant Coatings', *Life Assessment of Hot Section Gas Turbine Components*, IOM Communications Ltd, London, 2000.

25. J.R. Nicholls and D.J. Stephenson. *Corrosion Science*, 1992, **33**, (8), 1313–1325.

26. J.R. Nicholls and P. Hancock. 'The Analysis of Oxidation and Hot Corrosion Data – A Statistical Approach' in *High Temperature Corrosion*, R.A. Rapp, NACE-6, NACE, Houston, TX, 1983, 198–210.

27. T.E. Strangman. 'Turbine Coating Life Prediction Model' in *Proc. Workshop on Coatings for Advanced Heat Engines*, Washington, DC USA, 1990.

28. J.R. Nicholls and D.A. Triner. 'Parametric Equations for Valve Life Prediction' in *Diesel Engine combustion Chamber Materials for Heavy Fuel Operation*, Institute of Marine Engineers, London, 1990, 121–130.

29. J.R. Nicholls. *Mater. at High Temp.*, 1994, **12**, (1) 35–46.

30. N.J. Simms, J.E. Oakey, D.J. Stephenson, P. Smith and J.R. Nicholls. 'Erosion/corrosion of Gas Turbine Materials for Coal-Fired Combined Cycle Power Generation', *Wear*, 1995, 185–187.

31. J.R. Nicholls, P. Smith and J.E. Oakey. 'Prediction of Hot Salt Corrosion within Utility Gas Turbines', *Materials for Advanced Power Engineering* Part II, D. Coutsouradis *et al.* eds, Kluwer Publishers, 1994, 1273–1289.

32 P. Smith and J.R. Nicholls. 'Predicting Hot Corrosion in Utility Gas Turbine Engines'.
33 W. Tabakoff. *Surf. Coatings Tech.*, 1992, **52**, 65–79.
34 W. Tabakoff. *Wear.*, 1995, **186–7**, 224–229.
35 J.E. Restall and D.J. Stephenson. *Mater. Sci. Eng.*, 1987, **88**, 273–282.
36 J.R. Nicholls and D.J. Stephenson. *Wear.*, 1995, **186–7**, 64–77.
37 S. Majumdar, K. Natesan and Sarajedini. Argonne National Lab Report ANL-FE-81–8, Argonne, Ill., 1987.
38 N. Birks and F.S. Pettit. Report ARO-17421, 1984.
39 C.T. Kang, F.S. Pettit and N. Birks. *Metall. Trans.*, 1987, **18A**, 1785.
40 A.J. Markworth, V. Nagarajan and I.G. Wright. *Oxid. Met.*, 1991, **35**, (1/2) 89–106.
41 W. Wei and D. Deffenbough. DoE Report MC-22077–2022, 1986.
42 P. Hancock, J.R. Nicholls and D.J. Stephenson. *Surf. Coating Tech.*, 1987, **32**, 285–304.
43 D.J. Stephenson and J.R. Nicholls. *Corrosion Sci.*, 1993, **35**, 1015–1026.
44 D.J. Stephenson and J.R. Nicholls. *Wear*, 1995, **186–7**, 284–290.
45 J.R. Nicholls. *Mater. at High Temp.*, 1997, **14**, (3) 289–306.
46 J.A. Nesbitt and R.W. Heckel. *Thin Solid Films*, 1984, **119**, 281–290.
47 S.P. Cooper and A. Strang. In *High Temperature Alloys for Gas Turbines*. R. Brunetaud *et al.* ed., 249–260; Dordrecht, D. Reidel Publishing Co, 1982.
48 P. Mazars, D. Maresse and C. Lopvet. In *High Temperature Alloys for Gas Turbines*. R. Brunetaud *et al*, ed., 1183–1192; Dordrecht, D. Reidel Publishing Co, 1986.
49 J.A. Nesbitt and C.E. Lowell. *M.R.S. Symposium Proceedings*, 1993, **288**, 107.
50 J.A. Nesbitt and E.J. Vinarcik. 'Predicting the Oxidation Lifetime of β-NiAl-Zr Alloys' in *Damage and Oxidation Protection*, Haritos K. and O.O. Ochos, ed., Vol. 25–1, ASME, 1991.
51 J.A. Nesbitt J. A. 'Diffusional Aspects of the High Temperature Oxidation of Protective Coatings' in *Diffusion Analysis and Applications*, A.D. Romig and M.A. Dayanada, ed., TMS-AIME, Warrendale, PA, 1989, 307.

Combined Effects of Temperature Gradient and Oxidation on Thermal Barrier Coating Failure

Y.C. ZHOU[a,b] and T. HASHIDA[a]

[a]*Fracture Research Institute, Tohoku University, Sendai, Japan*
[b]*Institute of Fundamental Mechanics and Material Engineering, Xiangtan University, Xiangtan, Hunan, 411105, P.R. China*

ABSTRACT

The present article reviews the investigation about the coupled effects of temperature gradient and oxidation on the failure of thermal barrier ceramic coatings in the past ten years in our laboratory. The investigations include both experimental and theoretical studies. On the experimental investigation, the heating method, non-destructive evaluation and interface properties are mainly reviewed. An attempt has been made to develop four different heating methods which are furnace heating, burner heating, plasma heating and laser beam heating. The non-destructive evaluation such as temperature test, acoustic emission (AE) signals detected and impedance spectroscopy (IS) method have been successfully used in the study of TBC systems degradation. The interface properties such as oxidation and fracture toughness have been studied by test and observation. On the theoretical investigation, the temperature fields and the related thermal stresses fields have been analytical solved where the coupling effect of temperature gradient, oxidation, thermal fatigue, creep, morphology of the TBC system as well as the cooling rate were considered. The delamination cracking in thermal barrier coating system with ceramic coating deposited on a substrate has been studied analytically. Due to the complication of TBCs degradation, a real physical map of TBCs system failure is not obtained up to now. Further investigation is proposed.

1   INTRODUCTION

For fossil fuel fired power generation, the higher thermal efficiency is most effective for reducing the carbon dioxide emission. 'Development of the Advanced Gas Turbine' will create the next generation electric power system.[1, 2] The program included three development areas: (1) high-performance and high-reliability turbine vanes and blades; (2) hot parts cooling technology; (3) high temperature and low-NO_x combustor. The materials of high-performance and high-reliability include the high heat-resistance of new alloys and thermal barrier ceramic coating (TBCs). A TBC provides performance, efficiency, and durability benefits by reducing turbine cooling air requirements and lowering metal temperatures. Before the TBC system is used the following questions must be answered: How long can the TBC system survive in operational service?

Previous work has demonstrated that there are some important effects on TBCs operating. The first is thermal fatigue.[3, 4] The second is thermal growth oxidation (TGO) between bond coat and thermal barrier ceramic coating.[5, 6] The third is the roughness of the surface of bond coat.[7, 8] The fourth is oxygen and sulphur penetration along the grain boundaries.[9]

This paper reviews an attempt for the study of TBCs failure mechanism in the past ten years in the laboratory. The investigations include the experimental and theoretical study. Further study is proposed in the paper.

2 EXPERIMENTAL INVESTIGATION

There are many papers concerning life prediction and failure mechanism of TBCs system.[10–15] As we know, the TBC system is used to provide thermal insulation to critical air-cooled components by overlaying a strain-tolerant ceramic top coating. Therefore, there must be a temperature gradient in the thickness direction. TBC systems should begin to operate and stop operating. This means that the TBCs must survive many cycles of heating, heat hold and cooling. The TBC systems may encounter some special condition, for example, the system may stop operating and the system is suddenly cooled. The TBC systems are not a flat plate. There must be a curvature in some place, for example, the leading edge of a gas turbine blades. In a word, the failure mechanism is governed by the combined effect of temperature gradient, oxidation, thermal fatigue, creep, morphology of the TBC system as well as the cooling rate.

2.1 THERMAL SHOCK METHOD

In order to considered the above effects of TBCs system failure, one needs to design an experimental method for simulating the operating state which has the temperature gradient, fast or slow heating, fast or slow cooling, many cycles of heating/cooling as well as the curvature of TBC systems.

First is the burner heating method in which hydrogen and oxygen gases were used for the gas burner heating.[16] The flow rate of the fuel gas was regulated by needle valves. After predetermined surface temperatures were reached, the burner flame was shut off using a water-cooled chopper and the specimen was then allowed to cool to a room temperature. The specimen was internally gas-cooled to achieve various temperature drops within the coating layer. The burner method is very simple and convenient. But the relationship of temperature fields and the flow rate of fuel gas cannot easily be controlled.

Second is the plasma heating method in which the plasma torch is used to heat the sample.[17] The method is also simple and convenient. In this method, the plasma gun allows oxygen gas to be added in the plasma gas flow so that the oxidation resistance can be evaluated. Therefore, the great advantage of the method is to simulate the erosion and corrosion environments which TBCs system will encounter in addition to thermal loads.

The third method is the laser beam heating method in which heating of specimen surface was accomplished using a continuous CO_2 laser of 10.6μm output wavelength.[17–22] The unit is nominally specified as a 50W laser. The system was arranged so that the coated specimens can be exposed to a laser beam with a preset size (such as 8 mm or 6 mm in diameter), duration, and intensity. A chopper with a slot was rotated in front of the beam, allowing the beam to impinge on the specimen for various preset durations, either as a single pulse or as repetitive pulses. The outer surface of the specimen was heated by laser up to a predetermined temperature, and then the laser irradiation was stopped and the specimen cooled. The computer-controlled system shown in Fig. 1 allows heating/cooling processes to be automatically cycled for various duration.

The fourth method is the furnace and the method is very simple and convenient. In particular, the great advantage is that very long duration at elevated temperatures can be

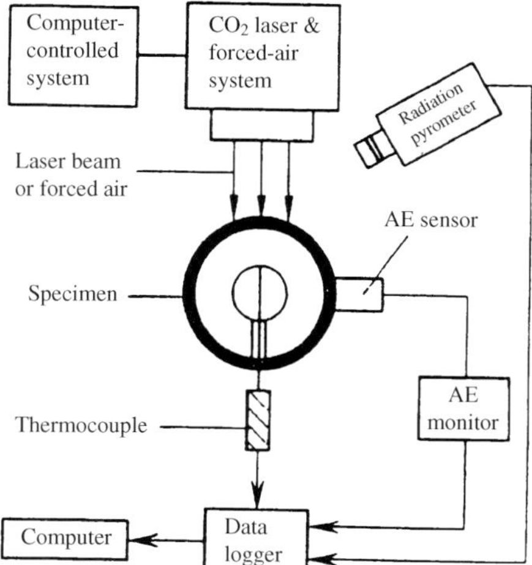

Fig. 1 Schematic of the experimental setup for CO_2 laser heating method.

achieved. It is very useful for investigating the oxidation processing of a bond coat. Therefore, the method has been used to study the thermal grown oxidation between bond coat and ceramic coating as described in Refs 23–25. However, the non-destructive test in the furnace is so simple as will be discussed below.

The above four heating methods are developed to study failure mechanisms in our laboratory. However, every method has its distinct advantage but at the same time has its own disadvantage. The common disadvantage is that the evolution of failure by mechanical and thermal loading is three-dimensional. This results in the impossibility of testing whole fields such as temperature, deformation as well as the crack initiation and propagation. The relation of experimental results and theoretical analysis is hardly possible. Therefore, the real physical concept for the failure mechanism is not really comprehensive. A new heating method needs to be developed. A new heating method is under development and the results will be published elsewhere.

2.2   Non-Destructive Evaluation

The evolution of TBCs system degradation is reflected by temperature fields, deformation, the change of materials structure, crack initiation and propagation as well as new materials formation.

The temperature on the ceramic coating surface is tested using a general infrared radiation pyrometer.[16–23] The method is very conventional. Although the response time is fast enough, about 40 ms for TBCs system failure study, only the average temperature on the focus size is tested. The focus is generally 6.0 mm in diameter. The temperature distribution is hard to determine. In the recent years, a new infrared radiation pyrometer with multi-infrared detectors has been available to test the temperature fields on the outer surface of TBC systems.[26, 27] The multi-point infrared radiation pyrometer was firstly developed to test the temperature fields near the crack tip. The pyrometer has high temporal response and high spatial resolution.

Concurrently with the above heating tests, acoustic emission (AE) monitoring was performed to detect the microfracture process of the coating materials.[16–25, 28] The disposition of the two or three transducers enabled the determination of the linear location of the AE sources emitted from the sample. Furthermore, AE signal analysis can be used to provide a way to predict the long term behavior of the TBC under thermal-cycling conditions. AE signals were detected using a broad band piezoelectric transducer with a resonant frequency close to 1MHz. The square of signal peak voltage was used as a measure of AE energy. The electric signals from the transducer were amplified with 80dB and bandpass filtered between 5 and 500 kHz.

Impedance spectroscopy (IS) was used to evaluate the formation kinetics and physical properties of the reaction layer between bond coat and ceramic coating in our laboratory for the first time.[23–25] IS method has been conventionally used to study the electrical behaviour of systems in which the overall system behaviour was determined by a number of strongly coupled processes.[29] In the evaluations of TBCs degradation, the IS method was developed to detect various defects such as delamination, spalling and cracking and other material damages. A schematic illustration of the used impedance measurement apparatus is shown in Fig. 2.

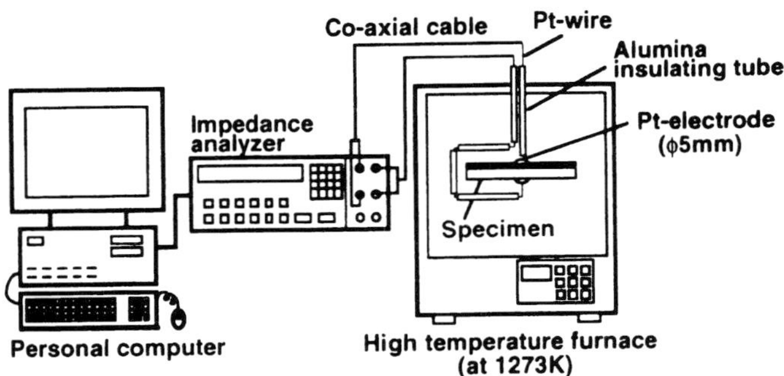

Fig. 2   Schematic layout of the impedance measurement apparatus.

The blister test was used to measure adhesive fracture toughness. A coating bonded to a substrate was debonded by applying a hydrostatic pressure which is a oil pressure as shown in Fig. 3. The basic ideas proposed by Jensen[30] have been adopted to test the interface fracture toughness of PSZ/Ni-superalloy. It was assumed that the delaminated ceramic coating is a thin plate. The non-delaminated coating was assumed to be bonded on the substrate with fully clamped conditions of thin plate at its edges. By analysing numerically the non-linear Karman plate equations, one can obtain the relation of external loads such as oil pressure and the radius of delaminated ceramic coating with the membrane stress N and bending moment M. Therefore, when the test of the radius of delaminated coating and the external loads, i.e., the oil pressure was done the energy release rate G_0 for interface

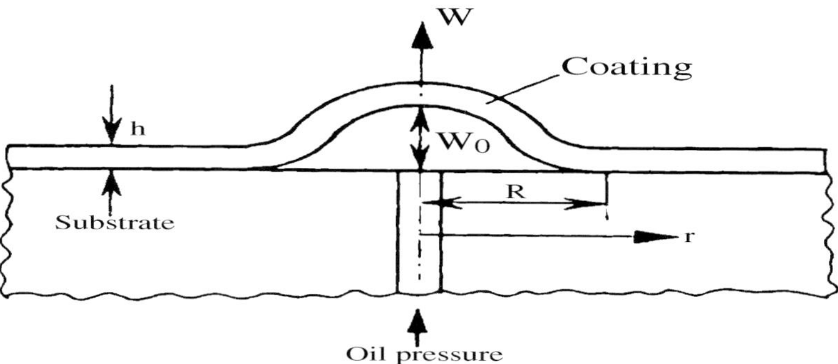

Fig. 3   Configuration for the test of interface fracture toughness.

crack was known. The radius of the delaminated ceramic coating was tested by ultrasonic test method and by SEM observation method.

2.3   MATERIALS AND TEST SPECIMENS

Two types of materials samples were prepared. One was the conventional coating which is a two-layer coating system consisting of PSZ layer (partially stabilized ZrO_2 by 8 wt%Y_2O_3) over a NiCrAlY bond coat. The coatings were air-plasma-sprayed onto a substrate of the specimens. The material of the substrate was SUS340 stainless steel or Ni base superalloy. Recently, the multilayer coating system, in particular, a functionally graded material (FGM) coating system has been proposed. This FGM coating was a five layer coating system of PSZ and NiCrAlY, and the composition was designed to have the same thermal shielding performance as that of the non-FGM coating. The coating was air-plasma-sprayed onto a substrate of the specimens. The material of the substrate was SUS340 stainless steel. Schematics of the cross-sections of the FGM coating are shown in Fig. 4.

The plan plate shape of the samples are used to detect the properties of interface such as fracture toughness, oxidation. The cylindrical shape of the specimens is used to simulate the curvature of the leading edge of gas turbine blades.

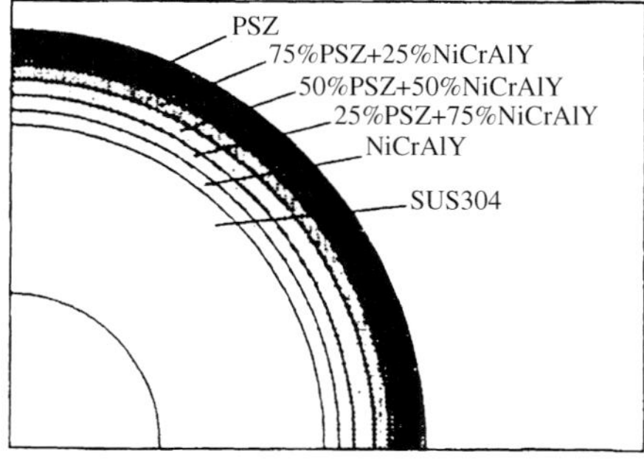

Fig. 4   Schematics of the cross-sections of FGM coatings.

2.4   SOME IMPORTANT RESULTS

A   The evaluation of degradation of TBCs system

The comparison of the AE behavior and SEM observation indicates that burner heating, plasma heating and laser heating tests all produced two types of coating damage: vertical cracking and interface delamination in non-FGM. The histories of temperature on the ceramic coating surface and substrate surface for the first cycle are shown in Fig. 5. The AE results are also shown in the figure. The corresponding SEM micrographs of the specimens' cross-section and the surface correspondence Fig. 5(a) are shown in Figs 6(a) and (b). AE signal detected at the beginning of the cooling stage corresponds to the vertical cracking shown as in Fig. 6(a) or surface crack shown as in Fig. 6(b). Figures 7(a) and (b) show the delamination damage correspondence Fig. 5(b) and typical SEM photo of ceramic coating completely failure. In Fig. 7(a), the delamination is observed in addition to vertical cracking and that the high AE activity during the cool down is associated with the

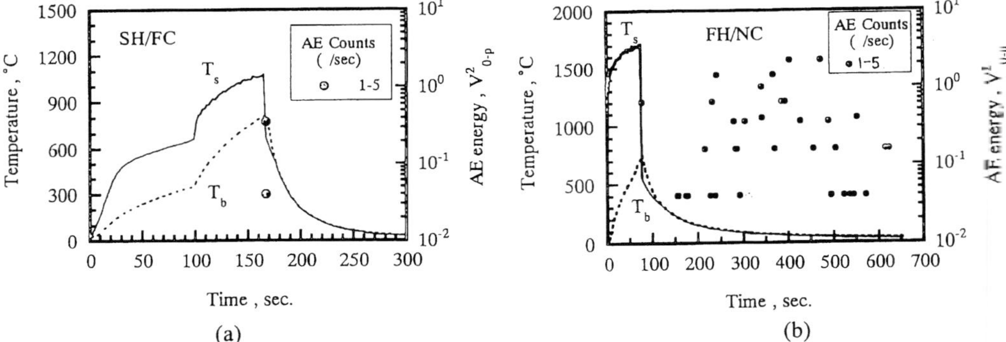

Fig. 5   Temperature histories and AE activity in non-FGM coating during the laser heating, (a) slow heating and fast cooling, (b) fast heating and natural cooling.

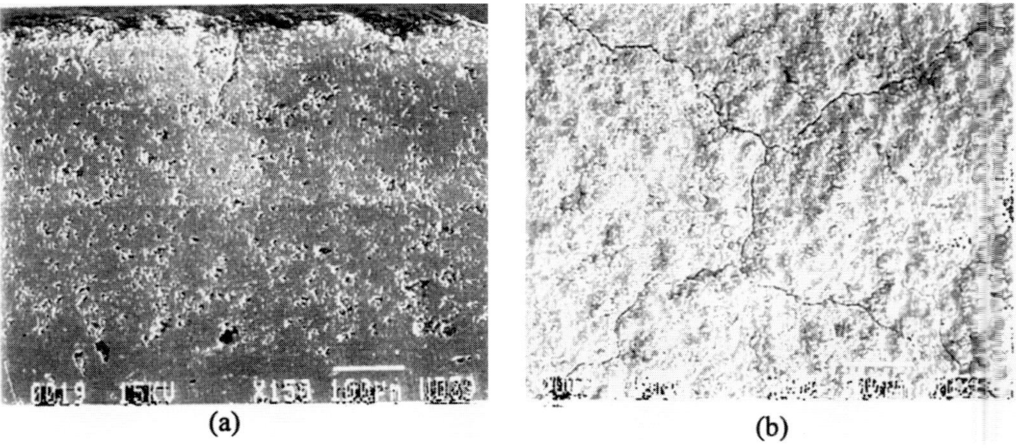

Fig. 6   SEM micrographs corresponding to Fig. 5(a) at the end of cooling, (a) cross-section micrograph, (b) surface micrograph.

delamination growth. The spallation shown in Fig. 7(b) is conventional ZrO_2 coating layer exposed to 5 thermal fatigue cycles with laser power of 34W. It can be seen that complete failure is the comprehensive result of radial crack, surface crack or interface crack and delamination.

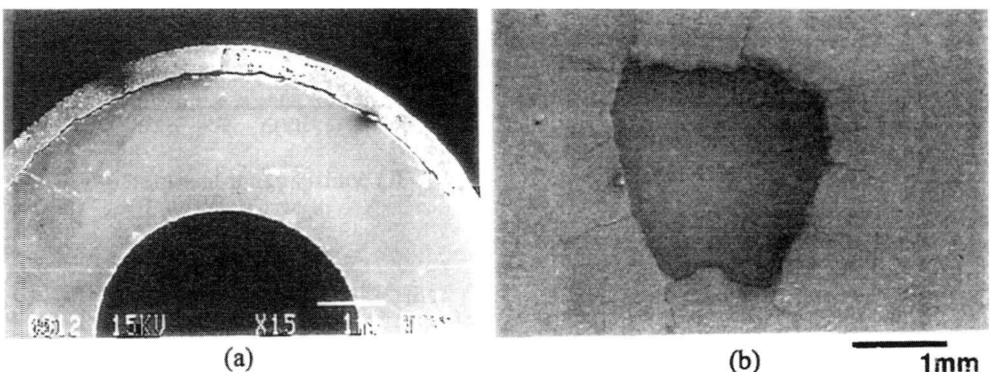

Fig. 7   Comprehensive failure, (a) cross-section of SEM micrographs corresponding to Fig. 5(b) at the end of cooling, (b) spallation of conventional ZrO_2 coating layer exposed to 5 thermal fatigue cycles (laser power = 34W).

The temperature drop through the specimen thickness primarily controls the thermal stresses. It is convenient to use the plane of temperature T_b on the substrate surface and temperature T_s on the ceramic coating to study the failure threshold for TBCs system. The physical concept is also clear and the optimum state for the range of advanced gas turbine can also be obtained on the T_s–T_b plane. In the plane, one can distinguish three regions which are no crack region, vertical crack region and delamination region as discussing by Hashida *et al.*[16–22]

In FGM failure experiment, the surface crack or vertical crack was visualised. No delamination growth was observed for the FGM within the tested temperature range where the conventional ceramic coating showed extensive delamination. The delamination growth was not obvious even under the higher laser power. From the T_s–T_b plane for FGM, one can conclude that the FGM has much higher thermal fatigue resistance compared to the conventional coating system through thermal cycling tests conducted under the simulated advanced gas turbine blade conditions. Therefore, the use of FGM is an effective way to enhance the thermal shock and thermal fatigue resistance of the thermal barrier coating.

The experiment results for fast or slow heating and fast or slow cooling lead to the conclusion that fast heating increases the damage to the coatings compared with natural slow heating. Forced air cooling also increases damage to the coatings compared with natural cooling. The thermal cycle consisting of fast-heating/forced-air-cooling is a suitable thermal cycle for an accelerated fatigue testing method.

B   Interface Properties

The mechanical and thermal properties of ceramic and metal are very different. The mismatch must produce interface problems in the material. Although in the new materials of FGM the composition and microstructure are varied continuously from place to place the thermal and mechanical mismatch exist. The properties of interface in TBC systems are studied on the respect of thermal growth oxidation and interface fracture toughness.

The oxidation properties were studied with IS method by test the frequency response of the impedance behaviour of specimens. The IS method was successfully applied to analyse the reaction layer formed at the interface between the TBC and the Ni base superalloy substrate. The combination of IS test and SEM observation show that the reaction layer does exist and the layer contains two layers of dark alumina and grey mixed oxide. Figure 8 shows the typical SEM image of the cross-section around TBC/NiCrAlY interface of the specimen aged for 3000h at 1000°C. YSZ and the two oxide layers formed can be presented by three impedance elements which are specific resistance, dielectric constant and thickness of each layer. It is possible to estimate the impedance behaviour using sensitivity analysis. The thickness of the alumina layer is calculated from the largest impedance value obtained at a region with saturated phase angle. It is possible to estimate the physical properties and the thickness of each layer by IS method, which allows to nondestructively evaluate the detrimental effects of the alumina layer in TBC integrity assessment.

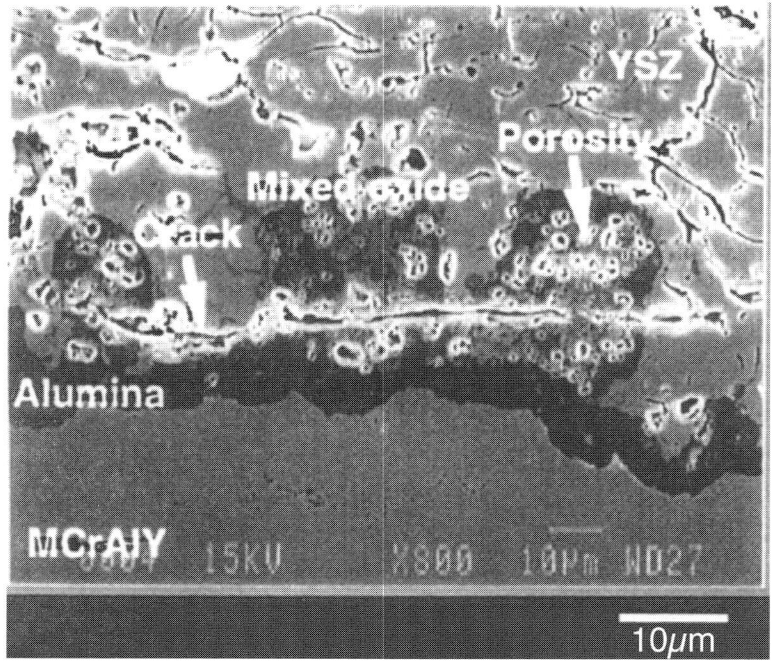

Fig. 8 Typical SEM image of the cross-section around TBC/McrAlY interface of the specimen aged for 3000h at 1000°C.

The interface fracture toughness K_{ic} is obtained by the following relation

$$K_{ic} = \sqrt{\frac{16 \cosh^2 \pi\varepsilon}{C_1 + C_2} G_0} \tag{1}$$

where ε is bimaterial constant, C_1 and C_2 are the compliance parameters of material 1 and material 2. Two kinds of ceramic coating sample are tested. One is dense and another is porous. The results for interface crack toughness are given in Fig. 9. The average interface crack toughness for dense and porous samples are, respectively, 0.9–1.2 MPa m$^{1/2}$ and 0.5–0.7MPa m$^{1/2}$. The average mode mixity measure ψ is −27°.

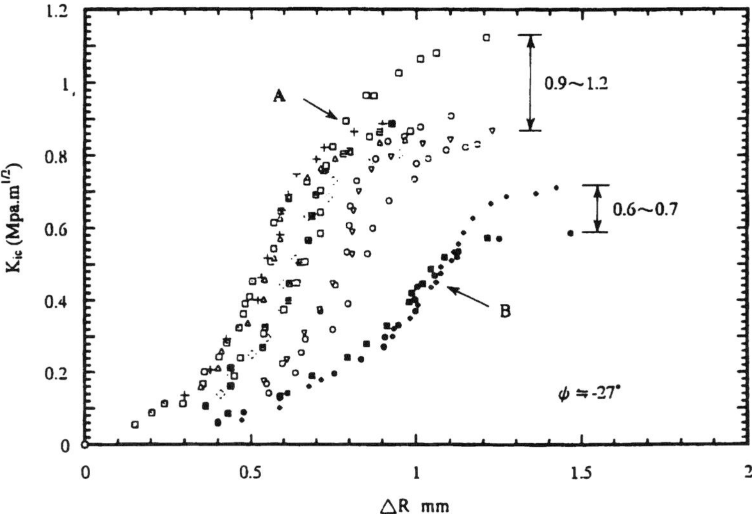

Fig. 9 Interface fracture toughness for PSZ/Ni-superalloy TBC systems with two kinds of sample, where A and B denote dense and porous materials, respectively.

3 THEORETICAL AND NUMERICAL INVESTIGATIONS

As previous stated, the failure mechanism is governed by the combined effect of temperature gradient, oxidation, thermal fatigue, creep and morphology of the TBC system as well as the cooling rate. In order to understand the mechanism of TBC systems operating at high temperature conditions, the combinations of experimental study with theoretical or numerical study are very important. On the theoretical investigation, the thermal stress fields in TBCs system by analytical and numerical method, the delamination cracking by analytical method are mainly reviewed.

3.1 Thermal Stress Fields in TBCs System

A *Theoretical Model*

For the study of TBC failure mechanism the physical map of temperature field and related thermal stress fields should be the first concerned problem.[31] The TBC system is assumed to be partially stabilised ZrO_2 by 8 wt%Y_2O_3 or mullite over a NiCrAlY bond coat sprayed on the nickel superalloy or steel substrate. In order to investigate the effect of different material combinations on the failure mechanism of TBC systems, four different combinations of TBC systems are studied. The TBC systems is a composite medium with four layers in cylindrical coordinate system. First the temperature fields for the non-homogeneous problem with energy generation in medium are solved analytically using Taylor transformation and Green's function approach. Secondly the analytical solutions for the thermal stress fields in the composite medium are obtained when the eigen strain rate is taken into consideration. The displacement fields can be written as

$$u_i(r,t) = A_i r + \frac{B_i}{r} + rC_i(r,t) + \frac{D_i(r,t)}{r} \qquad (2)$$

where A_i, B_i, C_i and D_i are determined by boundary conditions. Thirdly, the constitutive equations such as the creep of ceramic coating (PSZ and Mullite) and substrate (Ni-superalloy), plasticity of bond coat are given by a general formula. Fourthly, the thermal grown oxidation and the temperature dependence of thermal–mechanical parameters are taken into consideration.

B Some Important Results

A typical operating state for TBC systems is studied. The highest temperature for the coating surface is 1000°C and the lowest temperature for substrate surface is 700°C. Figure 10 shows the spatial distribution of temperature without and with considering thermal grown oxidation at time of 14000s. When TGO is not considered, the highest temperature difference for substrate exists in Mullite/Steel system and the least temperature difference exists in PSZ/Ni-alloy system. The highest temperature encountered to substrate is 709.57°C for PSZ/Ni-alloy system, however, that is 761.97°C for Mullite/Steel system. Therefore, the PSZ/Ni-alloy system is the optimum combination only if temperature field is considered for TBC systems operating at above typical condition. TGO does not affect the temperature fields in PSZ coating systems. But it effects temperature fields in mullite coating systems. It makes the temperature difference high in substrate.

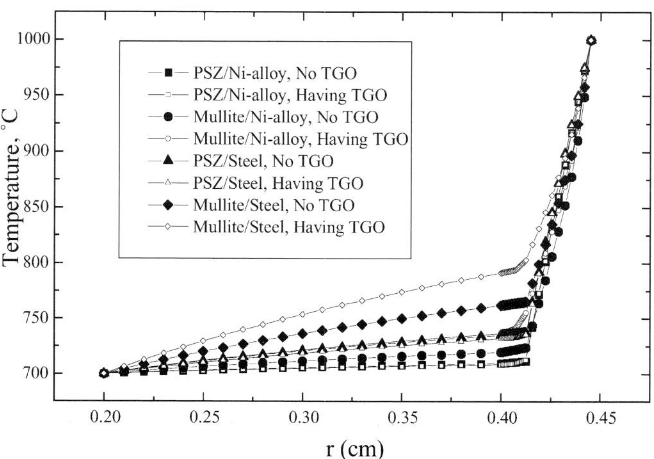

Fig. 10 Spatial distribution of temperature without and with considering thermal grown oxidation at time of 14 000 s.

The elastic stress fields for PSZ/Ni-alloy show that the radial stress is compressive and tangent stress is tensile in the period of heating and heat hold. The maximum values for compressive or tensile radial stresses exist at interfaces. The residual stresses are very small when TGO is not considered. The residual stress with TGO considered is larger than that with TGO ignored. It is very interesting to speculate that TGO may make the residual tangent stress from tensile change to compressive. Figure 11 gives the comparison of spatial distribution of tangent stresses at different time for PSZ/Ni-alloy system with TGO ignored and TGO taken into account.

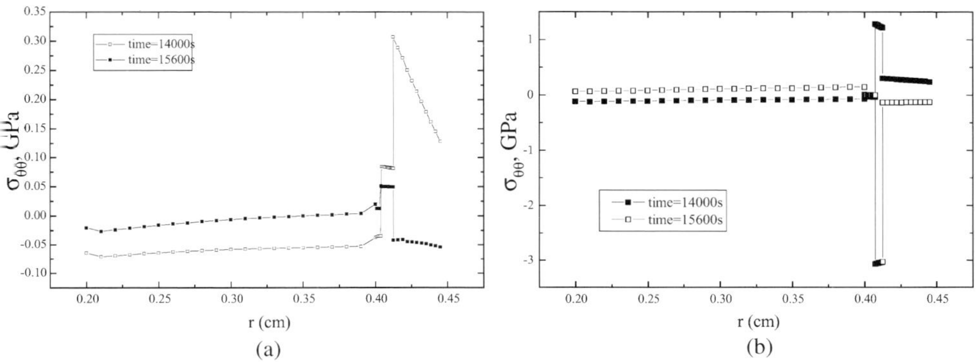

Fig. 11 Spatial distribution of tangent stresses at different time for PSZ/Ni-alloy system, (a) TGO ignored, (b) TGO considered.

The characterisation of thermal stresses in PSZ coating system is very different from that in mullite coating system. The residual radial stresses become more and more low for many cycles of heating/cooling in PSZ/Ni-alloy system. After some finite cycles of heating/cooling, the residual stress becomes compressive and not tensile. However, the residual radial stresses become higher and higher for many cycles of heating/cooling in a mullite coating system. The residual stress is permanently tensile. In the PSZ coating system, the residual tangent compressive stresses become more and more elevated for many cycles of heating/cooling. But in the mullite coating class, the residual tangent is almost constant for many cycles of heating/cooling. The difference of stress characterisation in PSZ coating system and mullite coating system is discussed. It may be due to the difference of mechanical behavior such as creep, thermal mismatch such as thermal expansion coefficients as well as mechanical mismatch which is reflected by Dundurs' parameters defined by Dundurs[32] as

$$\bar{\alpha} = \frac{\Gamma(\kappa_2 + 1) - (\kappa_1 + 1)}{\Gamma(\kappa_2 + 1) + (\kappa_1 + 1)} \quad , \quad \bar{\beta} = \frac{\Gamma(\kappa_2 - 1) - (\kappa_1 - 1)}{\Gamma(\kappa_2 + 1) + (\kappa_1 + 1)} \tag{3}$$

where $\Gamma = \mu_1/\mu_2$, the subscripts 1 and 2 denote ceramic and substrate, respectively.

The investigation of geometrical radius of TBC systems on the thermal stress shows that the effect is very obvious. The geometrical radius not only changes the quantum of thermal stress but also changes the characterisation of thermal stress. The compressive tangent stress for the system in a small radius may become tensile stress for the system in a large radius.

The change of compressive tangent stress to tensile tangent stress may greatly affect the mode of crack formation and propagation such as vertical crack or interface crack as observed in failure experiment in TBCs by laser beam thermal shock as reviewed in the above.

The effect of cooling rate on the residual stress was studied. The effect is due to the high creep rate of ceramic coating operating at high temperature. The effect of cooling rate on the residual stress is not large for systems operating at relative low temperature.

In the conclusion, the failure of TBC systems is the comprehensive effect of radial stress and tangent stress and temperature gradient in TBC systems. Generally, it is the compressive load in ceramic coating that lead the ceramic coating delaminating for typical TBCs system such as PSZ/Ni-alloy. This also means that the failure of TBCs syetem is the coupling effect of temperature gradient, oxidation, thermal fatigue, creep, morphology of the TBC system as well as the cooling rate.

3.2   DELAMINATION CRACKING IN TBCS SYSTEM

A   *Theoretical Model and the Analytical Solution*

The delamination cracking in thermal barrier coating system was studied.[33] The system consists of a ceramic coating of material No. 1 with thickness of h deposited on a super-alloy substrate of material No. 2 with thickness of H. Each material was taken to be isotropic and linearly elastic. A crack parallel to the interface is pre-existing in the interface. The problem is asymptotic in that the two material layers are infinitely long and the crack is semi-infinite. The structure was loaded by the membrane stress P and bending moment M. The temperature gradient along the radial direction was considered as a thermal loading. In addition, the constant stress of $\sigma_{yy} = \sigma$ is also considered. The conventions and geometry for the analysis of delamination cracking in TBC systems are schemed in Fig. 12. The energy release rate G, mode I and mode II stress intensity factors k_1 and k_2 as well as mode mixity measure ψ are derived as

$$G = \frac{h}{C_1} \left[\Pi_1 \left(\frac{P_0 C_1}{h}\right)^2 + \Pi_2 \left(\frac{M_0 C_1}{h^2}\right)^2 + \Pi_3 \left(\frac{P_0 M_0 C_1{}^2}{h^3}\right) + \Pi_4 \left(\frac{P_0 C_1}{h}\right) + \Pi_5 \left(\frac{M_0 C_1}{h^2}\right) \right.$$
$$\left. + (2Z_p \Pi_1 + Z_m \Pi_3)\left(\frac{P_0 C_1}{h}\right)(\sigma C_1) + (2Z_m \Pi_2 + Z_p \Pi_3)\left(\frac{P_0 C_1}{h}\right) (\sigma C_1) \right] \qquad (4)$$

$$k_1 = \sqrt{\frac{h}{b_{11} C_1} \frac{\lambda^{1/4}}{\pi \varsigma}} \left[\sqrt{\Pi_1}\left(\frac{P_0 C_1}{h}\right)\cos\phi + \sqrt{\Pi_2}\left(\frac{M_0 C_1}{h^2}\right)\cos(\phi = \alpha_0) \right] \qquad (5)$$

$$k_2 = \sqrt{\frac{h}{b_{11} C_1} \frac{\lambda^{1/4}}{\pi \varsigma}} \left[\sqrt{\Pi_1}\left(\frac{P_0 C_1}{h}\right)\sin\phi + \sqrt{\Pi_2}\left(\frac{M_0 C_1}{h^2}\right)\sin(\phi = \alpha_0) \right] \qquad (6)$$

The relative amount of mode II to mode I at the crack tip is measured by the angle ψ as

$$\psi = \tan^{-1} \left[\lambda^{-1/2} \frac{\sqrt{\Pi_1}\left(\frac{P_0 C_1}{h}\right)\sin\phi + \sqrt{\Pi_2}\left(\frac{M_0 C_1}{h^2}\right)\sin(\phi = \alpha_0)}{\sqrt{\Pi_1}\left(\frac{P_0 C_1}{h}\right)\cos\phi + \sqrt{\Pi_2}\left(\frac{M_0 C_1}{h^2}\right)\cos(\phi = \alpha_0)} \right] \qquad (7)$$

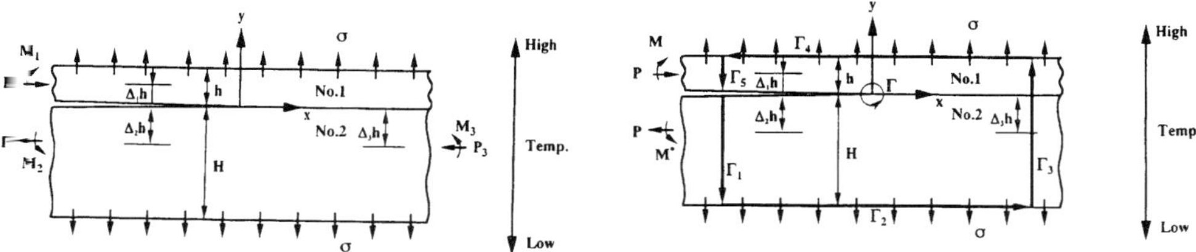

Fig. 12 Conventions and geometry for the analysis of delamination cracking in TBCs system.

where P_0, M_0 are the equivalent membrane stress and equivalent bending moment, but the contribution of constant stress σ is taken out. ϕ and α_0 are angle which are defined in Ref. 33. Other constants such as Π_i, λ et al. are defined in Ref. 33.

E Some Results

The theoretical model is used to study the failure characterizations of TBCs system operating at high temperature. Based on the results of thermal stress fields induced by the combined effect of temperature gradient, oxidation, thermal fatigue as well as creep,[31] the original loads such as P_1, M_1, P_3 and M_3 are easily obtained. The characterisations of original and equivalent loads are studied. Generally, ceramic coating is in tensile state or in compressive state for the system during heating, heat hold or cooling, respectively. The mechanical properties such as creep, plastic behaviours dominate the equivalent loads. Higher creep rate makes the residual loads be higher. The constant stress σ has definite effect on the equivalent loads P but has little effect on the loads M.

Temperature gradient has an important effect on the failure mechanism of TBCs system. The effect is complex and it depends on the Dundurs' parameters $\bar{\alpha}$ and $\bar{\beta}$. The characterisations of Dundurs' parameters $\bar{\alpha}$ and $\bar{\beta}$ depending on the interface temperature lead to the distinct characterizations of comprehensive stress intensity factors for PSZ coating and mullite coating. The comprehensive stress intensity factors may decrease, zero or increase with the increase of interface temperature for PSZ coating system. The comprehensive stress intensity factors increase with the increase of interface temperature for mullite coating system. The relation of K_i with Dundurs' parameters $\bar{\alpha}$ and $\bar{\beta}$ is not known and it may be obtained by non-dimensional analysis that will be done on the further.

The comprehensive stress intensity factors depend on constant stress σ. During heat hold, the comprehensive stress intensity factors become more and more high for mullite/Ni-alloy system when the stress σ changes gradually from high compressive stress to high tensile stress. The comprehensive stress intensity factors become more and more low for other TBC systems when the stress σ changes gradually from high compressive stress to high tensile stress. Figure 13 shows the comprehensive stress intensity factors for different TBC systems in the period of heat hold. On the other hand, the comprehensive stress intensity factors with TGO considered are higher than that with TGO ignored for every TBC system.

The delamination damage of ceramic coating may not mainly be in the form of mode I cracking and it may mainly damage in the form of mode II cracking for the system in the

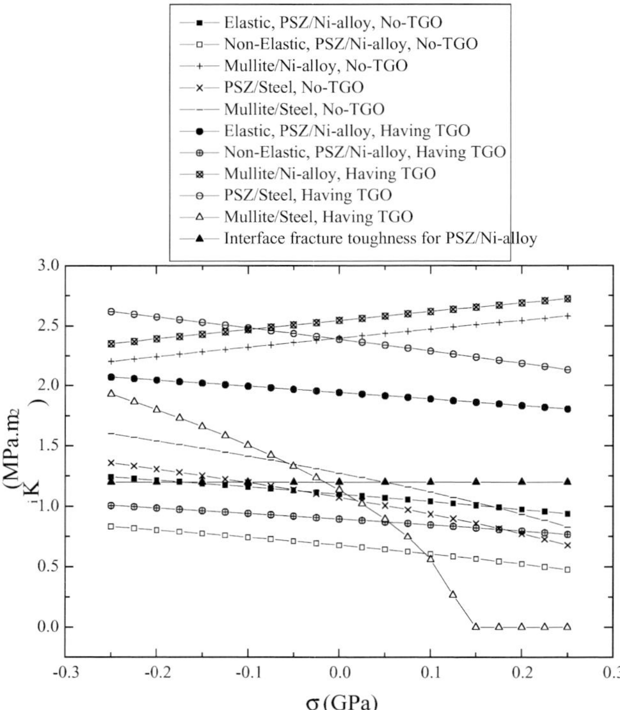

Fig. 13 Comprehensive stress intensity factors for different TBCs system in the period of heat hold.

heat hold. The delamination failure of ceramic coatings may mainly be in the form of mode I cracking and it may not be mainly failure in the form of mode II cracking for the system in the cooling or second and after second heating period.

The investigation of PSZ/Ni-alloy reveals that the TBC system may not fail in the form of coating delamination in the period of heat hold, but it may fail in the form of coating delamination during cooling or in heating period for the second or subsequent cycle. Figure 14 shows the histories of comprehensive stress intensity factors for PSZ/Ni-alloy TBC system with different constant stress σ. The conclusion is consistent with the experimental observations as reviewed above. The delamination of ceramic coating is mainly induced by the compressive load in the coating.

4　CONCLUDING REMARKS

The present article reviews the investigation about the coupled effects of temperature gradient and oxidation on the failure of thermal barrier ceramic coatings in the past ten years in our laboratory. The investigations include experimental and theoretical study.

On the experimental investigation, the heating method, non-destructive evaluation and interface properties are mainly reviewed. In order to study the effect of temperature

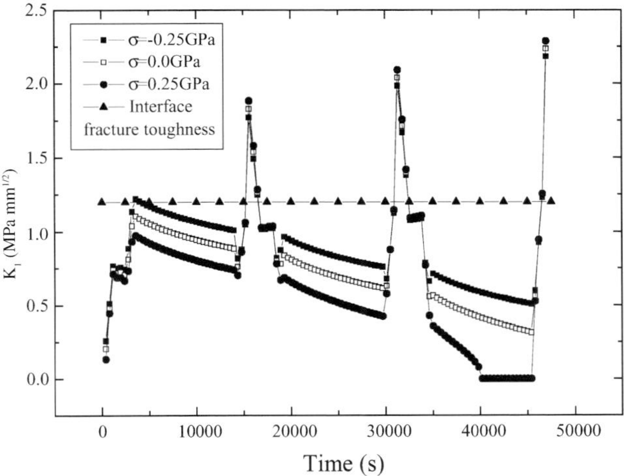

Fig. 14 Histories of comprehensive stress intensity factors for PSZ/Ni-alloy TBCs system with different constant stress σ.

gradient that exists in real TBC systems during service life, one must consider the simulation of thermal shock method. In the last ten years, an attempt has been made to develop four different heating methods, which are furnace heating, burner heating, plasma heating and laser beam heating. The non-destructive evaluation such as temperature test, acoustic emission (AE) signals detected and Impedance Spectroscopy (IS) method can be successfully used in the study of TBCs system degradation. Especially, IS method has been used to test the thickness of thermally grown oxidation in TBCs system. The interface properties such as oxidation and fracture toughness have been studied by test and observation.

With regards to the theoretical investigation, the temperature fields and the related thermal stresses fields have been solved analytically where the coupling effect of temperature gradient, oxidation, thermal fatigue, creep, morphology of the TBC system as well as the cooling rate were considered. The delamination cracking in thermal barrier coating system with ceramic coating deposited on a substrate was studied analytically.

Theoretical and experimental investigations all obtained some important results. However, due to the complication of TBC degradation, a real physical map of TBC system failure has not been obtained up to now. The following further investigation may be proposed.

(1) Experimental Research

The experiments are conducted by designing a system that really simulates the operating state of TBC system. In order to obtain the information of oxidation thickness, micro-cracks or macro-crack formation as well as the evolution of degradation some advanced methods should be developed. New samples should be prepared. The coatings may be divided into three kinds. The first is the non-FGM(functionally graded material) that consists of a PSZ layer (partially stabilized ZrO_2 by 8 wt%Y_2O_3) over a NiCrAlY bond coat. The second is qusi-FGM that consists of multi-layer coating system of PSZ and NiCrAlY. The third is the FGM that consists of continuous gradient chemical compositions of ceramic and NiCrAlY.

(2)   THEORETICAL RESEARCH

In the theoretical model, during heating, temperature hold and cooling, the micro-voids, micro-cracks may nucleate and grow. On the other hand, there must be an oxidation between the bond coating and ceramic coating. Using the experimental information, one model of the evolution of degradation is proposed. The degradation includes the micro-voids, micro-cracks and the oxidation thickness. The constitutive equations include the evolution of degradation. If the voids are large enough, the micro-cracks are long enough and the oxidation is thick enough, the macro-crack initiates to form and propagate. The result may induce the delamination or spallation of the ceramic coatings. With the help of experimental information two models should be proposed. One is failure criterion and another is the process of the crack propagation. For the future research a suitable code for predicting the service time of real TBCs system may be written if the failure mechanism of the TBCs system is clear based upon the above theoretical and experimental results. For writing the code the element finite method should be used for the complicatedly geometrical shape of turbine blades and vanes.

ACKNOWLEDGEMENT

The collaborative research for foreign researchers in Japan is provided to first author YCZ by JSPS (Japan Society for the Promotion of Science). This support is gratefully acknowledged.

REFERENCES

1. S. Amagasa, K. Shimomura, et al., 'Study on the Turbine Vane and Blade for a 1500°C Class Industrial Gas turbine', *Trans. ASME, J. Engng Gas Turbines Power*, 1994, **116**, 597–604.
2. I. Yuri, T. Hisamatsu, K. Watanabe and Y. Etori, 'Structural Design and High Pressure Test of a Ceramic Combustor for 1500°C Class Industrial Gas Turbine', *Trans. ASME, J. Engng Gas Turbines Power*, 1997, **119**, 506–511.
3. Y.R. Takeuchi and K. Kokini, 'Thermal Fracture of Multilayer Ceramic Thermal Barrier Coatings', *Trans. ASME, J. Engng Gas Turbines Power*, 1994, **116**, 266–271.
4. D.M. Zhu, R.A. Miller, 'Investigation of Thermal High Cycle and Low Cycle Fatigue Mechanisms of Thick Thermal Barrier Coatings', *Mater. Sci. Engng. A*, 1998, **245**, 212–223.
5. J. Cheng, E.H. Jordan, B. Barber and M. Gell, 'Thermal/Residual Stress in an Electron Beam Physical Vapor Deposited Thermal Barrier Coating System', *Acta Mater.*, 1998, **46** (16), 5839–5850.
6. V.K. Tolpygo, J.R. Dryden and D.R. Clarke, 'Determination of the Growth Stress and Strain in α-Al_2O_3 Scales During the Oxidation of Fe-22Cr-4.8Al-0.3Y Alloy', *Acta Mater.*, 1998, **46** (3), 927–937.

7   M.Y. He, A.G. Evans and J.W. Hutchinson, 'Effects of Morphology on the Decohesion of Compressed Thin Films', *Mater. Sci. Engng. A*, 1998, **245**, 168–181.

8   J.S. Wang and A.G. Evans, 'Effects of Strain Cycling on Buckling, Cracking and Spalling of a Thermally Grown Alumina on a Nickel-Based Bond Coat', *Acta Mater.*, 1999, **47** (2), 699–710.

9   H.L. Bernstein, J.M. Allen, 'Analysis of Cracked Gas Turbine Blades', *J. Engng Gas Turbines Power*, 1992, **114**, 293–301.

10   H.M. Tawancy, N. Sridhar, et al., 'Failure Mechanism of a Thermal Barrier Coating System on a Nickel-Base Superalloy', *J. Mater. Sci.*, 1998, **33**, 681–686.

11   G.M. Newaz, S.Q. Nusier and Z.A. Chaudhury, 'Damage Accumulation Mechanisms in Thermal Barrier Coatings', *J. Engng. Mater. Tech.*, 1998, **120**, 149–153.

12   K.F. Wesling and D.F. Socie et al., 'Fatigue of Thick Thermal Barrier Coatings', *J. Am. Ceram. Soc.*, 1994, **77** (7), 1863–1868.

13   T.A. Cruse, S.E. Stewart and M. Ortiz, 'Thermal Barrier Coating Life Prediction Model Development', *J. Engng Gas Turbines Power*, 1988, **110**, 610–616.

14   B.C. Wu, E. Chang, et al., 'The Oxide Pegging Spalling Mechanism and Spalling Modes of ZrO_2 8 wt% Y_2O_3/Ni–22Cr–10Al–1Y Thermal Barrier Coatings under various Operating Conditions', *J. Mater. Sci.*, 1990, **25**, 1112–1119.

15   R.A. Miller, 'Life Modeling of Thermal Barrier Coatings for Aircraft Gas Turbine Engines', *J. Engng Gas Turbines Power*, 1989, **111**, 301–305.

16   T. Hashida, H. Takahashi, et al., 'Burner Heating Method for Determining Thermal Shock Resistance of Ceramic Coatings for Gas Turbine Rotor Blades', *Proc. 3rd IUMRS Int. Conf. Advanced Materials '93, III/B: Composites, Grain Boundaries and Nanophase Materials*, M. Sakai et al. eds, 1994, 1291–1294.

17   H. Takahashi, T. Ishikawa, D. Okugawa and T. Hashida, 'Laser and Plasma-Arc Thermal Shock/Fatigue Fracture Evaluation Procedure for Functionally Gradient Materials', *Thermal Shock and Thermal Fatigue Behavior of Advanced Ceramics*, G. A. Schneider and G. Petzow eds, Kluwer Academic Publishers, 1993, 543–554.

18   T. Hashida and H. Takahashi, 'Laser Irradiation Thermal Shock and Thermal Fatigue Test Procedure', *Proc. 1st Int. Symp. On FGM*, Sendai, Japan, 1990, 365–373.

19   C.Y. Jian, T. Hashida, H. Takahashi, N. Shimoda, and M. Saito, 'An Accelerated Testing Method of ZrO_2-Based FGM Coating for Gas Turbine Blades', *Proc. 3rd Int. Structural and Functional Gradient Materials*, Lausanne, Switzerland, 1994, 419–424.

20   C.Y. Jian, T. Shimizu, T. Hashida, H. Takahashi and M. Saito, 'Development of Thermal Shock and Fatigue Tests of Ceramic Coatings for Gas Turbine Blades by AE Technique', *Proc. 12th Int. AE Symp.*, Sapporo, Japan, 1994, 369–374.

21   C.Y. Jian, T. Hashida, et al., 'Thermal Shock and Fatigue Resistance Evaluation of Functionally Graded Coating for Gas Turbine Blades by Laser Heating Method', *Composites Engng.*, 1995, **5** (7), 879–889.

22   C.Y. Jian, 'Study on Evaluation Method of Ceramic Coating System for Gas Turbine Rotator Blades', Doctoral Thesis, Tohoku University, 1996.

23   K. Ogawa, D. Minkov, T. Shoji, M. Sato and H. Hashimoto, 'NDE of Degradation of Thermal Barrier Coating by Means of Impedance Spectroscopy', *NDT & Int.*, 1999, **32**, 177–185.

24   K. Ogawa, T. Shoji, 'Non-Destructive Evaluation of Degradation in High Effeciency Gas Turbine Blades by an Impedence Spectroscopy Method 1st report – Evaluation of Ceramics Coating Thickness', *Proceedings of JSNDI Fall Conference*, 1997, 61–64.

25   K. Ogawa, 'In-Situ Measurement and Mechanistic Understanding of Degradation of Thermal Barrier Coating by Means of Impedance Spectroscopy Technique', Doctoral Thesis, Tohoku University, 1999.

26. A.T. Zehnder and A.J. Rosakis, 'On the Temperature Distribution at the Vicinity of Dynamically Propagating Cracks in 4340 Steel', *J. Mech. Phys. Solids*, 1991, **39**, 385–415.
27. Y.C. Zhou, Z.P. Duan and Q.B. Yang, 'A Thermal-Elastic Analysis on Laser-Induced Reverse Bulging and Plugging in Circular Brass Foil', *Int. J. Solids and Structures*, 1999, **36**, 363–339.
28. H. Niitsuma, M. Kikuchi, H. Takahashi, M. Suzuki and R. Sato, 'AE Classification and Micro Pop-In Cracking in Fracture Toughness Test', *Proc. 5th Int. Symp. Acoustic Emission*, Tokyo, NDI Japan, 1980, 411–420.
29. J.R. MacDonald, *Impedance spectroscopy*, New York, Wiley, 1987.
30. H.M. Jensen, 'The Blister Test for Interface Toughness Measurement', *Engng. Fract. Mech.*, 1991, **40** (3), 475–486.
31. Y.C. Zhou, T. Hashida, 'Coupled Effects of Temperature Gradient and Oxidation on the Thermal Stress in Thermal Barrier Coating', 1999, Submitted to *Acta Metall.*
32. J. Dundurs, 'Edge-Bonded Dissimilar Orthogonal Elastic Wedges', *J. Appl. Mech.*, 1969, **36**, 650–652.
33. Y.C. Zhou, 'Delamination Cracking in Thermal Barrier Coating System', 1999, submitted to *J. Appl. Mech.*

Investigation of Microstructural Development of PtAl Diffusion Coatings

JOHAN ANGENETE, EVA BAKCHINOVA and KRYSTYNA STILLER

Department of Experimental Physics, Chalmers University of Technology and Göteborg University,
SE-412 96 Göteborg, Sweden, email: angenete@fy.chalmers.se

ABSTRACT

Three commercial Pt modified aluminide diffusion coatings (RT22, SS82A and MDC150L) and one conventional aluminide diffusion coating (PWA73), deposited on a commercial single crystal Ni based superalloy (CMSX4) have been investigated. The samples were oxidised in still laboratory air and examined in Scanning Electron Microscope (SEM), Transmission Electron Microscope (TEM) and by Energy Dispersive Spectroscopy (EDS).

In the as-coated condition, several different precipitate families were found in all coatings, but after heat treatment only one (μ-phase) precipitate phase appeared. The presence of Pt, however, affected the precipitation behaviour during manufacture. It was found that the initial amount of Al has a profound influence not only on the diffusional behaviour of the coating, but also on the development of the interdiffusion zone.

INTRODUCTION

In the high temperature zones of gas turbines and aeroengines, Ni-based superalloys are used as base material in blades and vanes. These alloys are superior in terms of structural stability at high temperatures. However, their surface stability has to be improved by the use of coatings. Pt modified aluminide diffusion coatings (hereafter termed PtAl coatings) are mainly used to increase the high temperature oxidation resistance of Ni-based superalloys, but have also found application as bond coats for ceramic Thermal Barrier Coatings (TBC).

The addition of Pt to conventional aluminide coatings has been shown to improve the high temperature oxidation resistance.[1–5] However, the exact mechanisms behind the beneficial effects of Pt addition are still under debate. The present study is a part of a larger project, where the long term goal is to (1) assess the life limiting processes by which a PtAl coating degrades and (2) to clarify the role(s) of Pt. By carefully investigating the coating microstructure and microchemistry and their development during high temperature oxidation and linking them to the performance of the coatings, it will be possible to reach these goals.

EXPERIMENTAL

Three commercial Pt modified aluminide diffusion coatings (RT22, SS82A and MDC150L) and one conventional aluminide coating (PWA73) were applied to the same Ni-based single crystal superalloy (CMSX4). The nominal composition of CMSX4 is shown in Table 1. Discs with 4.2 mm thickness were cut from a 13 mm diameter circular rod of the substrate, oriented with the <001> direction along the rod. The coating

Table 1   Nominal composition of CMSX4.

Composition CMSX4	Ni	Co	Cr	Al	Ti	Ta	Mo	W	Re	Hf
wt%	61.7	9.0	6.5	5.6	1.0	6.5	0.6	6.0	3.0	0.10
at%	63.7	9.3	7.6	12.6	1.3	2.2	0.4	2.0	1.0	0.03

procedures were carried out by Chromalloy UK (RT22), Howmet Turbine Components Coating (SS82A), Thermatec (MDC150L) and Volvo Aero Corporation (PWA73).

Three of the coatings (RT22, SS82A and PWA73) are inward grown coatings, i.e. the slowly diffusing elements from the substrate are present throughout the entire coating structure, while MDC150L is an outward grown coating and as such contains less of those elements in the outer parts. The Pt in RT22, SS82A and MDC150L is deposited by electroplating of Pt to a thickness between 2 and 9 μm, SS82A containing somewhat less than the other two.

The samples were oxidised in still laboratory air at 1050°C, the PtAl samples for 50, 100, 200 and 500 h, while PWA73 was oxidised for 100, 200 and 1000 h. Of the PtAl coatings, two specimens were oxidised simultaneously for all times except 50 and 100 h, while only one PWA73 sample was oxidised for each time.

Before and after oxidation, all samples were weighed in a Sartorius R160P balance with a maximum readability of 10 μg, and the net mass gain per area unit was calculated.

Cross-sections of the samples were studied in backscatter mode in a CamScan S4–80DV Scanning Electron Microscope (SEM), equipped with a Link eXL Energy Dispersive Spectrometry (EDS) system. The samples were prepared by sectioning, moulding into conductive bakelite and polishing down to 1 μm metallographical finish. Cross-sections of the samples were also investigated in a JEOL 2000fx Transmission Electron Microscope (TEM), operated at 200 kV, with a Link AN10000 EDS system. The TEM samples were prepared by the technique described by Skogsmo and thinned down to electron transparency in a Gatan Precision Ion Polishing System (PIPS).[6, 7]

From cross-section backscatter images, volume fractions of precipitates were calculated, using a model for correction of the sectioning effect.[8]

It should be noted that the quantitative results from TEM-EDS and SEM-EDS are not calibrated against each other and therefore compositions measured with the two different techniques may deviate slightly. This is mainly a result of absorption effects in the NiAl system.[9]

RESULTS

RT22

A cross-section of the RT22 coating in the as-coated condition is shown in Fig. 1. It had an Outer Zone (OZ) that extended about 40 μm below the surface. Between the OZ and the substrate, a 10 μm wide Interdiffusion Zone (IZ) had formed. The thickness of the OZ did

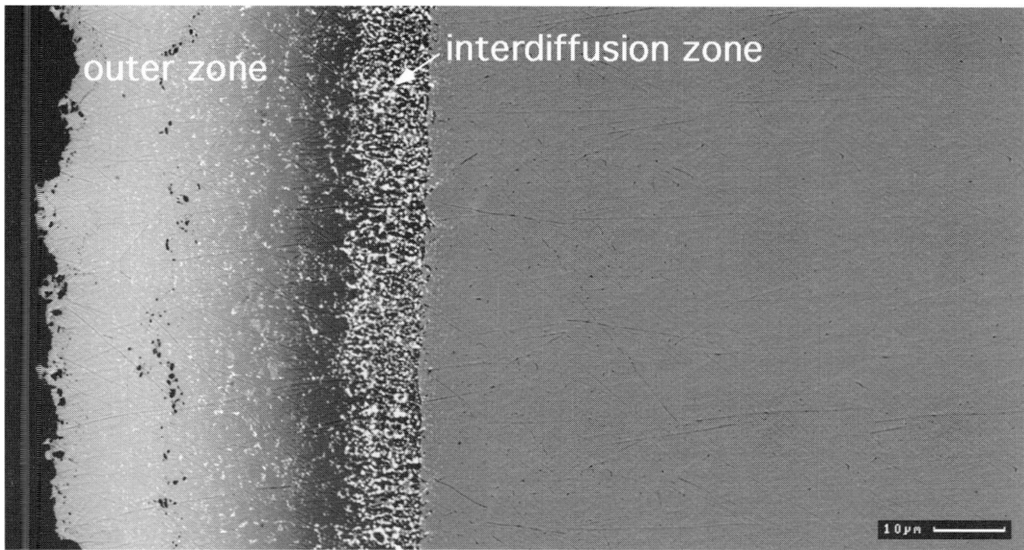

Fig. 1   Cross-section of RT22 as the as-coated condition.

however vary slightly between different samples and also at different positions on the same sample. On average, the OZ was about 41 μm thick and the IZ measured 10 μm.

In the as-coated condition, RT22 was found to contain several families of precipitates. Their average composition (measured by TEM-EDS) is shown in Table 2. In the outmost region, numerous small (about 100 nm) α-W and some σ-phase precipitates were found. In the IZ, both σ- and μ-phase and some small (about 100 nm) α-Cr precipitates were observed. The σ- and μ-phase precipitates were mainly blocky and their typical size varied between 0.2 and 0.5 μm. After heat treatment for 100 h, only μ-phase precipitates were present. The precipitate size increased with time but the volume fraction of the precipitates outside the IZ was found to be constant between 100 and 500 h. The thickness of the OZ was nearly constant during oxidation, but the total thickness of the coating structure increased by growth of the IZ into the substrate. After 500 h, the total coating thickness was about 65 μm, which means that the IZ grew about 200% in thickness.

In the IZ, the matrix consisted of two regions, one with β-NiAl and another, mainly with γ'-Ni$_3$Al. Figure 2 shows a Selected Area Electron Diffraction (SAED) pattern from a

Table 2   Average composition of the precipitate phases found in RT22 in the as-coated condition, as measured by TEM-EDS. The error limits represent the standard deviation of several measurements.

Phase	Co	Ni	Al	Cr	Mo	Re	Ta	Pt	W	Ti
α-W	2.6±0.4	18.3±5.7	22.8±8.5	3.1±0.8	2.2±0.6	4.1±3.4	6.4±4.3	4.4±2.7	34.2±9.1	0.3±0.5
μ	17.2±1.3	16.6±3.3	6.1±2.8	25.5±2	2.2±0.6	5.4±1.4	4±1	0.4±0.7	21.7±2.9	0.3±0.2
σ	17.6±4.9	30±12.7	11.7±6	19.7±5.7	1.7±1	1.5±1	5.4±4.6	1.2±1.2	9.4±4	0.5±0.4
α-Cr	11.7±9.2	19.5±7	8.7±3.1	48.5±1.7	1.2±0.7	3.6±0.3	0.8±0.1	0±0.1	5.1±1.6	0.6±0.3

γ'-Ni$_3$Al grain in the IZ with characteristic diffraction points, typical for the superlattice of the γ'-Ni$_3$Al structure. The boundary between the β-NiAl and γ'-Ni$_3$Al regions moved outwards through the IZ during heat treatment and after 500 h, protrusions with γ'-Ni$_3$Al reached the limit between the IZ and the OZ.

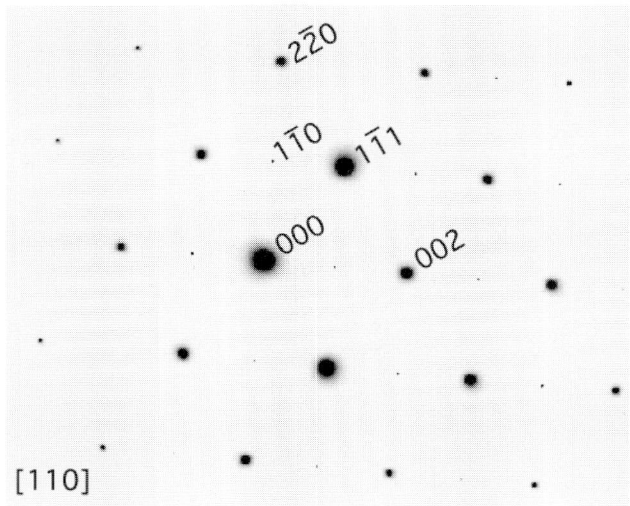

Fig. 2 SAED pattern of γ'-Ni$_3$Al grain in the IZ of RT22 after 500 h oxidation, showing characteristic superlattice reflections.

Around the μ-phase precipitates in the lower part of the IZ, the surrounding matrix was transformed to β-NiAl. These regions became larger after 200 h, but after 500 h, they were transformed to less Al rich phase, Fig. 3.

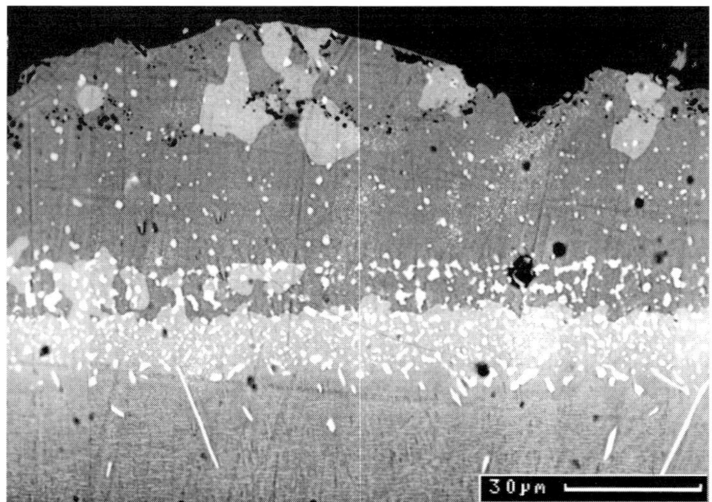

Fig. 3 Cross-section of RT22 after 500 h oxidation.

Typical TEM-EDS profiles of the matrix across RT22 in the as-coated condition and after 500 h heat treatment are shown in Fig. 4. From these, a rough estimate of the ratio between the average amounts of Al and Ni in the OZ can be done, giving about 1 in the as-coated condition and about 0.6 after 500 h.

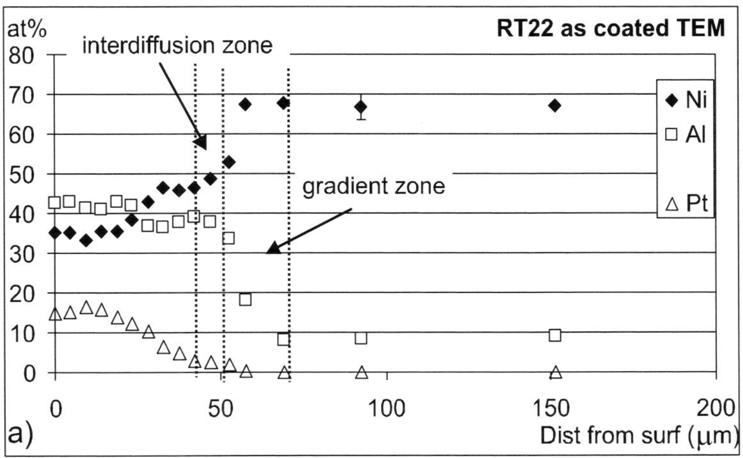

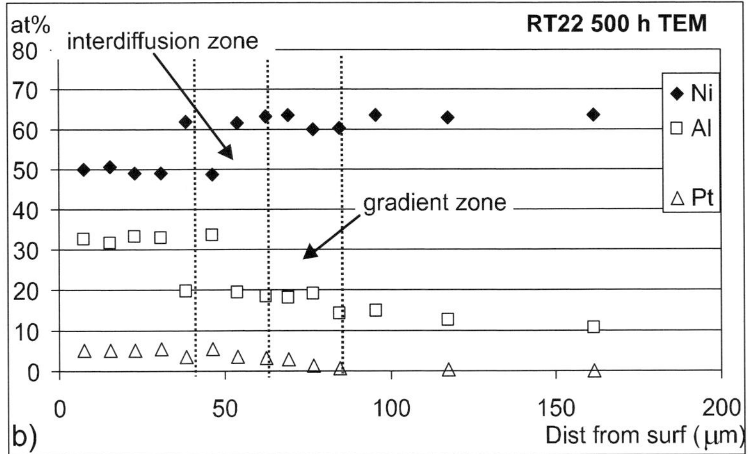

Fig. 4   TEM-EDS profiles of the matrix of the RT22 coating (a) in the as-coated condition and (b) after 500 h heat treatment. The profiles reach from the surface of the coating and into the substrate.

It is clear that considerable interdiffusion occurred during heat treatment, i.e. inward diffusion of Al and Pt, while Ni diffused through the IZ into the OZ. Moreover, an Al gradient was observed below the IZ, forming a zone termed the Gradient Zone (GZ). After 500 h heat treatment, the typical distance from the surface at which the amount of Al was higher than the substrate composition was about 80 µm. It should be noted that some occasional needle-shaped µ-phase precipitates were observed in the GZ after 500 h.

SS82A

In the as-coated condition, SS82A had a two-phase structure in the outmost 20 μm, consisting of PtAl$_2$ and β-NiAl, Fig. 5. The composition of these two phases, measured by TEM-EDS is shown in Table 3. As in RT22, numerous α-W precipitates were observed in the OZ, but the outmost 10 μm of the coating was free of precipitates. In the two-phase region, the α-W precipitates were predominantly observed in the PtAl$_2$ grains, but also some precipitates were found in the β-NiAl grains, Fig. 5. Occasional σ-phase precipitates were found in the outmost part of the OZ and also in the IZ. Moreover, μ-phase particles were observed in the inner parts of the OZ and in the IZ. The composition of the precipitate phases in this coating was very similar to those found in RT22. In the as-coated condition, the thickness of the OZ was around 80 μm (20 μm two-phase zone plus 60 μm intermediate zone) and the IZ was about 10 μm thick.

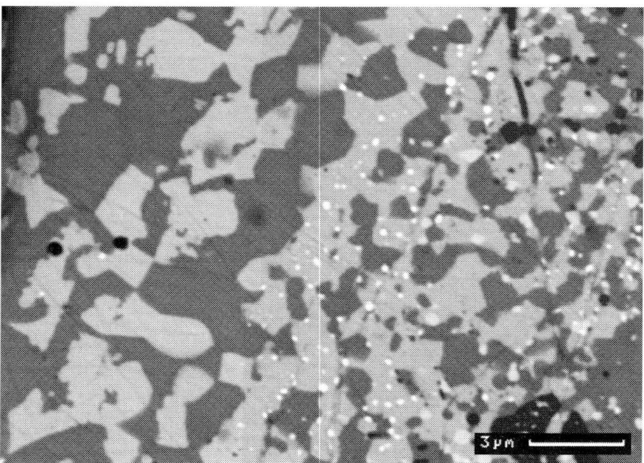

Fig. 5   Cross-section of the outmost part of SS82A in the as-coated condition.

Table 3   Composition of the matrix phases in SS82A in the as-coated condition, as measured by TEM-EDS. The error limits represent the standard deviation of several measurements.

Phase	Co	Ni	Al	Cr	Mo	Re	Ta	Pt	W	Ti
PtAl$_2$	2±0.1	11.4±0.3	54.6±0.7	7.4±0.4	0.6±0.2	−0.1±0.3	−0.2±0.2	22.2±0.9	0.3±0.1	1.4±0.1
β-NiAl	5.5±0.3	36.3±1.7	53.8±2.4	1.1±0.4	0.1±0.1	0±0.1	−0.1±0	3±0.5	0.1±0.1	0.1±0

After 50 h heat treatment, the PtAl$_2$ grains were dissolved, leaving a β-NiAl matrix enriched in Pt. As in RT22, only μ-phase precipitates were present after heat treatment. A zone near the surface remained virtually precipitate-free, showing that SS82A contained less of the precipitate-forming elements than RT22 in the vicinity of the surface. As for RT22, the volume fraction of precipitates in the OZ was constant, but the precipitate size increased, during heat treatment.

As in RT22, β-NiAl regions formed around the μ-phase precipitates in the inner parts of the IZ, but the transformation back to the Al-deficient phase was not complete after 500 h heat treatment.

Another similarity with RT22 was that during heat treatment, the thickness of the OZ was nearly constant, while the thickness of the IZ increased and that a GZ grew into the substrate. The thickness of the IZ increased however much more than in RT22, about 500% after 500 h, giving a total coating thickness of about 130 μm. No needle-shaped precipitates were observed in the GZ.

TEM-EDS profiles of the coating in the as-coated condition and after 500 h are shown in Fig. 6. Estimates of the average Al/Ni ratio in the OZ gave about 1.7 in the as-coated condition and about 0.5 after 500 h. After 500 h, the Al profile extended about 130 μm from the surface.

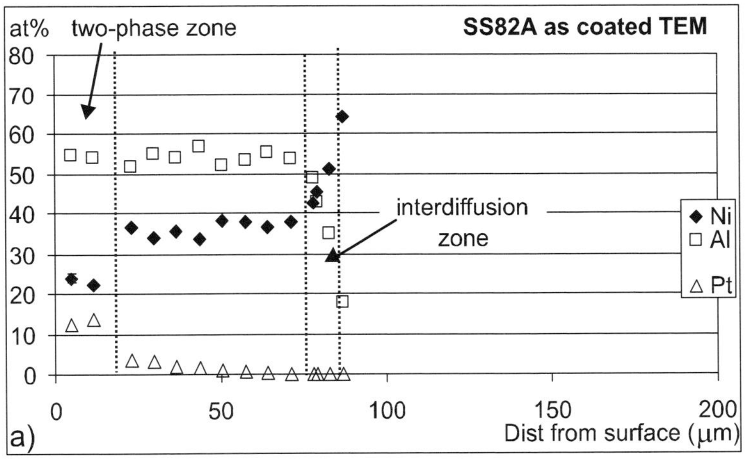

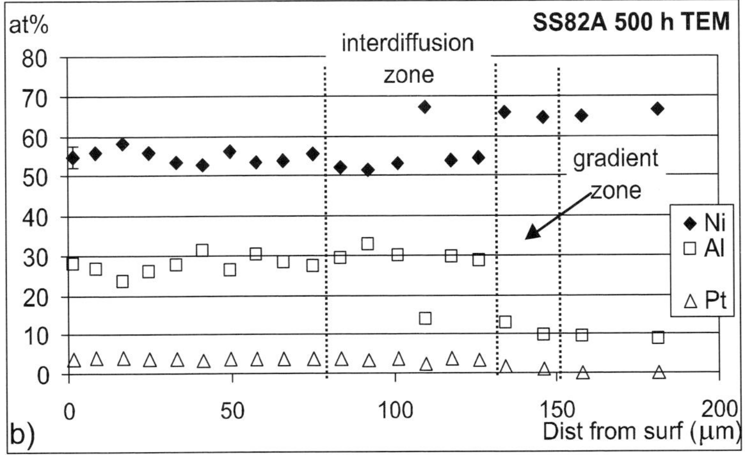

Fig. 6   TEM-EDS profiles across SS82A (a) in the as-coated condition and (b) after 500 h.

MDC150L

The MDC150L coating did not contain any precipitates at all in the about 20 μm thick OZ, Fig. 7. The 20 μm thick IZ consisted of σ- as well as μ-phase precipitates with composition similar to the other two PtAl coatings. At the boundary between the IZ and the OZ, an array of Al_2O_3 grains was observed. Most likely, these were remnants from the grit blasting process prior to the coating cycle and can as such be considered as inert markers of the original component surface.

During heat treatment, the thickness of the OZ and the IZ remained fairly constant, apart from some few needle shaped μ-phase precipitates that protruded as far as about 20 μm from the IZ into the substrate after 500 h, Fig. 8. Only μ-phase precipitates were found after heat treatment, and again, they had the same composition as was observed in RT22.

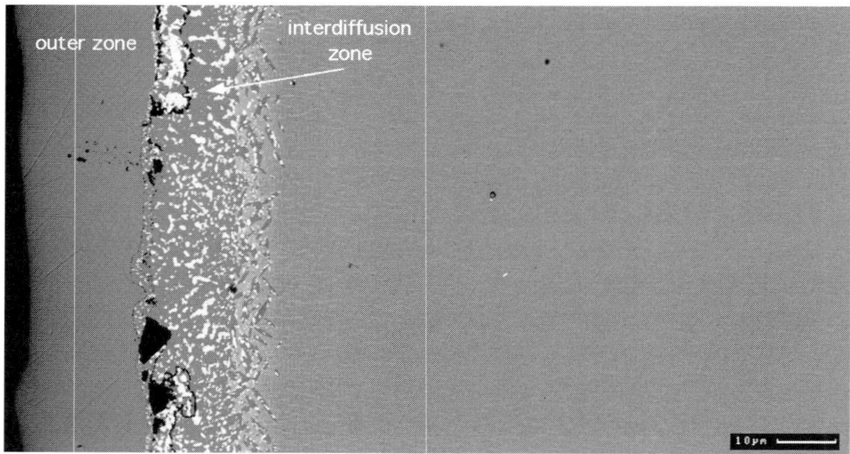

Fig. 7   Cross-section of MDC150L in the as-coated condition.

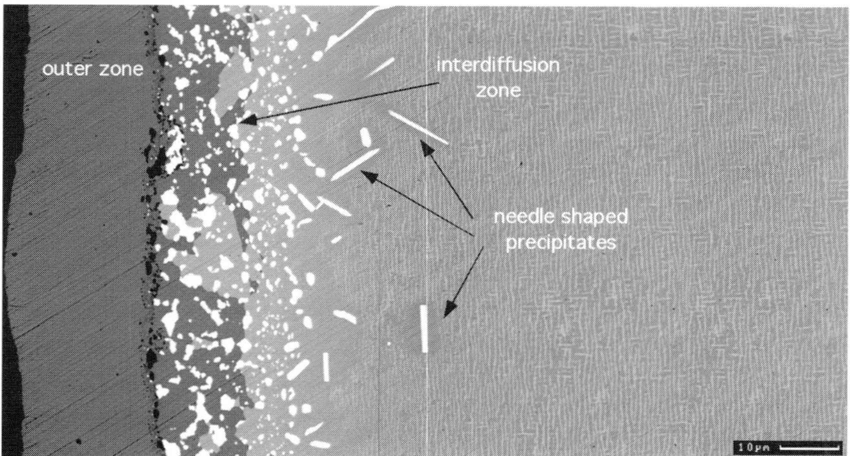

Fig. 8   Cross-section of MDC150L after 500 h heat treatment.

As in the other two PtAl coatings, β-NiAl grains formed around precipitates near the substrate during heat treatment up to about 100 h. After 500 h, these had been transformed to the less Al-rich phase.

The Pt profile in MDC150L was much smoother than in the other two PtAl coatings in the as-coated condition, Fig. 9. This coating also had the lowest Al content in the OZ. The average Al/Ni ratio was found to be about 0.8 in the as-coated condition and decreased to about 0.5 after 500 h.

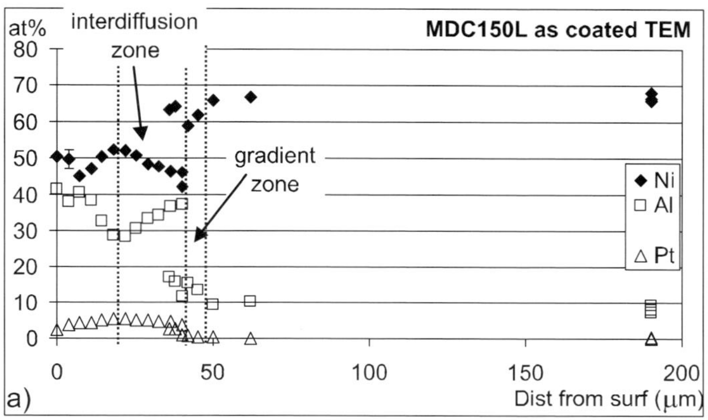

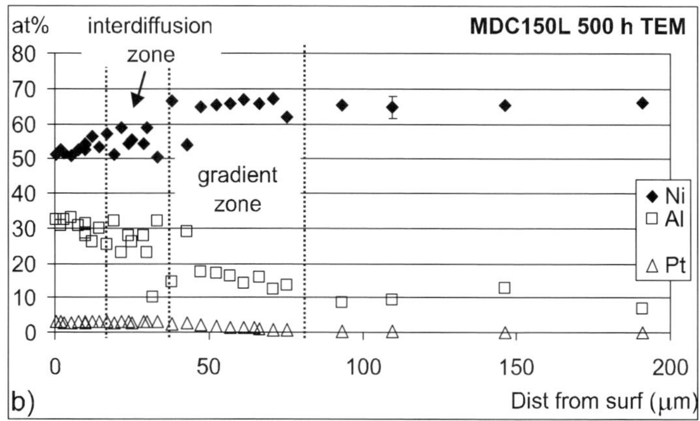

Fig. 9 TEM-EDS profiles across MDC150L (a) in the as-coated condition and (b) after 500 h.

A wide GZ formed below the IZ after 500 h heat treatment, Fig. 9, reaching about 30 μm below the IZ.

PWA73

A cross-section of the PWA73 coating in as-coated condition is shown in Fig. 10. The OZ was about 55 μm thick, and was decorated with numerous precipitates. The thickness of the IZ was about 18 μm. In the as-coated condition, most of the precipitates in the OZ were found to be α-Cr, but also σ-phase precipitates and a few α-W particles were observed. In the IZ, only μ-phase precipitates could be detected. Again, the μ-phase composition did not differ largely from the composition observed in the other coatings.

Fig. 10   Cross-section of PWA73 in the as-coated condition.

After heat treatment for 200 h, the thickness of the OZ remained nearly constant, but the thickness of the IZ increased to 30 μm, see Fig. 11. Moreover, as in the PtAl coatings, only μ-phase precipitates were observed both in the OZ and in the IZ.

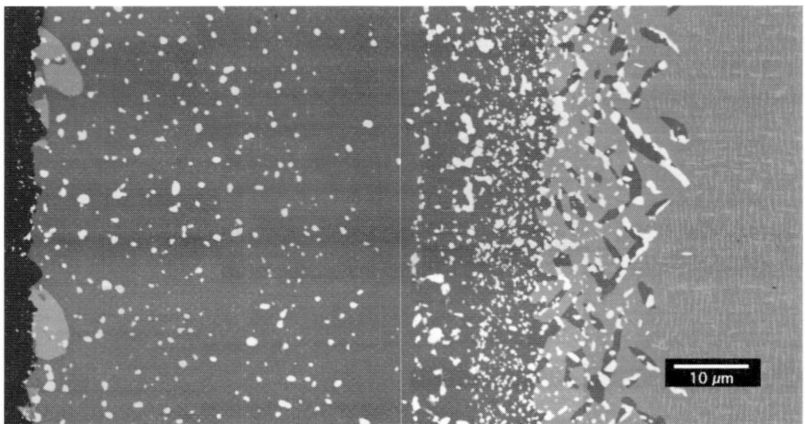

Fig. 11   Cross-section of PWA73 after 200 h heat treatment.

No compositional TEM-EDS profiles of this material have been acquired at this stage.

DISCUSSION

In the as-coated condition, the Al concentration in the outer zone in the PtAl coatings was highest for the SS82A coating, intermediate for RT22 and lowest for MDC150L. Comparison of these observations with the measured growth of the IZ of the PtAl coatings indicates that a high initial Al content in the OZ causes a large growth of the IZ. This implies that the increased Al concentration in the substrate due to the inward Al diffusion reduces the solubility of precipitate-forming elements such as Cr, Co and W. Correlation between the position of the IZ/substrate interface and the Al profiles suggests that there is a critical Al concentration (lower than 20 at%) at which μ-phase precipitates start to form. In the case of RT22 and MDC150L, an about 30 μm long tail of the Al profile, formed the GZ. In this region, the Al content was too low for extensive precipitation of μ-phase, and only some occasional needles were formed. In SS82A, on the other hand, the Al concentration in the GZ was lower and no needle-shaped precipitates formed.

The formation of the β-NiAl zones in the vicinity of μ-phase precipitates in the IZ is explained by the relatively high amount Ni incorporated into the precipitates. This causes local Ni depletion around the precipitates, promoting formation of β-NiAl. During prolonged heat treatment, Ni is supplied by the substrate and the β-NiAl regions are transformed to a less Al-rich phase. The large inward flux of Al in SS82A however delays this transformation.

The fact that the volume fraction of the precipitates in the OZ was constant over time (from 100 h to 500 h) in both RT22 and SS82A shows that the flux of precipitate-forming elements as Cr, Co and W to (or from) the OZ was balanced by a change of their solubility in the β-NiAl matrix. This solubility change is naturally connected to the change in β-NiAl composition, as Al and Pt diffuse towards the substrate.

The location of the numerous α-W precipitates in RT22 and SS82A in the as-coated condition coincides with high concentrations of Pt, which indicates that Pt has an effect on the stability of this phase. This is supported by the rare occurrence of α-W in the sample without Pt, PWA73. In this latter coating, large amounts of α-Cr precipitates were found in the as-coated condition. This supports the previously reported effect of Pt to reduce precipitation of this phase.[10] Instead of forming α-Cr precipitates, Cr is incorporated into the Cr-rich σ-phase in as-coated PtAl coatings.

The fact that only μ-phase precipitates exist after heat treatment in the conventional aluminide as well as in all the PtAl coatings shows that the substrate composition has a strong influence on which precipitate phase that is stable during heat treatment. It also shows that at Pt concentrations of about 3 to 5 at%, Pt does not affect the stability of μ-phase precipitates. The composition of the μ-phase precipitates varies within very small limits in all samples, indicating that this phase had reached a stable configuration with respect to composition.

SUMMARY

- The development of PtAl coatings during high temperature exposure is largely dependent on the initial Al concentration in the coating. High Al levels promote a rapid growth of the IZ.

- There appears to be a critical Al level at which precipitation of μ-phase in the substrate is caused by inward diffusion of Al. This is the mechanism by which the IZ grows inwards during heat treatment.

- The presence of Pt affects the precipitation characteristics of the coating during manufacture, in terms of promoting α-W and suppressing α-Cr.
- The substrate composition dictates which precipitate phase that is stable at 1050°C. Low amounts of Pt (around 3 to 5 at%) appear not to have any effect on the stability of μ-phase.

ACKNOWLEDGEMENTS

The High Temperature Corrosion (HTC) centre, Volvo Aero Corporation and ABB Stal AB are acknowledged for financial support and supply of samples.

REFERENCES

1. K. Bungardt, G. Lehnert and H. Meinhardt, US Patent #3,677,789, 1972.
2. K. Bungardt, G. Lehnert and H. Meinhardt, US Patent #3,692,554, 1972.
3. K. Bungardt, G. Lehnert and H. Meinhardt, US Patent #3,819,338, 1974.
4. M. Goebel, A. Rahmel, M. Schuetze, M. Schorr and W. Wu, 'Interdiffusion between the platinum-modified aluminide coating RT 22 and nickel-based single-crystal superalloys at 1100 and 1200°C', *Mater. High Temp.*, 1994, **12**(4), 301–309.
5. P. Korinko, M. Barber and M. Thomas, 'Coating characterization and evaluation of directionally solidified CM 186 LC and single crystal CMSX-4', 1996 International gas turbine and aerongine congress and exhibition, Birmingham, UK 1996.
6. S. Vuorinen and J. Skogsmo, in *Surface modification techniques*, T. Sudarshan and D. Bhat eds. TMS, 1988.
7. S. Vuorinen and J. Skogsmo, 'Characterization of α-Al$_2$O$_3$, κ-Al$_2$O$_3$ and α-κ multioxide coatings on cemented carbide', *Thin Sol. Films.*, 1990, **193–194**, 536–546.
8. J. Angenete, Licentiate Thesis, Chalmers University of Technology and Göteborg University 1999.
9. D. Williams and C. Carter, *Transmission electron microscopy*, Plenum Press, 1996.
10. H. Tawancy, N. Abbas and T. Rhys-Jones, 'Role of platinum in aluminide coatings', *Surf. Coat. Technol.*, 1991, **49**(1–3), 1–7.

Diffusion Barriers for Gas Turbine Superalloy Blades

M. UUNONEN, P. KASKI, P. HENTTU and P. KETTUNEN

Institute of Materials Science, Tampere University of Technology, Tampere, Finland

ABSTRACT

Nickel base superalloys are protected against high temperature oxidation by, for example, MCrAlY type overlays. However, the diffusion of Ni from the superalloy and that of Al from the overlay causes detrimental structural changes in the overlay as well as in the sections of superalloy close to the interface. In a European concerted action COST 501/II programme, a diffusion barrier of the type TiN+AlN+TiN was developed for preventing this detrimental diffusion up to 1100°C. In this programme the bonding of the developed diffusion barrier to the superalloy remained insufficient. The work of solving the bonding problem is continued in a BRITE-EURAM project. On the basis of laboratory scale tests, suitable candidates for bonding layers have been found, and they will be tested during the second part of the project on simple test pieces as well as on real blades in conditions simulating the real gas turbine conditions.

1 INTRODUCTION

The inlet temperature is one of the most important factors determining the efficiency of a gas turbine – the higher the inlet temperature, the better the efficiency. The inlet temperature, however, depends on the temperature that the blades of the hottest zone tolerate during continuous operation of the turbine. This is why a continuous search for more temperature-resistant materials for the blades of the hottest zone has been common in the development of gas turbines during the last decades. Today, nickel base superalloys permit the metal temperature in the hottest zone to reach values of about 950°C. By using gas cooling and thermal barriers, the inlet temperature can be raised somewhat above the metal temperature of the blades, as is done in aeroturbines, but still the maximum metal temperature of the blades decisively determines the inlet temperature and thus also the efficiency of the turbine.

Superalloys are exceptional materials in the sense that their strength stays reasonably good up to temperatures close to their melting point. This is due to their microstructure containing large γ' precipitates of the type Ni_3Al which in many alloys start to dissolve first at temperatures well above 1100°C. Thus, possibilities to raise the metal temperature of the blades of the hottest zone from the present values seem to exist.

One of the problems, appearing at temperatures at and above 1000°C, is the diffusion between the superalloy blades and their overlays protecting the blades against high temperature corrosion and acting as a bonding layer between the blades and their thermal barriers. The overlays are of the type MCrAlY, where M means Ni, Co or their alloys. The protective capability is due to the formation of an Al_2O_3 layer in oxidizing atmospheres. At and above 1000°C, the diffusion of Al from the overlay into the superalloy causes Kirkendall cavities at the intermediate boundary, diluted zone of precipitates close to it, and precipitates in the upper section of the superalloy. These phenomena may reduce the strength properties in the vicinity of the boundary. Diffusion of Ni from the superalloy into the overlay reduces, in turn, the protective capability of the overlay. Thus, diffusion of Al and Ni at these temperatures damages both of the structures; therefore an effective diffusion barrier is needed between the superalloy and its protective overlay at metal temperatures of 1000°C and above.

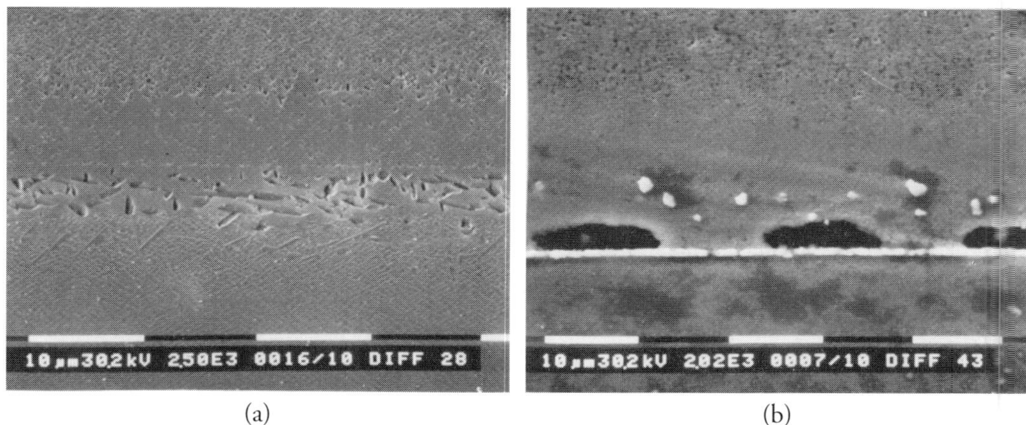

(a)   (b)

Fig. 1 Phenomena appearing at the intermediate boundary between the superalloy MA 6000 and the overlay CoNiCrAlYTa: (a) diluted zone of precipitates above and large precipitates below the boundary after an annealing for 50 h at 1000°C, (b) diluted zone and Kirkendall cavities after an annealing for 50 h at 1100°C.

The basic requirement for such a barrier is that it is capable of diminishing the diffusion of Al and Ni at the required metal temperature. The second requirement is that the barrier can be bonded sufficiently well to the superalloy as well as to the overlay. The bonding must also be capable of surviving the thermal shocks, temperature variations and loading that the gas turbine blades will be exposed to.

2 DEVELOPMENT OF DIFFUSION BARRIERS

The development of suitable diffusion barriers was taken as one of the tasks in the European concerted action COST 501/II programme. First the goal was set at 1000°C. At this temperature, a single dense TiN layer bonded by Ti was capable of reducing the detrimental phenomena due to the diffusion of Ni and Al, as shown in Fig. 2. At 1100°C, however, the capability of the single barrier was found insufficient to stop the diffusion of Al.

After this, the goal was set at 1100°C. Also the superalloy and the overlay were changed, the former to SRR99 and the latter to LCO22 (Amdry 995). A triple layer of TiN+AlN+TiN bonded with Ti appeared in laboratory tests to be capable of reducing the diffusion phenomena sufficiently. Figure 3 illustrates the barrier and its bonding more closely.

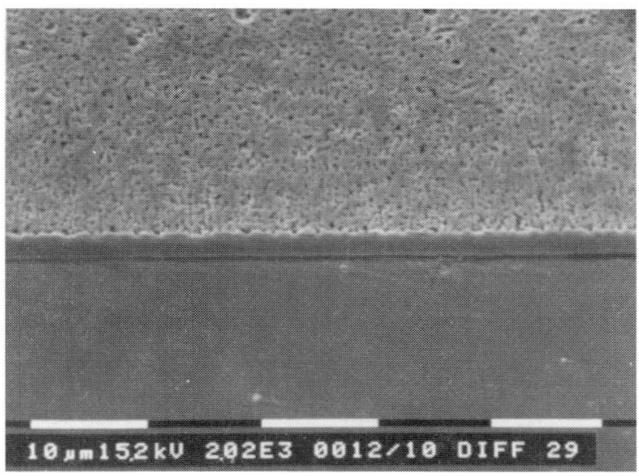

Fig. 2 The boundary between superalloy MA 6000 and the overlay CoNiCrAlYTa with a diffusion barrier of TiN after annealing for 50 h at 1000°C.

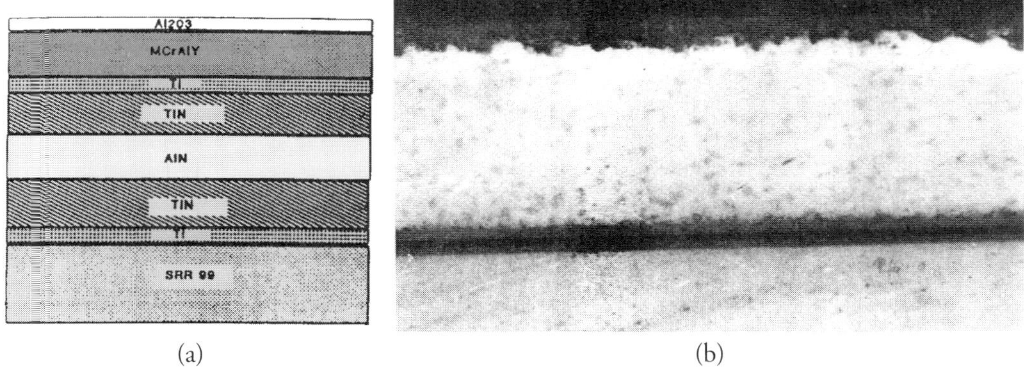

(a)   (b)

Fig. 3 The developed multilayer diffusion barrier: (a) schematically, (b) as seen by scanning electron microscopy.

In burner rig tests used for the blades of aero gas turbines, the diffusion barrier together with the overlay spalled off due to the used bonding. This indicated that a better bonding is required at this higher temperature.

3 DEVELOPMENT OF THE BONDING

The development of a suitable bonding for the TiN+AlN+TiN multilayer diffusion barrier at the temperature of 1100°C has been the subject of the present BRITE-EURAM project being active for three years. The basic ideas of the development can be characterised as follows:

- no phases melting at or below 1200°C are allowed to be formed due to the used bonding layer(s).
- the thermal expansion and lattice parameters of the adjacent layers should not differ by more than 15–30% from one another.

- limited amount of solubility between adjacent layers is preferred.
- the bonding layer(s) should be as thin as possible in order to increase their strength.

3.1 SEARCH FOR THE BONDING LAYERS

Thermodynamic calculations as well as experimental results showed that a continuously changing graded bond layer from the composition of the superalloy substrate to that of the TiN layer of diffusion barrier was not possible without forming of phases melting below 1200°C. For this reason, the idea of graded bonding was abandoned, and the effort was directed towards developing layers where all the above mentioned ideas are fulfilled.

Figure 4. illustrates the thermal expansion behavior of the used superalloys, SRR99 and IN 738, of the components of the diffusion barrier, and of some of the developed candidates for bonding layer(s). The expansion values are measured from small samples produced by HIPing and subsequent annealing. The actual sputtered bond layers may expand slightly differently, but directionally the values of Fig. 4 are applicable. Table 1 correspondingly shows the lattice parameters of the alloys in question.

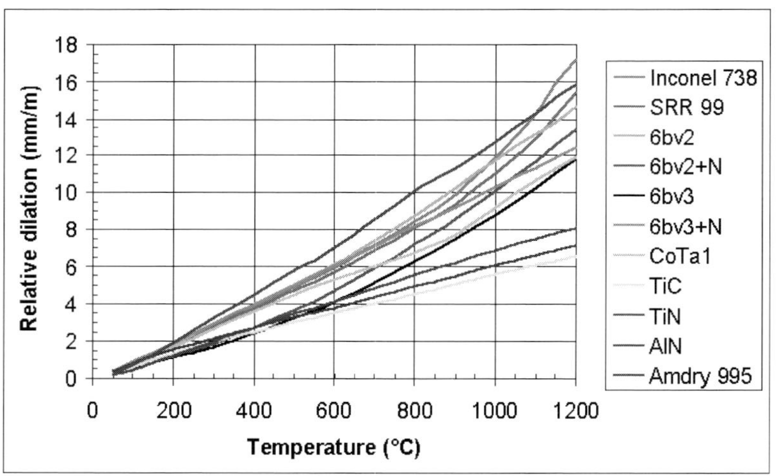

Fig. 4   Thermal expansion of the studied materials as a function of temperature.

For bond strength measurements, coated samples were produced from each of the bond layers so that the substrate was always a superalloy and the bond layers were between the superalloy and TiN and between TiN and the overlay. Bond layers, identified as 6bv2 and 6bv3, were used in two different conditions; as such and as nitrided on the side against TiN. The latter condition was produced by changing the sputtering atmosphere from argon to nitrogen. Table 2 describes examples of such coated samples.

3.2 TESTING OF BONDING STRENGTH

The bonding ability of bond layers was tested in two conditions, after production of coated samples and after annealing in purified argon for 24 h at 1100°C, first at the room

Table 1 Melting points, thermal expansion coefficients, and lattice parameters of the studied materials.

Materials	Melting point (°C)	TEC at RT (μm m^{-1} K^{-1})	TEC at 1150°C (μm m^{-1} K^{-1})	Lattice system	Lattice parameter (Å)
Superalloys:					
SRR99	>1200	9.05	23.25	cubic	$a = 3.6$
IN738	>1200	9.33	28.51	FCC	$a = 3.6$
DB's:					
TiN	2949	8.07	9.31	cubic	$a = 4.2$
AlN	2300	5.54	5.65	cubic	$a = 3.11$
Bond layers:					
6bv2	~1200	10.06	19.27	cubic	$a = 2.2$
6bv3	>1200	8.95	14.92	cubic	$a = ?.?$
CoTa1	>1200	7.45	15.43	CPH	$a = 2.4$ $c = 4.1$
CoTa2	>1200				
TiC	3067	4.82	7.09	FCC	$a = 4.3$
Top coatings:					
LCO22	>1200	13.0	17.8	CPH	$a = 2.5$ $c = 4.2$
Amdry 995	>1200	11.7	15.66	CPH	$a = 2.6$ $c = 4.2$

Table 2 Examples of the results from tensile tests on coated samples.

Specimen	Coating type	Fracture strength (N mm^{-2})	Place of rupture
9E12 HT	6bv2/TiN/6bv2/LCO22	69	glue
9E13 HT	CoTa1/TiN/CoTa1/Amdry	58	glue
9E14 NHT	CoTa1/TiN/CoTa1/Amdry	18	Interface CoTa/TiN
9E15 NHT	6bv2/TiN/6bv2/Amdry	68	glue
9E16 NHT	6bv2/TiN/6bv2/N+Amdry	68	glue
9E17 HT	CoTa1/TiN/CoTa1/N+Amdry	57	glue
9E18 NHT	6bv2/TiC/TiN/TiC/6bv2/Amdry	67	glue
9E18 HT	6bv2/TiC/TiN/TiC/6bv2/Amdry	66	glue
9E19 NHT	6bv2/TiN/AlN/TiN/6bv2/Amdry	53	coating
9E21 NHT	6bv2+N/TiN/N+6bv2/Amdry	46	glue
9E23 NHT	6bv2+N/CoTa1/TiN/CoTa1/N+6bv2/Amd	16	coating
9E23 HT	6bv2+N/CoTa1/TiN/CoTa1/N+6bv2/Amd	46	glue
9E26 NHT	6bv3/TiN/6bv3/Amdry	45	glue

NHT = not heat treated; HT = heat treated

temperature by means of tensile tests. The coated specimens were glued under slight pressure to tensile test rods and accurately aligned in tension direction, at a temperature specified for the hardening of the glue. After hardening of the glue, the fracture strength was measured by a tensile test and the rupture place surface was identified microscopically. If the fracture occurred along the glue, the bonding was accepted for further testing. Table 2 illustrates some of the results from these tests.

After tensile tests, specimens similar to those showing fracture strength higher than the glue were subjected to thermal cycling in air between 500°C and 1100°C. In this test, the specimens were moved first from room temperature to 1100°C, then cycled between 500°C and 1100°C so that the holding time after stabilization of the temperature at the maximum and minimum temperature was one hour. After reaching the desired number of cycles, the specimens were cooled from 1100°C to room temperature. The average heating rates were 4.5 Ks^{-1} and 5.5 Ks^{-1} during the first heating to 1100°C and the subsequent heating periods from 500°C to 1100°C, respectively. Correspondingly the average cooling rates during the cycling of temperature was 6.7 Ks^{-1}. The samples were studied after 10, 100 and 1000 cycles. No spalling off phenomenon was observed in any of the studied samples. The two bond layers of CoTa type and that of TiC reacted, however, with the diffusion barrier, and therefore they seem not to be promising as bonding layers.

In the burner rig tests, the samples were first heated to 1100°C, then subjected to thermal cycling between 850°C and 1100°C, and after a desired number of cycles cooled from 1100°C to room temperature. Figure 5 illustrates the arrangement of these tests. The specimens were studied after 10 and 100 cycles. All the specimens passed the tests without any spalling or reaction with the diffusion barrier.

Fig. 5   The arrangement of burner rig tests.

In the remaining period of the project, coated samples will further be tested by thermo-mechanical tests, where the samples are stressed with a constant tensile stress of 200 MPa while temperature is cycled between 850°C and 1100°C. In other tests, temperature is kept constant at 1100°C and the stress is cycled between 10 MPa and 200 MPa.

Bondings passing all the above mentioned tests will then be applied to real turbine blades together with diffusion barrier, overlay and thermal barrier, and tested in conditions simulating those in real gas turbines.

4 DISCUSSION

The above results are from laboratory scale tests of an ongoing project, not from final full size tests. In spite of this, they provide an important basis on how the bonding for actual blades should be realised.

First, a continuous graded bonding from the composition of superalloy to that of TiN is not possible without forming of low melting phases hazardous to the bonding at the temperature of 1100°C. Thus, solution is sought from separate intermediate layer(s) matching reasonably with the superalloy as well as with TiN.

From the properties of the bond layer, thermal expansion seems to be very important for the bonding capability. Beside this, a slight solution possibility between adjacent interfaces may be beneficial for the bonding. The thermal expansion of TiN is very much smaller than that of metallic alloys, and therefore one of the main questions is the finding of a suitable metallic bond layer fitting in thermal expansion sufficiently with TiN.

From the developed bonding layers, the layers 6bv2 and 6bv3 appear most promising. Nitriding of these alloys on the interface against TiN may be positive. Both of these alloys have shown durable bondings in tensile tests as well as in the different thermal cycling tests.

Final decisive results will be received from the full size tests on real blades in simulated turbine conditions, which will be the final stage of the total project.

5 CONCLUSIONS

A successful bond layer between TiN+AlN+TiN diffusion barrier and the superalloy substrate cannot be achieved on the basis of a continuously changing graded bond layer. This is why the solution must be sought from intermediate layers. From the developed layers, 6bv2 and 6bv3 as such or in suitable combinations seem the most promising. Final tests on the best candidates will be carried on real turbine blades.

Thermomechanical Fatigue of Coated Superalloys

M.I. WOOD, D. RAYNOR and R.M. COTGROVE

ERA Technology, Leatherhead, Surrey, UK

ABSTRACT

The thermomechanical fatigue of IN738 coated with either a 'ductile' CoNiCrAlY coating or a 'brittle' Pt–Al are reported for 180° and 135° out of phase cycles. The CoNiCrAlY coating has no effect on the TMF life of the IN738 substrate, although the life of the coating can be quantitatively described using a hysteretic energy approach. The Pt–Al coating had a detrimental effect at high strain ranges, but no effect at low strain ranges. This effect has been related to the coating's DBTT, in conjunction with the effect that coating cracking has by eliminating any crack initiation period for the uncoated alloy.

INTRODUCTION

Transient operation of a gas turbine imposes a thermal fatigue loading on the hot section components. The subsequent damage to these components can be one of the main factors governing overhaul intervals, either because the integrity of the component is threatened, or because further operation would render the part unrepairable. Whilst thermal fatigue damage in vanes is commonly repairable, that in rotating blades is more problematical if it is located away from the tip or upper part of the aerofoil.

In this context, it is important to appreciate that first row rotating blades are coated in all modern units, and that thermal fatigue damage in such components almost invariably initiates at the surface, i.e. in the coating. This state of affairs applies to both aero and industrial gas turbines. From a design perspective, the performance of the coating-substrate system under thermal fatigue conditions ought therefore to be an important consideration. However, most designs are based on isothermal low cycle fatigue (LCF) data, sometimes of uncoated material.

The shortcomings of this approach is evidenced by the increasing, although still small, quantity of published work on thermomechanical fatigue properties of uncoated, and sometimes coated, superalloys. Most of this work is within the last 10 years, although some of the earliest work does date back to the 1960s. Within this body of work there is a strong bias towards aeroengine materials and test conditions, as opposed to those for industrial units (the ratio of effort being approximately 5:1). Only when one looks at the results for single crystal alloys is the division of effort between coated and uncoated material found to be roughly equal.

Some of the important parameters and properties affecting coating behaviour were set out in the 1970s (e.g. Ref. 1, 2). Ideas such as ductile to brittle transition temperature (DBTT), and its importance in governing cracking tendencies, were used with a fair degree of success to modify coating composition to reduce in-service thermal fatigue problems (e.g. Ref. 3, 4). However, it became apparent from service behaviour (e.g. Refs 3, 5) that undue reliance on DBTT properties could be misleading, and that a more comprehensive understanding was required, treating the coating as an engineering material in a composite structure.

Whilst more data on coated material is now available, the actual behaviour of the coating is often still not explicitly (mechanistically) incorporated into model of the behaviour of the coated system. However, there are studies where the behaviour of the coating itself, rather than just its effect, has been one of the objectives of the work[6, 7].

The work described here considers the TMF behaviour of IN738LC with two different coatings, one 'ductile' and one 'brittle' (using DBTT type parlance). These two coatings effectively represent opposites in their balance of properties. Particular attention has been paid to distinguishing between the behaviour of the coating itself, and the effect of the coating on the coated substrate.

MATERIALS AND TEST PIECES

Test blanks were made in conventionally cast IN738LC. This was hot isostatically pressed (HIP) using industry standard conditions (1204°C for 4h). After machining they were coated with either a ~75 or 150 μm thick low pressure plasma sprayed CoNiCrAlY coating or a ~75μm thick Pt–Al coating. In both cases these were applied by commercial vendors. The CoNiCrAlY coating was surface finished to give a roughness of 2.5 μm (RMS). After coating, the test pieces were given the standard IN738 heat treatment of 2h at 1120°C then 24h at 845°C.

All the TMF tests on coated or uncoated material used the same standard machined all over LCF type test piece with a machined (and honed) hollow bore.

TMF TESTING

MACHINE

All TMF testing was carried out on a specially configured Instron 8500 series servo electric machine fitted with water cooled hydraulic grips. Side contact extensometry was employed for strain controlled testing. Heating was carried out using a radio frequency induction unit. Cooling of the testpiece took place by a combination of radiation and convection from the testpiece, coupled with conduction into the water cooled grips. No forced cooling was used in the test programme. The specimen temperature was measured using an infra-red optical pyrometer.

After a standardised sequence of initiation cycles, the TMF tests commenced by adding the required mechanical strain cycle to the free thermal expansion cycle generated by the specimen. As in all test of this type, it was the total strain which was measured and used to control the TMF cycle.

TMF TESTING

TMF tests were carried out using two basic cycles types, classified either as 180° linear out-of-phase, or 135° out-of-phase (OP) cycles, with $R = -1$, Fig.1. A two minute strain control dwell occurred at both maximum and minimum cycle temperatures. In the former case this was to allow some coating relaxation to take place, whilst in the latter case this was to ensure the stability of the test cycle.

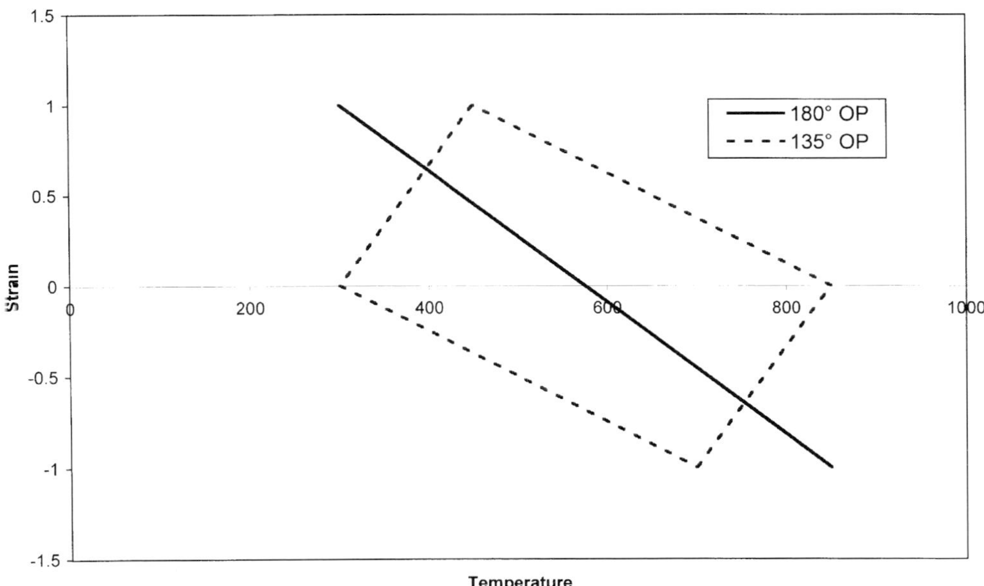

Fig. 1 TMF cycle types.

Temperatures were cycled between 300°C and 850°C or 900°C, with a loading rate of 10°C/sec. Since the cooling rates were natural, the rate varied as a function of the temperature. At the higher temperatures it reached ~10°C sec⁻¹.

Test piece endurances were derived from the tangent point for the start of the sustained tensile stress drop in the evolutionary maximum stress-cycle plot. This should not be confused with the life for crack initiation in the coating which was established through repeat surface examination, as set out in the next section

SURFACE EXAMINATION

Most tests were continued to failure, but some were terminated early for sectioning and metallography. Many tests were periodically interrupted so that acetate replicas of the surface could be taken, after which the tests were restarted. This sequence was often repeated a number of times throughout the life of a specimen to produce a set of replicas which could be used to follow the development of surface damage. Examination using both optical and scanning electron microscopy were used to characterise and quantify the evolution of surface damage.

TMF RESULTS

UNCOATED MATERIAL

Base line tests were carried out for both cycle types and peak temperatures. The behaviour is shown in Fig. 2, together with a best fit line incorporating the data from both

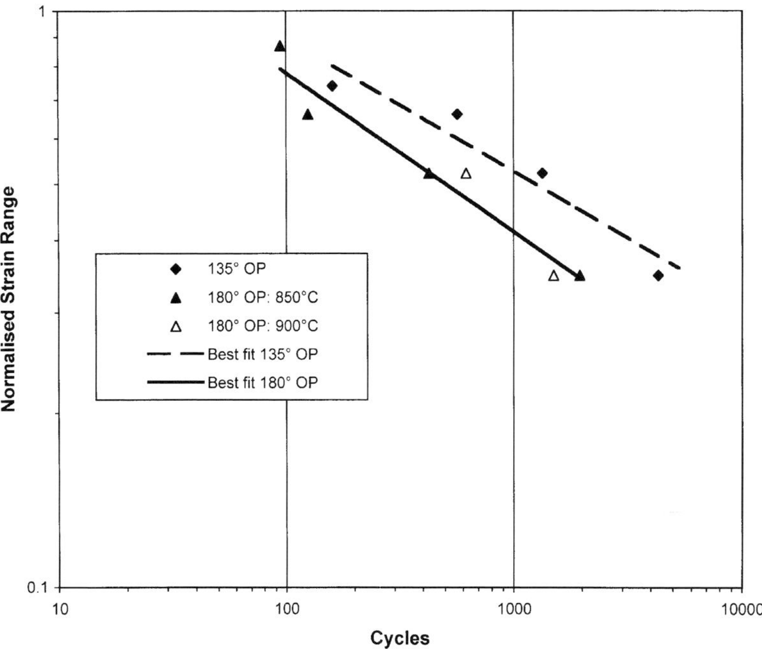

Fig. 2   Life of uncoated IN738 for 180° and 135° OP cycles.

temperatures. As can be seen, the endurances were longer under the 135° OP cycle conditions. Cracks initiated at grain boundaries and then propagated in a transgranular manner.

CoNiCrAlY COATED MATERIAL

TEST PIECE ENDURANCE

The lives are shown in Figs 3 and 4 for the 180° and 135° OP cycles respectively. Overall these indicate that the coating has had little or no effect on the endurance life of the substrate. It is also apparent that the small differences in coating thickness and peak temperature employed have similarly not had a significant influence on the overall behaviour.

COATING ENDURANCE

The as processed coating contains a large number of surface features carried over from the original plasma sprayed surface (Fig. 5). During cycling, cracks initiate and grow from these residual features (Fig. 6). This development of this progressive cracking is qualitatively illustrated in Fig. 7, based on the surface examinations carried out during the periodic interruptions of the tests. This multiple initiation and growth of cracks is also evident in metallographic sections taken either during or at the end of tests (Fig. 8). Based on destructive examinations on interrupted tests, it is also possible to mark on Fig. 7 the

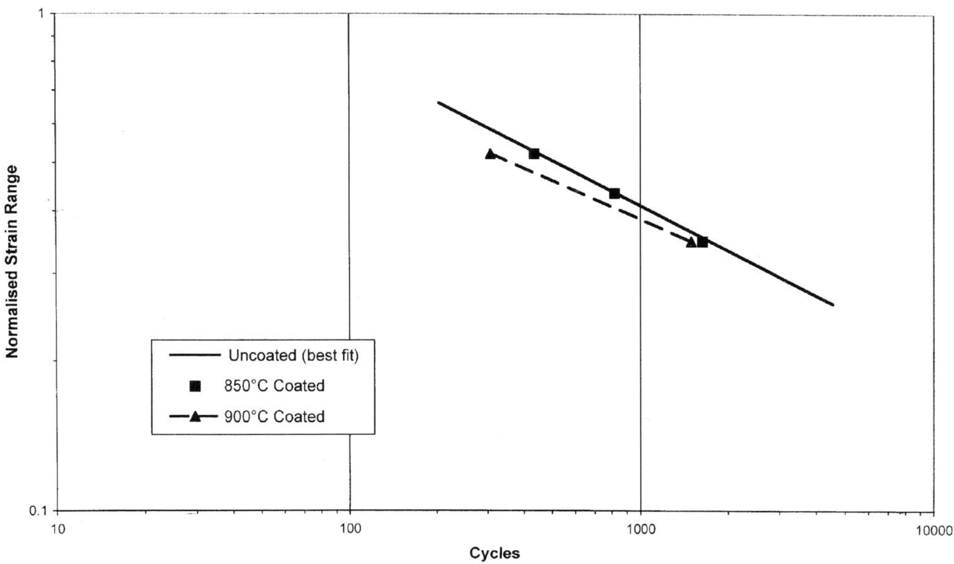

Fig. 3   Life of CoNiCrAlY coated IN738 for 180° OP cycle.

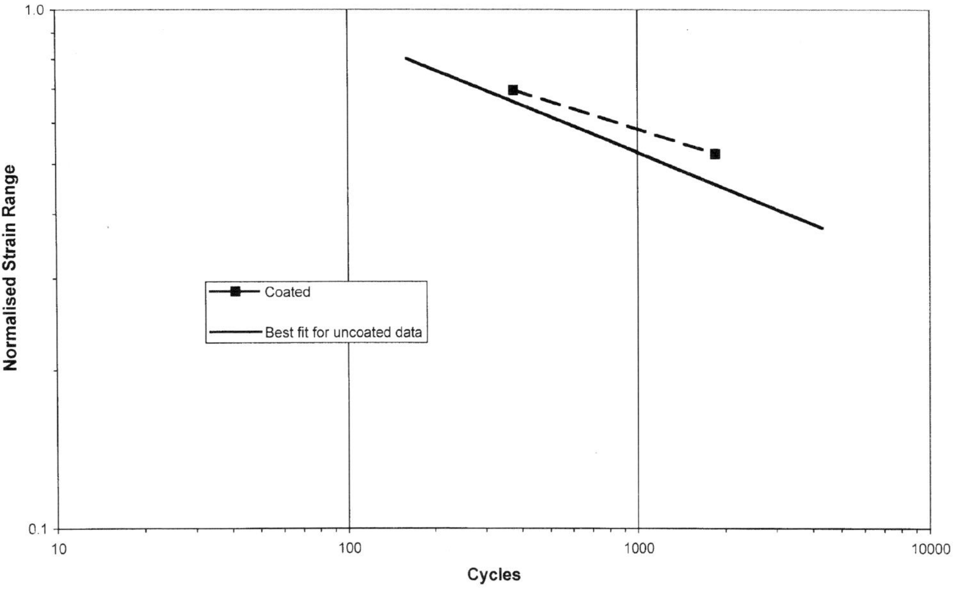

Fig. 4   Life of CoNiCrAlY coated IN738 for 135° OP cycle.

boundary when the surface initiated cracks reach the coating-substrate interface. Whilst it is marked as a discrete line, in reality it can only represent a certain probability of the cracks having reached the interface, given the large number and distribution of initiation sites. On reaching the coating substrate interface, a fairly high percentage of the cracks continue to grow into the substrate in a transgranular manner (Fig. 9).

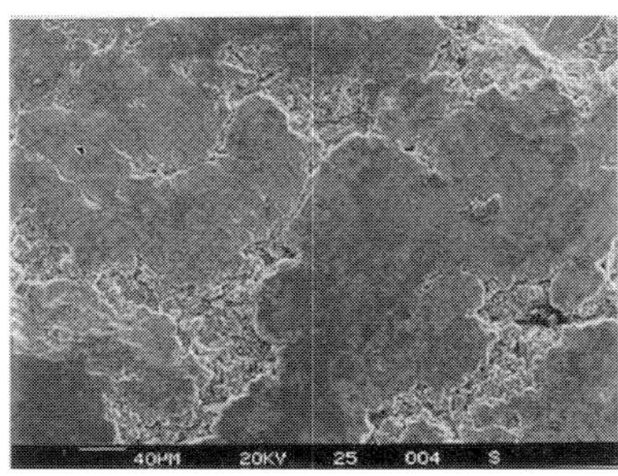

Fig. 5 CoNiCrAlY coating as processed.

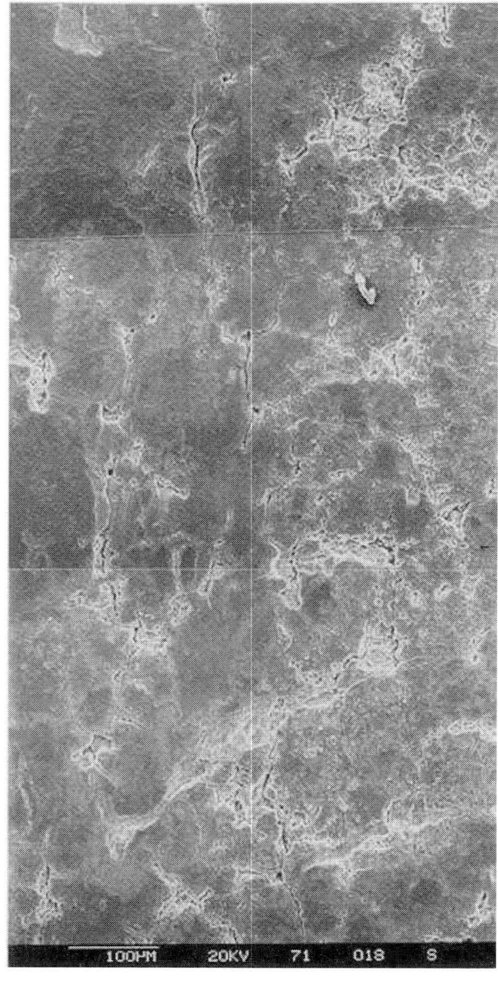

Fig. 6 CoNiCrAlY coating after 1200 cycles. (stress axis horizontal).

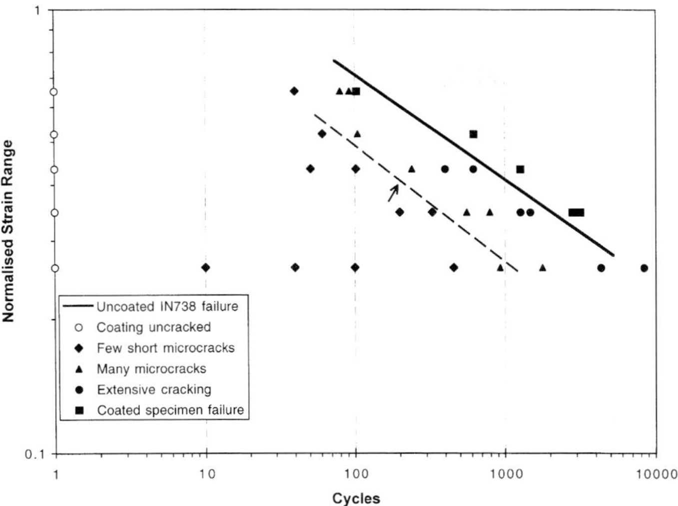

Fig. 7   Qualitiative damage development map for the CoNiCrAlY coating.

Fig. 8   Cross section of CoNiCrAlY coating at end of test.

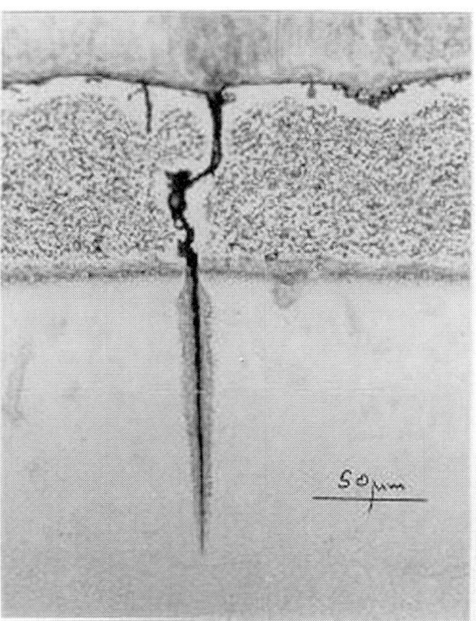

Fig. 9   Crack growth beyond the CoNiCrAlY coating.

Pt–Al COATED MATERIAL

Test Piece Endurance

Testing was carried out only for 180°OP cycles. The lives for both maximum temperatures are shown in Fig.10. At the higher strain ranges all tests showed a reduction in the total endurance lives. However, at the lowest strain range, whilst the test to 900°C behaved in the same manner as those at the higher strain ranges, the test to 850°C did not shorten the life at all.

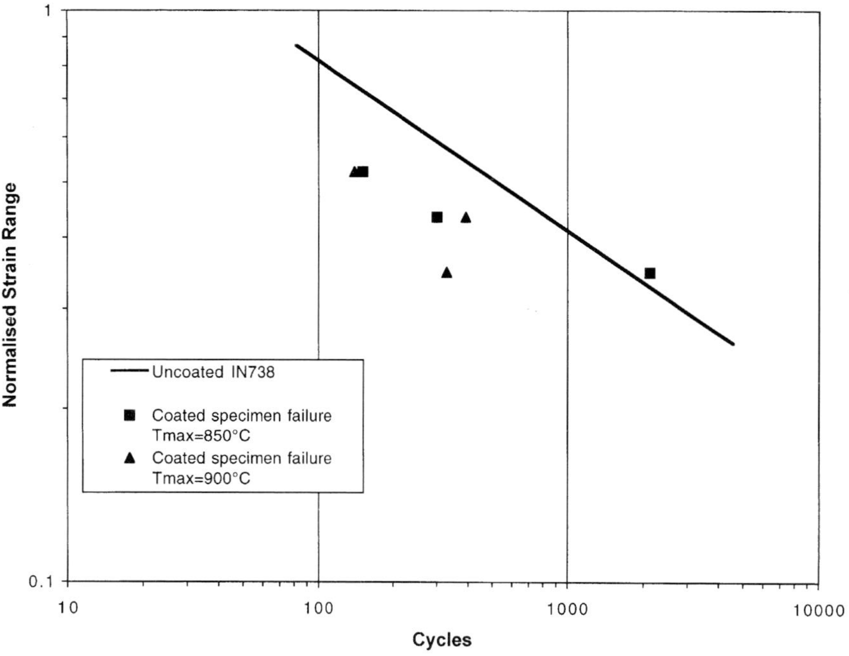

Fig. 10 Life of Pt–Al coated IN 738 for 180° OP cycle.

Coating Endurance

Surface examination during the tests to 850°C showed a clear difference in behaviour between the tests at the two higher strain ranges and that at the lowest (where no endurance penalty was observed). In the former two cases, many long straight cracks formed extremely rapidly, probably during the first TMF cycle (Fig.11). In contrast, no cracks were observed on the surface during any of the examinations (the last being at 600 cycles, with test piece failure at ~2000 cycles). Metallography showed that at the higher two strain ranges many coating cracks had propagated deep into the substrate, whereas in the case of the lowest strain range tests, only a few secondary cracks were seen, and these had only small amounts of growth into the substrate.

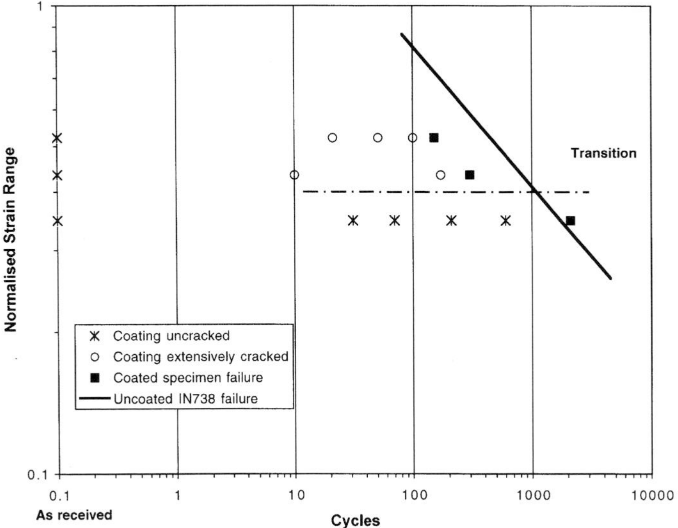

Fig. 11   Qualitative development of cracking in Pt-Al coated IN 738.

OTHER COATING WORK

Freestanding creep and stress relaxation test pieces were machined from thick low pressure plasma sprayed CoNiCrAlY deposits (after full heat treatment) or from hollow test pieces where the substrate had been removed by back machining. A variety of tests were undertaken in the 650°–800°C range to generate a stress – strain rate power law relationship. Applying this model to predict the relaxation of the coating at higher temperatures underlines the rapidity of the stress offloading: Fig. 12.

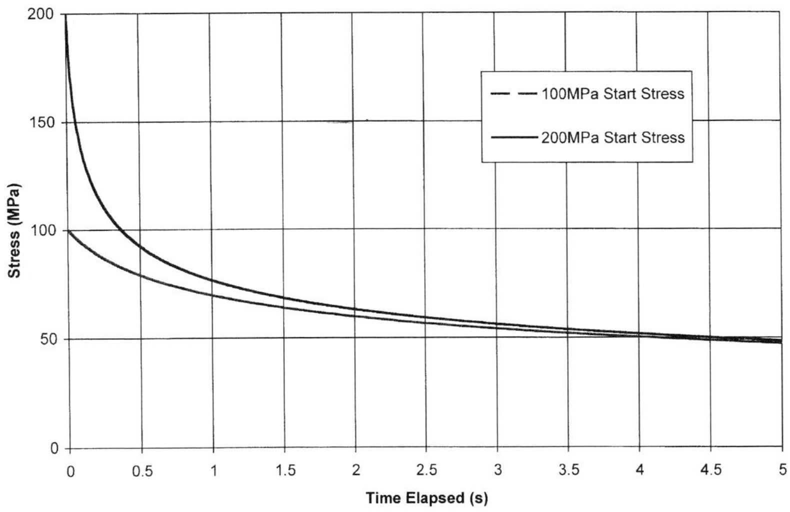

Fig. 12   Relaxation of CoNiCrAlY at 800°.

This material model, combined with tensile, elastic modulus and thermal expansion data generated elsewhere in the program (Ref. 8) was then used to calculate the hysteresis loop for the CoNiCrAlY coating during TMF cycles. An illustrative stabilised stress-temperature loop after shakedown is shown in Fig. 13 for the 180° OP cycle at the minimum and maximum strain ranges used. As may be observed, the coating is stress free at the peak cycle temperature.

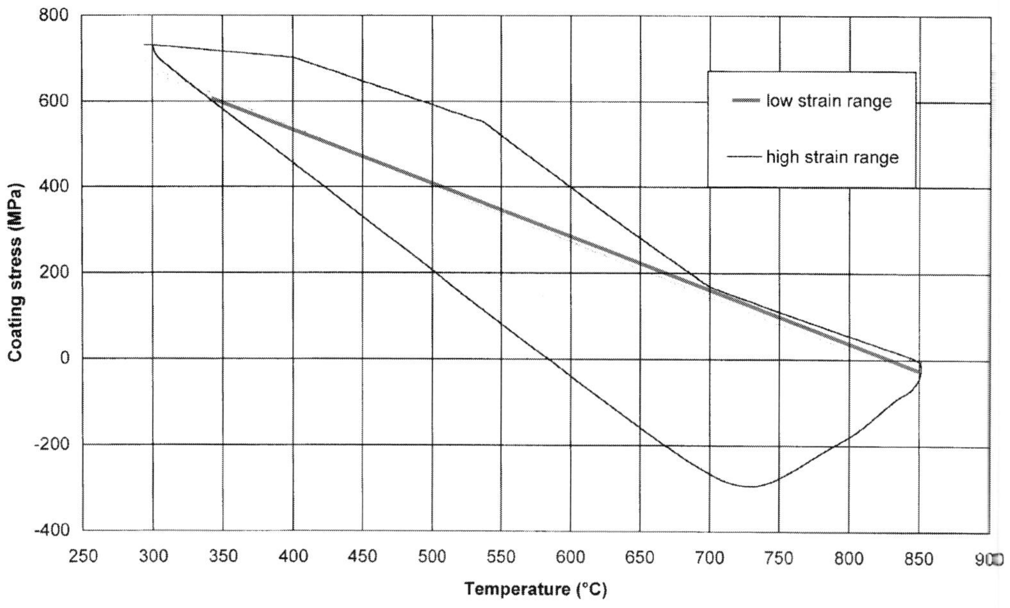

Fig. 13 Calculated hysteresis loop for the CoNiCrAlY coating in a 180° OP cycle.

DISCUSSION

Because the two coating systems considered here were known to have different mechanical properties ('ductile' v. 'brittle'), it should not be surprising that their behaviour under TMF conditions is also different. For the 'ductile' CoNiCrAlY coating there is progressive cumulative damage accumulation at all strain ranges, whereas for the 'brittle' Pt–Al coating there appears to be a marked transition in behaviour above a particular strain range.

So far as the 'ductile' CoNiCrAlY coating is concerned, the modelling route which has been used is the hysteretic energy approach (Refs 9,10). A general form of the relationship is

$$N_f = A(\Delta\varepsilon_t)^\alpha(\sigma_{max})^\beta$$

where

N_f	= number of cycles to failure
$\Delta\varepsilon_t$	= total mechanical strain range
σ_{max}	= maximum tensile stress
A,α,β	= constants

The model emphasises the role of the tensile stresses in governing life, which is appropriate given the large tensile stresses generated in the coating, sometimes exceeding yield at the minimum test temperature. It is worth noting that there are no dwell terms in this formulation (i.e. terms describing the cycle frequency or the amount of time spent at the maximum temperature). This is somewhat different to the usual form of this (and other) descriptive equations, but it arises directly from the modelling of the coating deformation where it can be seen that relaxation at above ~800°C is effectively instantaneous, and well within the 2 min dwell used in this work. Hence there are no purely mechanical consequences from the length of the dwell period (of interest here). However this does not preclude different dwell time dependencies arising from oxidation or structural change within the coating which have not been specifically included in the model at this stage. This is discussed later.

In applying the model, two differing approaches have been used. One is 'conventional', in that the strain range input into the model is the total mechanical strain range (modified to incorporate the thermal expansion mismatch for the coating). The second approach has been to use just the tensile strain range (again modified for the thermal expansion mismatch), on the basis that the coating always starts the cool down cycle from zero stress. These two approaches are the same for 180° OP cycles, but differ slightly for 135° OP cycles.

To implement the model required the following inputs:

(i) $\Delta\varepsilon_{coat\ total}$: mechanical strain range defined in the test conditions, but modified to incorporate the thermal expansion mismatch (i.e. total strain range applied to the coating),

(ii) $\Delta\varepsilon_{coat\ tensile}$: the peak tensile strain range imposed on the coating, again allowing for thermal expansion mismatch effects,

(iii) σ_{max} : from the calculated hysteresis loops,

(iv) N_f : Based on the interrupted surface examination data.

The constants in the hysteretic energy-life equation have then been calculated through a multiple linear regression fit. The calculated cyclic life of the coating (based on coating strain ranges) has been plotted for comparison with the overall test piece life – Fig. 14 (but expressed in terms of the substrate strain range imposed during the test). To illustrate the standard of the fit between the analysis and the input data, the two sets are plotted against each other in Fig. 15, along with a ± x2 scatterband. It can be seen that both approaches can describe the input data to within this standard of accuracy. This level of 'goodness of fit' has been found in several other TMF studies of coated material (when dealing with total lives) – e.g. Ref. 11.

The overall neutral affect of the CoNiCrAlY coating on the TMF life of the coated IN738 can be understood by noting that the life of the coating (i.e. the time for the cracks to grow through to the interface) is very similar to the crack initiation life for the uncoated IN738 (based on the surface examination work on the uncoated alloy). Since the effective crack initiation lives are therefore equal, and crack growth into the substrate occurs by the same transgranular mode, the life of both types of test piece would be expected to be equal.

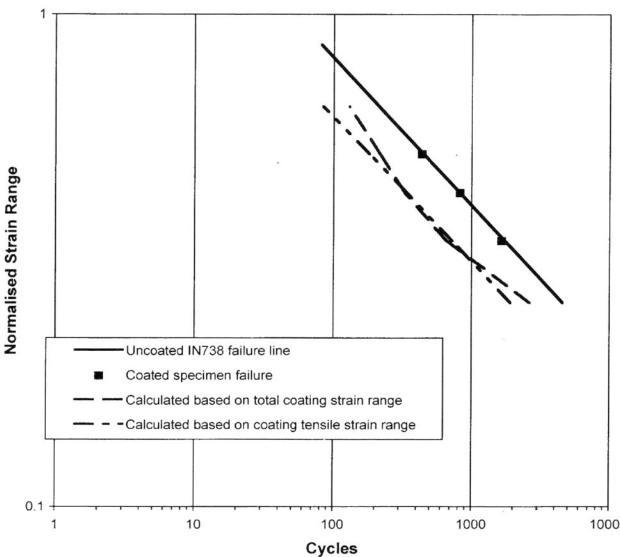

Fig. 14 Comparison of model predictions and experimental data for CoNiCrAlY coating: 180° OP cycle.

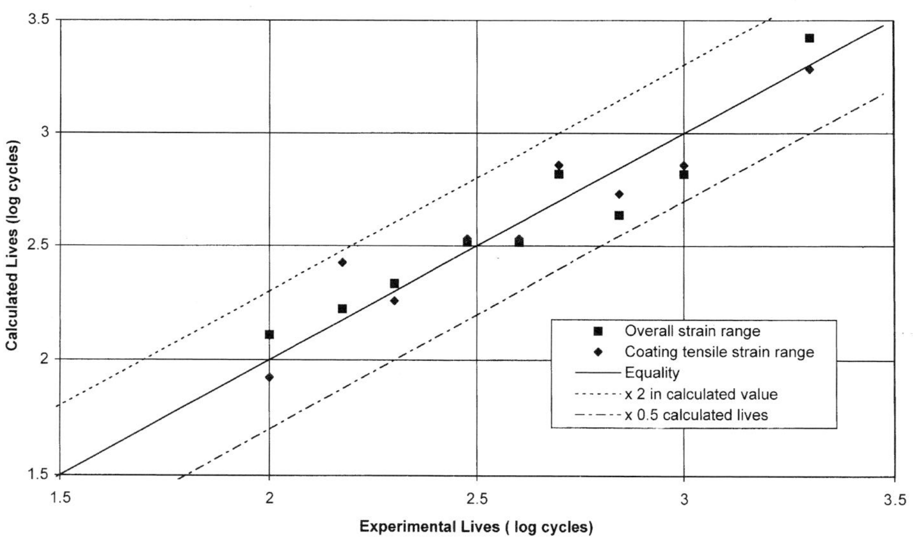

Fig. 15 Comparison of model predictions and experimental data for CoNiCrAlY coating: 180° and 135° OP cycles.

For the 'brittle' Pt-Al coating, a very different approach has been taken. The surface examination results, together with other observations on similar coatings[8] indicates that there is a strain range above which the coating cracks instantaneously: if tests are carried out below this strain range then cracking occurs very late in life. This is effectively the original DBTT type explanation of the behaviour.

The consequence of this early crack initiation is that the test duration is then governed by crack propagation. Based on the surface examination work on the uncoated alloy, it is possible to subtract the observed life to crack initiation (for the uncoated alloy) to determine the crack propagation period. This propagation period has been plotted in Fig. 16, and it can be seen that it compares well with the failure lives of the Pt-Al coated IN738 under conditions when the coating cracks. If the coating did not crack then no detrimental effect was observed.

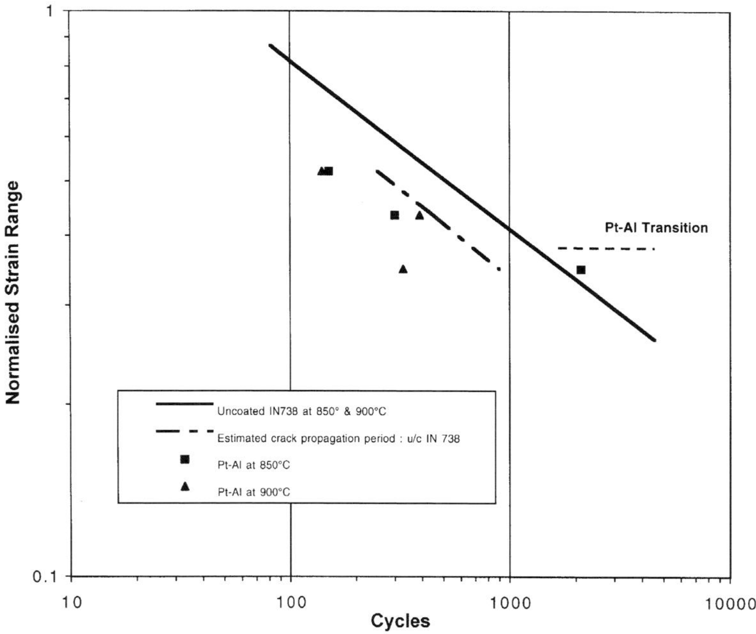

Fig. 16 Comparison of life of Pt–Al coated IN 738 and crack propagation life for uncoated IN 738.

The DBTT behaviour of the coating was not determined specifically, but the transition strain observed here is very much in accord with the values found in the literature for this generic type of coating.[12] However, as with all 'brittle' coated systems of this type, it is inexact to characterise the failure behaviour as a single value. Rather, one should consider the statistical aspect of failure in such systems.[13]

One shortcoming of all approaches based on the properties of new, undegraded material is that the properties and structure of the coating do alter during the test. As with the mechanical properties of coating in general, there is little available information on this subject, and some of it is apparently contradictory. For MCrAlY type coating there is information from small punch work on exposed CoNiCrAlY coatings.[14] This indicates that the coating ductility decreased and the DBTT increased. Whilst the model employed in the current work does not explicitly use either of these properties, it is indicative that the mechanical response of the coating is altering with exposure. It is noteworthy that some of the industrial original equipment manufacturers are carrying out TMF tests with much longer dwell cycles to included more exposure effects into the TMF test.[15] For 'brittle' coatings, whilst the current work indicates a transition in behaviour, other studies on aluminide coated alloys have shown that some of the time dependent coating changes affect properties of relevance, even within a DBTT type approach. DBTT studies on aged aluminised IN100 have shown that exposure lowered the DBTT, possibly by reducing the aluminium content of the coating.[16] A different aspect of the effects of exposure can be seen in Ref.17, on aluminised SRR99. In this work, the composition change in the coating allowed the formation of a $L1_0$ phase. On cooling this phase could be expected to form martensitically, inducing locally high stresses, possibly acting as a crack initiation site.

CONCLUSIONS

For the relatively ductile CoNiCrAlY coating:

(a) The TMF behaviour is governed by the formation and slow incremental growth of cracks from residual processing features and defects in the surface of the coating.

(b) The cracking of the coating occurs at about the same time as crack initiation for uncoated IN738. Whilst the cracks continue to grow from the coating into the substrate, they do not affect the overall life of the testpiece as the crack initiation period for the testpiece as a whole has not been significantly altered.

(c) The high temperature deformation of the coating has been experimentally determined and incorporated into a deformation model for use under TMF conditions. This has shown that the coating relaxes to zero stress at the peak temperature within seconds.

(d) The life of the coating has been modelled using an hysteretic energy approach. This can predict coating life within a factor of $\pm$ x2.

For relatively homogeneous coatings such as Pt–Al which are 'brittle' at low temperatures:

(a) The coatings can fail immediately in a brittle manner if the effective tensile strain imposed on the coating is above a critical value at low temperature. This is the same as the DBTT approach.

(b) If the coating cracks then the life of the test piece is shortened. This is because the coating cracks propagate into the substrate, eliminating the crack initiation period for the uncoated material.

(c) If the effective peak tensile strain is less than this critical value than the coating does not crack. The life of the test piece is then unaffected.

ACKNOWLEDGEMENTS

The author gratefully acknowledges the permission of the Directors of ERA Technology to publish this paper.

The financial sponsorship and technical input of the sponsors of the ERA joint industry project 'The TMF of Coated Superalloys' is gratefully acknowledged.

REFERENCES

1. G.R. Leverant, T.E. Strangman and B.S. Langer, 'Parameters controlling the thermal fatigue properties of conventionally cast and directionally solidified turbine alloys', *3rd Int. Symp. Superalloys: Superalloys: Metallurgy and Manufacture*, Claitors Publishing Division, 1976, 285–295.
2. G.W. Goward, 'Protective coatings for high temperature alloys: State of Technology', in Symposium on properties of high temperature alloys with emphasis on environmental effects, Las Vegas, NV, 1976, 806–823.
3. G.W. Meetham, Use of protective coatings in aero gas turbine engines', *Material Science and Technology*, 1986, **2**, 290–294.
4. I. Linask and J. Dierberger, 'A fracture mechanics approach to turbine aerofoil design', ASME 75-GT-79, American Society of Mechanical Engineers, 1974.
5. J.W. Fairbanks and R.J. Hecht, 'The durability and performance of coatings in gas turbines and diesel engines', *Material Science and Engineering*, 1987, **88**, 321–330.
6. D.M. Nissley, 'Thermomechanical fatigue life prediction in gas turbine superalloys: A fracture mechanics approach', *AIAA Journal*, 6 June 1995, **33**, 1114.
7. J. Bresses, J.M. Martinez-Esnaola, A. Martin-Mezizoso, J. Timm and M. Arana-Antelo, 'Coating effects on crack growth in a single crystal nickel based alloy during thermomechanical fatigue', *Thermomechanical fatigue behaviour of materials*, Vol. 2, ASTM 1263, American Society for Testing and Materials, Philadelphia, 1996, 82–95.
8. M.I. Wood and D.R. Raynor, Unpublished work, ERA Technology 1996.
9. W.J. Ostergren, A damage function and associated failure equations for predicting hold time and frequency effects in elevated temperature LCF', *J. of Testing and Evaluation*, 1976, **4**, 327–339.
10. D.M. Nissley, T.G. Walker and K.P. Walker, 'Life prediction and constitutive models for engine hot section anisotropic materials program, NASA CR-189223, National Aeronautics and Space Administration, 1992.
11. H.L. Bernstein, T.S. Grant, R.C. McClung and J.M. Allen, 'Predictions of thermomechanical fatigue for gas turbine blades in electric power generation', *Thermomechanical Fatigue Behaviour of Materials*, ASTM-STP 1186, American Society for Testing and Materials, 1993, 212–238.
12. D. Vogel, L. Newman, P. Deb and D.H. Boone, 'Ductile to brittle transition temperature of platinum modified coatings', *Mat. Sci. and Eng.*, 1987, **A88**, 227–231.
13. T.C. Totemeier and J.E. King, 'Isothermal fatigue of an aluminide coated single crystal superalloy: Part 1', *Met. and Mater. Trans.*, 1996, **27A**, 353–361.
14. Ito, *High temperature degradation of coating and substrate in gas turbine blades*, ASME Conference Preprint 95-GT-358, American Society of Mechanical Engineers, 1995.
15. N. Czech, F. Schmitz and W. Stamm, 'Thermal Mechanical fatigue behaviour of advanced overlay coatings', *Mat. and Manufacturing Processes*, 1995, **10**, 1021–1035.
16. H. Bernard and L. Remy, 'Ductile brittle transition of an aluminide coating in IN100 superalloy', *High temperature materials for power engineering*, Kluwer Academic Press, 1990, 1185–1194.
17. K.M. Ostolaza K.M., et al., 'Evolution of nickel aluminide diffusion coating on a single crystal superalloy under TMF', *Surface Treatment*, 1995.

High Temperature Deformation and Crack Growth in γ-TiAl Base Intermetallic Materials

BILAL DOGAN

GKSS Research Centre, Institute for Materials Research, Max-Planck-Str., D-21502 Geesthacht, Germany.
bilal.dogan@gkss.de

ABSTRACT

Fracture toughness and creep crack growth of a light-weight γ-TiAl base intermetallic alloy, Ti48Al2Cr, with a duplex microstructure has been investigated at 700°C, that is the service temperature of components to be made of γ-TiAl. Single edge notch bend specimens of two sizes, and compact tension specimens were tested both with EDM slit notches and fatigue precracks. The loading mode was varied to study the applicability of different test techniques in high temperature crack growth characterisation of intermetallics. The present investigation showed that the inhomogenity in microstructure, specimen size and geometry effects need to be considered in fracture toughness characterisation of TiAl base intermetallics. The starter sharp crack may be introduced by fatigue precracking and also by EDM provided notch root radius ρ is ≤ 0.05mm. The fracture toughness data showed specimen size and geometry dependence. Brittle fracture directs attention to the crack length measurement method that requires improvement. Due to the low ductility of TiAl base intermetallics at 700°C the fracture toughness data were correlated with fracture parameters K and J-integral. The crack growth data from CT specimens at 700°C were correlated with K and $C^*(t)$, and the applicability of the existing method for testing and fracture toughness characterisation of TiAl intermetallics is discussed. Creep deformation data are included along with crack growth data in an attempt to pave the way to the life assessment methodology for components made of γ-TiAl base intermetallic alloys.

1 NOMENCLATURE

a	Crack length
a_e	Initial crack length
Δa	Crack growth
B	Specimen thickness
CCG	Creep crack growth
CT	Compact tension specimen
$C^*(t)$	Creep crack growth parameter[21]
CTOD	Crack tip opening displacement
δ_5	*CTOD* measured at the initial crack tip[23]
E	Young's modulus
EDM	Electric discharge method
ε	Strain rate
ε_c	Creep component of strain rate
F	Applied load
F_c	Critical load[17]
F_{max}	Maximum load

FPC	Fatigue precracking
HIP	Hot isistatic pressing
IC	Investment cast
J	J-integral
J_c	J calculated at F_c
$J_{0.2}$	J at 0.2 mm crack growth
J_i	J at crack initiation
K	Stress intensity factor
K_{max}	K calculated from F_{max}
K_c	K calculated from F_c
$K_{0.2}$	K at 0.2 mm crack growth
K_{Ic}	Plane strain fracture toughness
K_J	K calculated from J
K_{Jc}	K calculated from J_c
K_{tip}	Stress intensity at the crack tip
K_{fmax}	K_{max} in cyclic loading (Fatigue precracking)
$K_{applied}$	K calculated from applied load
ΔK	Cyclic stress intensity factor
n	Secondary creep exponent
ρ	Notch root radius
r_p	Crack tip plastic zone size
R	Stress ratio
$R_{p0.2}$	Yield stress
R_m	Tensile strength (UTS)
R_f	Fracture stress
SEN	Single edge notch specimen
σ	Stress
$\Delta\sigma$	Cyclic stress range
t	Test time
t_f	Fracture time
U_{pl}	Plastic work at fracture[17]
ν	Poisson's ratio
Vc	Creep component of deflection
Ve	Elastic component of deflection
V_t	Total deflection (Load line displacement)
W	Specimen width
γ	TiAl phase
α_2	Ti$_3$Al phase

INTRODUCTION

The industrial need for improvements in the performance of aircraft has been the main driving force for the progress made in the development of materials for turbine applications. Development of Ni-based alloys and conventional high temperature titanium based alloys via improved alloying chemistry and process technologies has reached its limit for industrial applications. The maximum operating temperature of some Ni-based super-alloys extends beyond 80% of their melting temperature. Therefore, significant increases in maximum service temperatures may no longer be expected. On the other hand, limited stiffness, creep strength, oxidation resistance and susceptibility to 'titanium fire' limits the application of titanium alloys below 550°C. However, industrial demand for higher operating efficiency and increased fuel economy as well as reduction of CO, CO_2, NO_x emissions prompted the search for light-weight structural materials with retained mechanical properties at higher temperatures such as intermetallics.[1–5]

Enormous effort has been made in the field of intermetallic materials during the past two decades.[6–10] The work concentrated on the light weight γ-TiAl base intermetallic alloys which are being considered for future high temperature engineering applications to replace the current Ni and Ti base alloys. Several potential applications have been identified for TiAl-based alloys in the aerospace, automotive and turbine power generation markets. The aircraft industry is pursuing the implementation of these alloys in aircraft engines. Potential components for turbine engines contain rotational parts such as low-pressure turbines, high-pressure compressor blades, and high-pressure turbine blade cover plates, and stationary parts such as transition duct beams, vanes, swirlers, various cases, and nozzle flaps and tiles.[11] The components made of TiAl-based intermetallics have successfully passed the extensive qualification tests, such as engine tests, bench tests and rig tests by several industrial companies, including MTU, Rolls–Royce, General Electric, Pratt & Whitney and IHI.[11,12] The automotive industry has also intensified its effort by qualification and introduction of exhaust valves and turbocharger turbine wheels made of TiAl based alloys. Recent engine tests performed with cast TiAl tubocharger rotor exhibited better acceleration response and higher maximum rotational speed than its counterpart Inconel rotor. This is followed by the use of TiAl turbocharger turbine wheels in commercial cars of a special type.[13] As this impressive progress in application of these materials continues, the industry concentrates on the remaining barriers such as the development of low cost, high volume manufacturing methods and development of life assessment methodology to apply to these new class of engineering materials.

The tremendous advance in production of components made of TiAl based intermetallics has been accompanied by advances in the understanding of the fundamental aspects of the alloys.

The sequence of phase transformations involving α-phase decomposition is qualitatively understood.[11] Compositional effects on the microstructure, and mechanical and creep properties have been studied intensively. Additions such as C, Si, N, Ta and W appear to improve the creep resistance. Higher creep resistance is achieved with a lamellar microstructure.[9] High cycle fatigue resistance is excellent up to 800°C.[11,14] Oxidation resistance becomes a critical factor at higher temperatures (i.e. 800°C) for a long time especially under thermal and/or loading conditions.[11] However, considering the potential application temperatures of about 700°C, the γ-TiAl base intermetallic alloys remain promising materials for engineering applications provided the problems of relatively low fracture toughness and fast crack growth rates are overcome. The concern is that component life will then be limited in the presence of relatively small existing or in-service initiated defects or flaws. Therefore, understanding the mechanisms of deformation, fracture and high temperature crack growth need be improved. This has been recognised and is also being worked on intensively.[9, 15–19]

The fracture mechanics approach has not been fully developed in the field of intermetallics. For this reason, the fracture toughness values reported in the literature reflect the various methods used for fracture toughness determination. Furthermore, during fracture mechanics testing the crack deviation at lamellae may result in an invalid fracture toughness test.[17] Creep crack growth (CCG) testing of creep brittle materials, such as intermetallics, has been subject to an international effort[20] that produced recommendations to be incorporated in the only existing test method for creep crack growth rate testing of metals, ASTM E 1457–92.[21] Hence, a study of fracture behaviour at service temperature, specimen size and geometry effect on toughness, and the applicability of the present test methods in fracture toughness and creep crack growth testing of intermetallics is needed.

The present paper reports on a systematic study of deformation, and crack initiation and crack growth in a γ-TiAl base intermetallic alloy with Cr addition. Creep and fracture toughness data are determined at 700°C. Effect of specimen size and geometry on fracture toughness are studied. This paper attempts to highlight some of the issues pertinent to fracture toughness and CCG testing of creep brittle materials using experimental data obtained on a TiAlCr alloy. The creep deformation and rupture resistance of these alloys depend on the microstructure. The fracture behaviour depends strongly on the microstructural constituents and the orientation of the lamellae in the process zone of the crack tip. The deformation by microcracking and crack branching increases with temperature up to 700°C. Therefore, there is a need for an in-depth investigation of its creep resistance and CCG behaviour. The applicability of the present test procedure for testing metals[21] in creep crack growth testing of intermetallics is discussed. Emphasis is placed particularly on the relevance of crack growth data for defect assessment.

EXPERIMENTAL PROCEDURE

MATERIAL AND METALLOGRAPHY

A γ-TiAl base intermetallic alloy, Ti 48at%Al 2at%Cr, referred to hereafter as TiAlCr, with O: 0.037–0.063, N: 0.008–0.016, C:0.011–0.013 was investment cast (IC). The cast billets were subsequently hot isostatically pressed (HIP) at 1200°C for 4hrs under 2.0kbar to remove porosity.

Metallographic specimens were prepared using conventional methods, and polarised light micrographs were taken of the polished specimens. The duplex microstructure of the cast and HIPped material consist of γ grains of 15μm and α₂+ γ lamellar regions of 320μm in size (Fig. 1). The test pecimens were sectioned from the IC+HIPped billets some of which were heat treated at 1300°C in air to increase ductility and achieve a balance of properties.

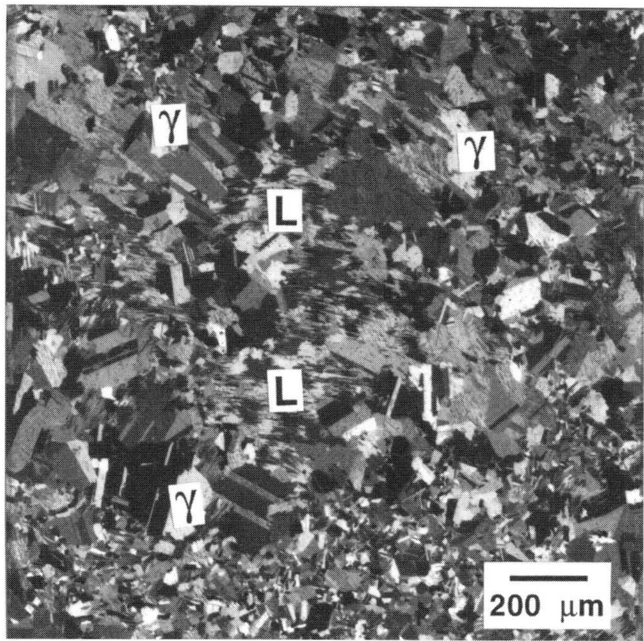

Fig. 1 Polarised light micrograph of microstructure of investment cast (IC) and HIPed Ti43Al2Cr. L: lamelae(γ+α₂), γ: TiAl.

SPECIMENS AND TESTING

Specimens

The tensile, creep, single edge notched (SEN) and compact tension (CT) specimens were spark eroded (Electric Discharge Method, EDM) from investment cast near net shape forms and ground to final specimen dimensions.

The SEN specimens were of two sizes; a) $5\times10\times50mm^3$ and b) $10\times20\times90mm^3$, with specimen width $W=2B$, where B is the specimen thickness. The CT specimens were 50mm wide and 10mm thick and side grooved 20% after precracking or machining of EDM notches. The slit notches in SEN and CT specimens were introduced by EDM using a 0.1 mm diameter wire where the notch root radius, ρ, was 0.05mm.

Tensile and creep tests

The tensile properties of the material were determined at 700°C in air at cross head speeds of 0.05 mm min⁻¹ and 5 mm min⁻¹. The testing rates were the same as those chosen to study the effect of loading rate on fracture toughness of the material using SEN specimens.

Creep tests were carried out under constant load conditions at initial stress levels to achieve test times covering a wide stress–strain rate range. The data were assessed to determine the creep properties at 700°C for both as cast+HIPped and HIPped+heat treated material at 1300°C.

Fracture toughness tests

A series of SEN specimens were fatigue precracked (FPC) to a_o/W=0.5 in four-point-bending (4PB) in a servohydraulic machine at 50Hz. A constant tensile cyclic stress intensity ΔK=3.2MPa$\sqrt{m}$ was chosen in order to control the crack growth as the crack growth rates in TiAl are extremely sensitive to the applied ΔK level.[22] The CT specimens were precracked under load control with a load ratio R=0.7 and initial maximum stress intensity K_{max}=9.1MPa$\sqrt{m}$ and ΔK=3MPa$\sqrt{m}$. Details of the fatigue precracking have been reported elsewhere.[16]

Fracture toughness testing of SEN specimens in 4PB were carried out at 700°C in air, in a servomechanical 10kN machine at cross-head speeds of 0.05 mm min^{-1} and 5 mm min^{-1}. Load and displacement data were recorded during the tests. The crack growth in SEN specimens were studied by unloading some of the specimens in the non-linear range of the load-displacement record prior to fracture. These specimens were heat tinted at 400°C to mark the crack length before final cracking.

Fracture toughness testing of CT specimens was conducted at 700°C at loading rates of 0.005 to 0.600 mm h^{-1} to determine the loading rate effect on deformation and fracture toughness values as in SEN specimens. The tested and heat tinted specimens were cracked open for crack length measurement on fracture surfaces. Fracture surfaces of the tested SEN and CT specimens were examined in an SEM for fracture mode and fracture sites.

Creep crack growth tests

Creep crack growth testing of CT specimens was carried out under displacement rate control at 1, 5 and 10 μm h^{-1}, and under constant load on an electro-mechanical machine. The specimens were heated to the test temperature in a four zone electric resistance furnace to provide uniform temperature distribution in the test section of specimens, with a temperature control of ± 3°C. The load, displacement in the load line, V_{LL}, and at the crack tip, δ_5, and the crack length were continuously monitored and recorded for further evaluation. Measurement of displacements were made by a non-contact laser scanner method designed in GKSS for high temperature fracture toughness testing.[23] The displacement measurements are obtained from the relative displacement of pins, welded on the specimen surface in the load line and at the crack tip at 5 mm distance across the crack plane, on which the laser beam is scanned.

The direct current potential drop (DCPD) method was employed to monitor crack initiation and crack growth in CT specimens. The crack length was evaluated from the electrical potential measurement and the crack growth rate, *da/dt*, was determined using 7 point incremental (second order) polynomial method, following the test standard.[21] The obtained crack length value from potential measurement was compared with the crack length measured on the fracture surface after completion of the test.

DATA ASSESSMENT

Fracture toughness data

Although the fracture mode was mainly brittle, the non-linearity in load-displacement behaviour of TiAlCr calls for consideration of non-linear fracture parameters such as K_c or non-linear fracture mechanics using the *J*-integral. Therefore, along with *K* data, the *J* values are calculated from the applied load and the displacement of loading points from $J=J_{el}+J_{pl}$, where J_{el} is calculated from the elastic stress intensity factor *K* and the J_{pl} is obtained from the plastic work at fracture, U_{pl}, calculated from the plastic part of the area under the load-displacement curve.[17]

$$J_{el}=K^2(1-v^2)/E, \quad \text{and,} \quad J_{pl}=h_{pl}/B(W-a)\int_0^{V_{pl}}PdV_{pl} \tag{1}$$

The fracture parameters K_c and J_c at F_c were determined for SEN specimens. The F_c is defined as the critical load at fracture or the load at significant pop-in.[17] The fracture parameters $K_{0.2}$ and $J_{0.2}$ were determined at 0.2mm crack growth for CT specimens.

The stress intensity factor K_J (K_{Jc}, and $K_{J0.2}$) was calculated from *J*, $K_J^2=J.E/(1-v^2)$, with $v=0.3$ and $E=152000$MPa. The validity of data is checked using the criteria given in the test procedure.[24]

Creep crack growth data

Due to the brittle fracture behaviour of TiAlCr at 700°C, the crack growth data is correlated with *K* and $C^*(t)$ following ASTM standards E399[25] and E1457,[21] respectively.

For the tests as in the present work, where the load, the load-line deflection rates, and the crack size measurements are available $C^*(t)$ is determined from,

$$C^*(t) = (F(dV/dt)/BW)\eta(a/W,n) \tag{2}$$

where *F* is the applied load, *B* and *W* are the specimen thickness and width, respectively, dV/dt is the measured load-line deflection rate, *n* is the creep exponent, and η is a geometric function whose value depends on the crack size and *n*.[21]

RESULTS

TENSILE AND CREEP TESTS

The 700°C tensile and creep data for the as-cast + HIPped, and HIPped + heat treated at 1300°C specimens are given in Table 1. The reported tensile data is an average value of two tests, which had a marginal scatter. Creep parameters in the table were determined from tensile creep tests where the data were evaluated from registered creep strain as a function of time for each stress level. The data were further assessed to determine the minimum (steady state) creep rate which is plotted as a function of stress in Fig. 2. Both materials showed transition in creep deformation behaviour with increasing stress. The transition is observed at σ=320MPa√m and 350MPa√m for heat treated, and as cast + HIPed conditions, respectively, leading to two creep exponents as seen in Fig. 2. However, the creep exponents reported in Table 1 are average values for the entire data set for each material condition.

Table 1   Tensile and Creep Data of TiAlCr at 700°C.

Material	$R_{p0.2}$ (MPa)	R_m (MPa)	E-Modulus (MPa)	A_5 (%)	D_1	m	n
As cast	372	522	152 450	0.03	8.0×10^{-5}	6.56	7.78
HT 1300°C	326	412	152 000	0.34	7.0×10^{-5}	9.13	9.19

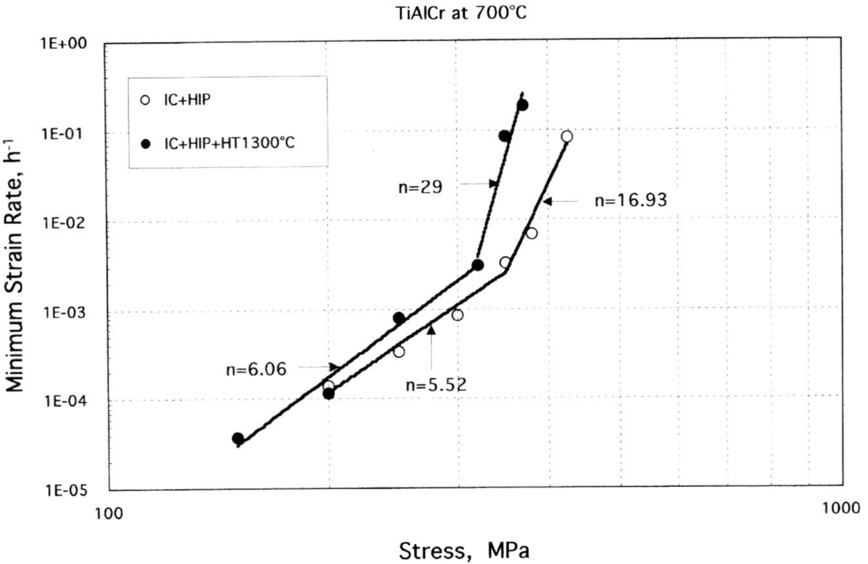

Fig. 2   Steady-state creep rate as a function of stress for IC + HIPed and heat treated (HT 1300°C) TiAlCr.

FRACTURE TOUGHNESS TESTS

The toughness data, K_c, J_c and K calculated from Jc, K_{Jc} for the two sizes of SEN bend specimens having slit notches and FPC starter cracks are given in Table 2. The CT50 specimen data, $K_{0.2}$, $J_{0.2}$ and J_i, are included in the table to illustrate the specimen geometry effect on the fracture toughness.

Table 2   Effect of specimen size and geometry on fracture toughness of SEN and CT specimens as-cast TiAlCr at 700°C.

Specimen size (mm)	$K_{c,slit}$ SEN (MPa√m)	$K_{c,FPC}$ SEN (MPa√m)	$K_{Jc,slit}$ SEN (MPa√m)	$K_{JC,FPC}$ SEN (MPa√m)	$J_{C,slit}$ SEN (N mm⁻¹)	$J_{C,FPC}$ SEN (N mm⁻¹)	$K_{0.2,FPC}$ CT50/10 (MPa√m)	$J_{i,FPC}$ CT50/10 (N mm⁻¹)	$J_{0.2,FPC}$ CT50/10 (N mm⁻¹)
SEN: 5 × 10 × 50	34.2	32.0	43.7	40.3	12.5	10.6	45.0	14.1	24.0
SEN: 10 × 20 × 90	34.2	32.0	39.3	44.1	10.1	12.7			

For the SEN specimens, the K_{Jc} values are higher than K_c values because the deformation is accounted for within the J determination. The J_c data from the SEN specimens compares with J_i from the CT50 specimens. It is also seen in the table that the K_{Jc} data from the SEN specimens agree well with $K_{0.2}$ for the CT50 specimens.

CREEP CRACK GROWTH TESTS

The scatter and pop-ins in the recorded experimental data may be related to the mixed fracture mode with secondary cracking and crack front tunneling as observed in an SEM on fracture surfaces of tested specimens (Fig. 3). The effect of loading mode on the crack tip deformation and crack extention observed in the SEM on sectioned specimens is depicted in Figs 4 and 5, for constant load and displacement rate control test specimens respectively.

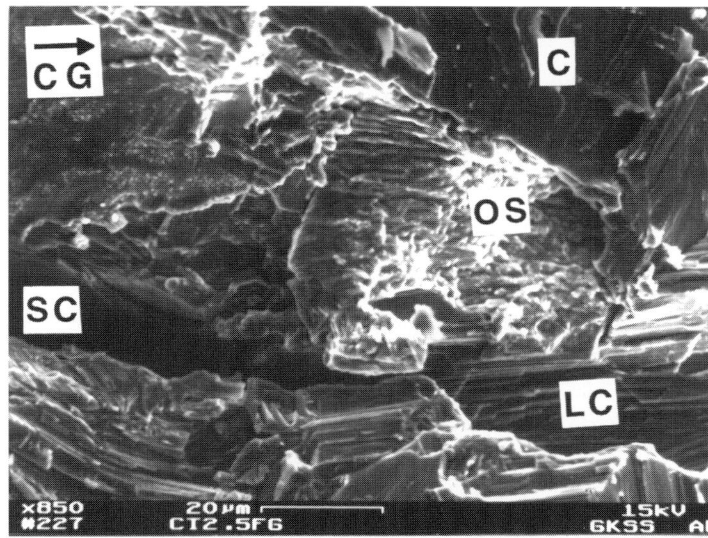

Fig. 3 Fracture surface of a CT specimen tested at 700°C showing a mixed fracture mode at the tip of a growing crack. OS: Oxidised fracture surface, C: Cleavage facet, LC: Lamellar cracks, SC: Secondary crack. Crack growth (CG) direction as indicated.

The complete set of creep crack growth data obtained from CT specimens for both as-cast +HIPped and HIPped+heat-treated materials at 700°C are correlated with crack tip parameters K and $C^*(t)$ in Figs 6 and 7, respectively. The scatter in correlated data is large, particularly in the K-correlation. Therefore, K data are reduced to transition range only, that is data for $\Delta a \leq 0.2$mm, showing a good correlation, as depicted in Fig. 8. On the other hand, the crack growth data from displacement rate controlled tests (Fig. 7) showed decreasing $C^*(t)$ due to crack growth rate effects beyond F_{max}. In Fig. 9, these data are reduced up to F_{max}, that give a linear crack growth correlation. The data from constant load test, however, show two parts that calls for study of effects of sharp starter crack and loading mode in CCG testing and validity of crack growth data. Therefore, the crack growth data are further analysed comparing the data

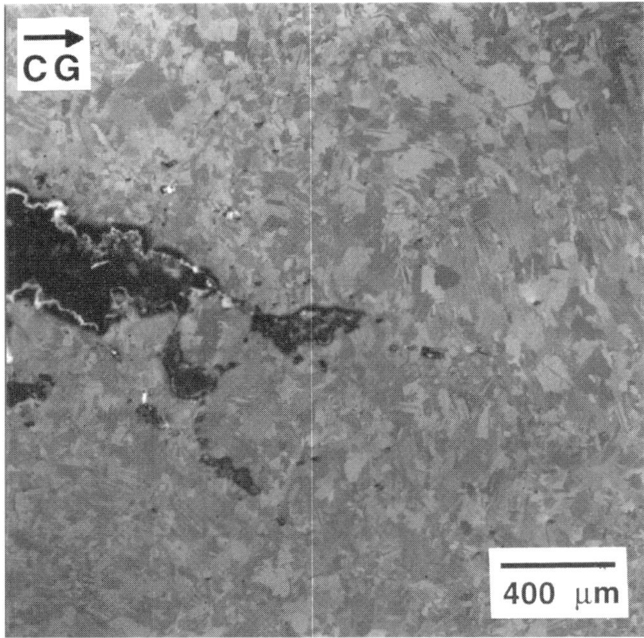

Fig. 4 Polarised light micrograph of a sectioned specimen of constant load tested HIPed + HT 1300°C material, showing the crack tip damage and crack growth.

Fig. 5 Polarised light micrograph of a sectioned specimen of displacement rate controlled tested HIPed + HT 1300°C material, showing the crack tip damage and crack growth.

from constant load and displacement rate control tests in Figs 10 and 11, respectively. Both tests showed the ratio of creep component of displacement rate (dVc/dt) to total displacement rate (dVt/dt) ≥ 0.9 for the whole range of tests (Figs 10(a) and (b)), satisfying the validity requirement of the test standard[21]. However, the rather brittle fracture mode observed in SEM (Fig 3) calls for analysis of deformation and deflection rate partitioning. The total load line deflection, Vt, together with elastic, Ve, and creep components of deflection, Vc, are shown in Fig. 11(a) and (b), for contant load and displacement rate control tests as a function of normalised time, t/t_f, respectively. A difference in the growth of creep deformation is seen from the variation of creep component of deflection, Vc, in two figures.

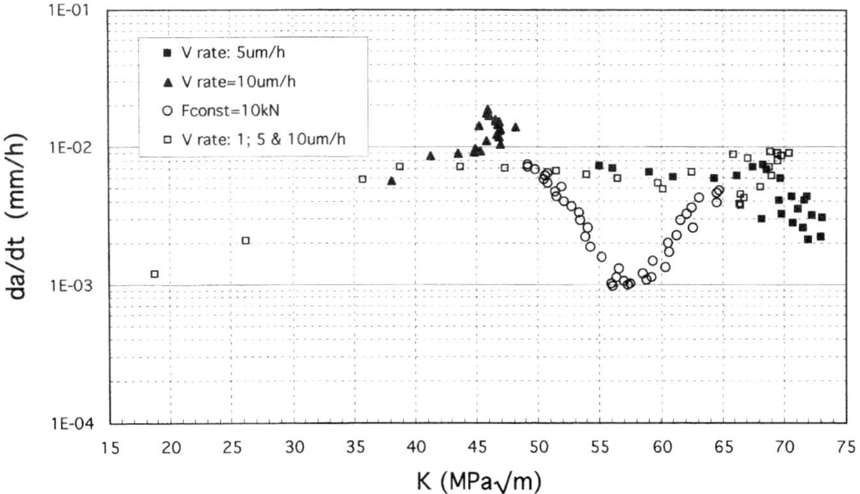

Fig. 6 Crack growth rate as a function of K at 700°C (complete set of data).

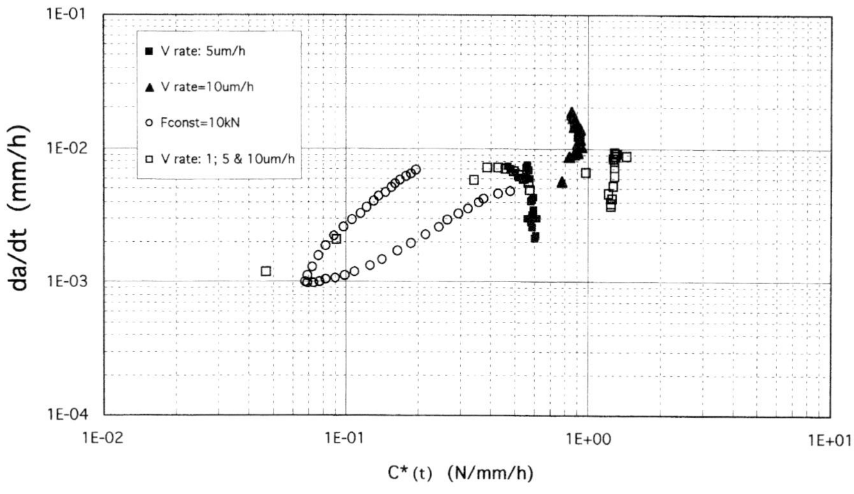

Fig. 7 Crack growth rate as a function of $C^*(t)$ at 700°C (complete set of data).

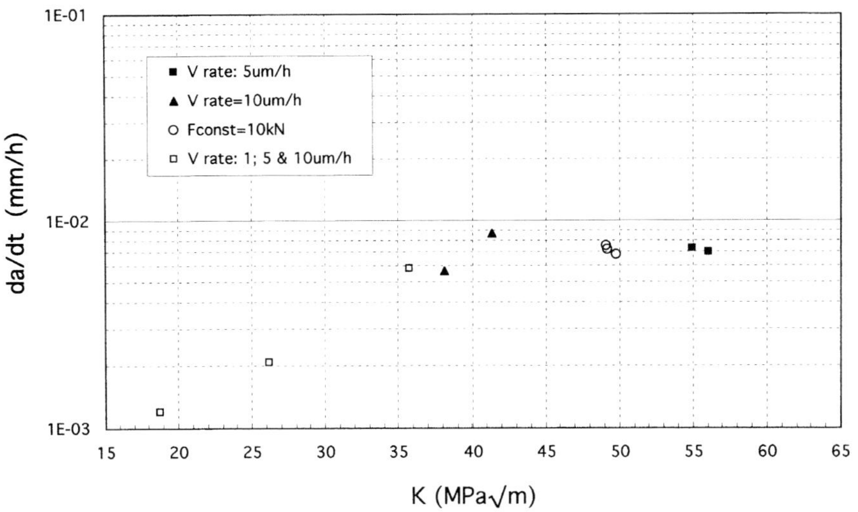

Fig. 8   Crack growth rate as a function of K (reduced data for $\Delta a<0.2$mm).

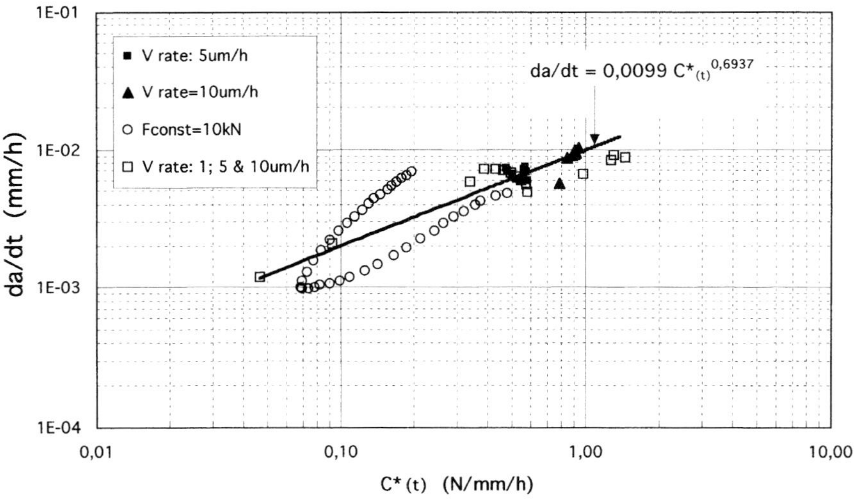

Fig. 9   Crack growth rate as a function of $C^*(t)$ (reduced data for $F<F_{max}$).

DISCUSSION

INTRODUCTION OF SHARP STARTER CRACK

During fatigue precracking of brittle materials at low fatigue loads subcritical crack may initiate and then stop requiring an increase in load for propagation. The fact that the fatigue strength of TiAlCr was of the same order of magnitude as its tensile strength,[7] fatigue precrack loads as high as the critical loads in toughness testing were needed. This

220

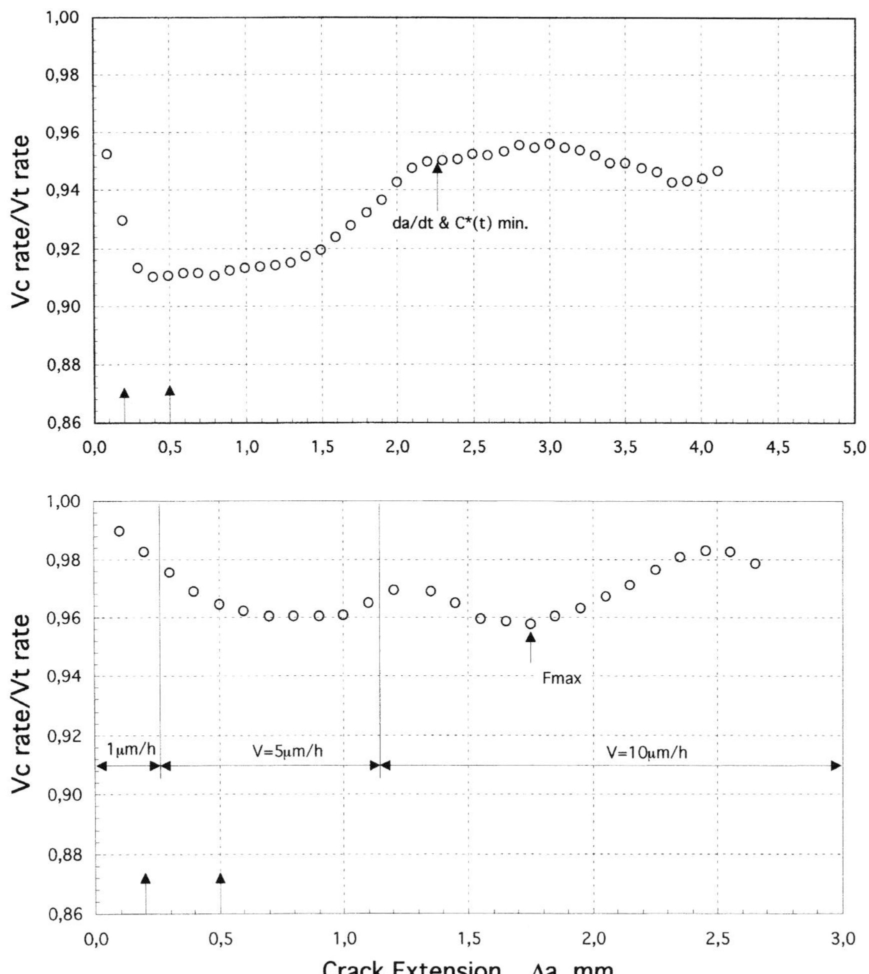

Fig 10 Ratio of load-line deflection rate due to creep (dVc/dt) to a total load-line deflection rate (dVt/dt) as a function of crack extension (Δa) for HIPed + HT 1300°C TiAlCr tested at 700°C (a) under constant load (*Fconst.* = 10 kN), (b) displacement rate control (*V* rate = 1, 5 and 50 μm h^{-1}).

see a narrow load range for fatigue precracking that was investigated intensively.[16,20] The microstructure, particularly lamellar grain size and orientation, affects the fatigue crack initiation and fatigue crack growth leading to multi-crack initiation and crack branching,[17] which are mainly affected by the lamellar orientation at the crack tip.

The FPC load calculated from formulas given in the test procedures[24] overestimate the precrack loads by a factor of three. Fatigue precrack forces for brittle materials with limited non-linear deformation prior to fracture are checked using a critical load[17] F_c. The FPC loads for TiAlCr were of the same order of magnitude as F_c. On the other hand, the ratio $Kmax/E \leq 1.5 \times 10^{-4} \sqrt{m}$ gives a good estimate of FPC force. Hence, if F_c is not available, the ratio $Kmax/E = 1.5 \times 10^{-4} \sqrt{m}$ can be used to estimate FPC load.

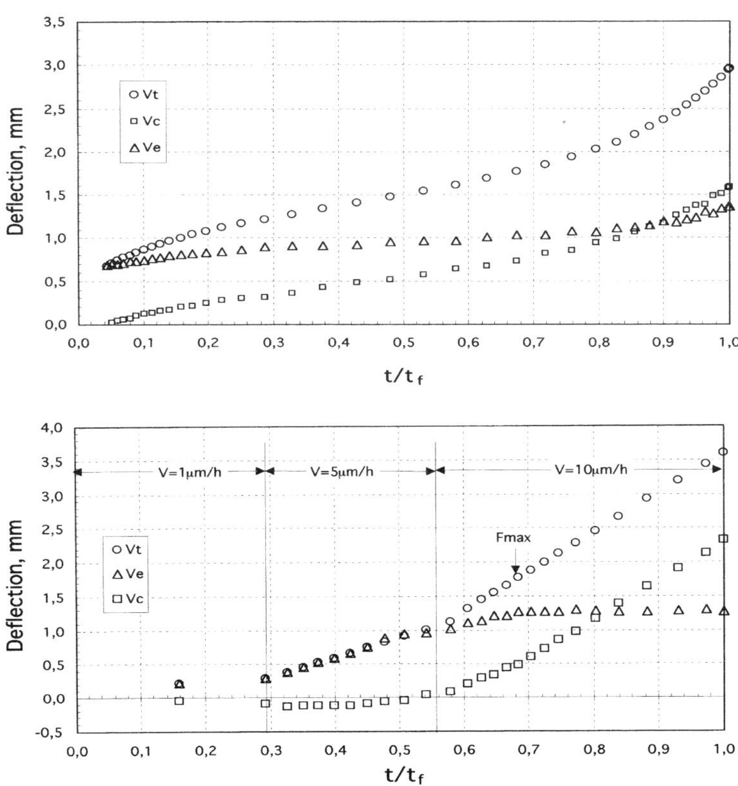

Fig. 11 Load line deflection (*Vc*: creep component, *Ve*: elastic component, *Vt*: total load-line deflection) as a function of normalised test time, t/t_f, for HIPed + HT 1300°C TiAlCr tested at 700°C (a) under constant load, (b) displacement rate control.

The experience with TiAl intermetallics[16,17] showed that the above discussion is valid for fine microstructures where the lamellar grains are small. However, coarse lamellar grains at machined notch tips may lead to starter cracks growing along lamellae oriented at an angle to the main crack plane which leads to crack deviation, and invalidates the test.

The difficulty faced with the introduction of a sharp starter crack may be overcome by spark erosion (EDM) of a fine (i.e. tip radius $\rho \leq 0.05$mm) starter slit notch.[17, 20]

CRACK LENGTH MEASUREMENT

The measurement of crack length in toughness testing of TiAlCr requires further study. Crack lengths measured on the side surfaces of specimens were smaller than those measured on fracture surfaces by up to 30 per cent. The crack fronts were uneven and the crack growth was nonuniform by more than 20 per cent or 0.15 mm, not satisfying the standard requirements.[21] Crack tunneling and unbroken ligaments on fracture surfaces contribute to the discrepancy in crack length measurements. Crack branching and bifurcation, as in Fig. 3, lead to violation of the validity criteria.

Out-of-plane cracking is particularly promoted by large grains and lamellae. An out-of-plane cracking limit needs to be specified for titanium aluminides similar to the one in Fig. 8 of ASTM E647. The crack deflection in excess of 20° may call for mixed-mode stress analyses to compute K as it is also suggested in ASTM E647.

The crack growth was monitored during testing of CT specimens using the DCPD method. Crack lengths determined from DCPD were compared with those determined from fracture surfaces. The difference was as high as 50 per cent, with those measured on fracture surfaces being higher, which also did not satisfy the standard requirement. In determining the crack lengths on fracture surfaces, the unbroken shear ligaments were ignored, whereas crack lengths from the DCPD were erroneous because of surface contacts and unbroken ligaments.

It is recommended here that the DCPD method be used for monitoring crack growth in TiAl base intermetallics with initial and final crack lengths corrected with those from the fracture surface.

FRACTURE PARAMETERS

The fracture parameters determined according to the test procedures characterise the fracture behaviour of homogeneous materials which are size-independent provided certain procedural validity requirements are satisfied.[21, 24, 25]. Certain combinations of material and test specimen geometry may lead to crack growth data which do not meet the validity requirements for size independence.

A noticable difference is seen between the K_{Jc} and K_c values. The J integral calculation, used to determine K_{Jc} takes into account the brittle deformation and crack growth by fracture of ligaments and linkage with microcracks in lamellae. It is important to note here that in SEN specimens the crack growth prior to fracture varied by up to 1.0mm, which is much higher than the allowed crack growth value of 0.2mm given in the test procedures for fracture toughness determination (i.e. $K_{0.2}$, $J_{0.2}$ at 0.2mm crack extension) from tests with stable crack growth. If the crack growth in SEN specimens could be determined reliably, the toughness data would be corrected for the crack growth. This is important because most of the toughness data published in the open literature are determined using the initial crack length, ignoring the crack extention in SEN specimens.

The present work showed that specimens of $B \geq 5$mm are required to obtain size and loading rate independent fracture toughness data with SEN specimens. Note that it is difficult to assess the size independence because of scatter in the J_c data arising partly from variations in the amount of stable crack growth prior to unstable fracture. The scatter may be reduced if the data are limited to crack growth less than 0.2 mm. More data are needed to clarify this point.

The non-linear deformation in the force-displacement diagram is accounted in calculating K_{Jc} for the SEN specimens. A good agreement is seen between K_{Jc} from tests on SEN specimens and $K_{0.2}$ from CT specimens that can be considered as the size and geometry independent crack growth initiation toughness of the material.

The available data suggest that EDM slit notches with tip radius of $\rho=0.05$ mm or smaller can be used as sharp starter cracks in fracture toughness testing of TiAlCr. The value of ρ is much smaller than the microstructural feature, e.g. lamellar regions of 320µm in size, that controls fracture.

The crack growth data from tests on CT specimens met the validity criteria[24, 25] except from the crack length determination. The fracture resistance increased slightly with crack extension.

CREEP CRACK GROWTH

Crack growth in TiAlCr at 700°C is brittle, as depicted in Fig. 3. Therefore, during creep crack growth in these materials, the effects of crack growth on the crack tip stress fields cannot be neglected. Attention needs to be paid to the applicability of the stationary crack tip parameter $C^*(t)$ for correlating creep crack growth behaviour. The concept of $C^*(t)$ was originally proposed for characterizing creep crack growth under extensive steady state creep conditions. Its validity has been extended for extensive primary and secondary creep. However, if any primary creep is present, the magnitude of $C^*(t)$ changes with time. It is reported that TiAl basis intermetallics show large primary creep at the present test temperature.[9] In the small scale creep regime, $C^*(t)$ is not path-independent and also does not uniquely characterize the crack tip stress fields. Further, K-fields will exist until the crack begins to extend at the incubation time t_i, which needs to be defined for experimental determination of the initiation phase.

In the light of above discussion, the complete set of creep crack growth data are correlated with both $C^*(t)$ and K (Figs 6 and 7), followed by the reduced data (Figs 8 and 9) to shed some light on the applicability of the concepts of K and $C^*(t)$ in these materials. The data from transition range correlate only with K. This is because immediately after loading, an elastic, or elastic–plastic stress distribution is generated ahead of a crack tip prior to the onset of creep. The criteria was set in the standard[21] to ensure the extensive creep stress distribution will be produced ahead of the crack for the data to correlate with $C^*(t)$. The complete CCG data set satisfied the transition time $t_{test} > t_T$, and deflection rate ratio $(\mathrm{d}Vc/\mathrm{d}t)/(\mathrm{d}Vt/\mathrm{d}t) \geq 0.9$ as seen in Fig. 10(a) and (b), although the material is brittle as seen in Fig. 11 and also in SEM observation (Fig. 3). The variation of deflection rate ratio in Fig. 10 does not follow the predicted behaviour.[26] Similar discrepancy is also reported for creep brittle Al and Ti alloys.[27] As numerical work was not done on the present material data, it may be noted that this aspect needs further clarification particularly if assessments of components made of TiAl are to be conducted.

The stationary stress state is achieved in the second part of constant load test (Fig. 9), the slope of which agrees well with displacement rate controlled tests. The decreasing $\mathrm{d}a/\mathrm{d}t$ and $C^*(t)$ part of the constant load test is attributed to the transition of crack growth mode of a fatigue precracked specimen where initial crack growth follows the transgranular fatigue crack path that changes to creep crack mode with accumulation of damage with time. This behaviour in TiAl is also reported by Fuji et al.[19] It is important to note that this transition occurs over a period of finite crack extension, as large as half of the test time, $t/t_f = 0.5$ (Fig. 10(a)).

The data from displacement rate controlled tests exhibit typical transition behaviour over a range of crack growth caused by a change of stress state via changing applied load under displacement rate control in creep brittle material. This discussion is supported by the partitioned deflection in Fig. 11(b), where the creep component of deflection is 0 for a long period of time, $t/t_f \geq 0.5$. It was followed by constant Ve and increasing Vc, however, crack growth rate effect dominates after F_{max}. On the other hand, the constant load test shows an increasing Vc from beginning of the test onwards as expected (Fig. 11a.) The difference in deformation mechanisms can also seen in the SEM at the crack tip of specimens (Figs 4 and 5), where a larger deformation zone in the constant load test specimen is compared with smaller deformation zone and straight crack growth in displacement rate controlled test.

The present work directs attention to the need for further study of creep crack behaviour of intermetallic alloys. Special emphasis needs to be placed on the creep deformation mechanisms at the crack tip and effect of growing crack on the deformation and crack growth behaviour associated with microstructural considerations. It is well known that microstructural variations strongly affect the creep deformation as well as fracture behaviour in intermetallics[9,17] particularly in lamellar structures. This directs attention to the need for more data on TiAl basis intermetallics that are considered for component production, for a sound conclusion and use of data in component assessment procedures.

CONCLUSIONS

The microstructure influences the deformation and creep behaviour of TiAlCr alloys. The creep deformation depends on the stress level and leads to a change in creep exponent with increasing stress.

The fracture mode in crack growth test specimens at 700°C is brittle with microcracking observed in the γ-phase and along lamellar interfaces.

A large discrepancy, up to 50 per cent, between crack lengths measured on fracture surfaces and those determined using DCPD method were caused by unbroken lamellar ligaments, crack branching and crack tunnelling. This causes difficulties in crack length measurements that requires further study.

EDM slit notches with a tip radius of 0.05mm can be used as sharp starter cracks in SEN and CT specimens.

The fracture toughness data are specimen size and geometry dependent. SEN specimens smaller than $5 \times 10 \times 50 mm^3$ show loading rate dependence.

Creep crack growth tests can be done both under constant load and displacement rate control provided the effects of crack growth on stress state is accounted for. Crack growth data from displacement rate controlled tests beyond a maximum load (F_{max}) may not be correlated with crack tip parameters.

The time dependent crack growth data obtained at 700°C assessed following ASTM standard[21] correlates with $C^*(t)$. However, much work is needed particularly for the starter sharp crack requirement, crack length measurements and validity of data, to assess creep brittle intermetallics. The data from the transition range i.e. $\Delta a \leq 0.2mm$ may be correlated with K.[25]

The creep crack growth data together with creep rupture data provides indispensible input for life assessment of components made of γ-TiAl based intermetallics.

ACKNOWLEDGEMENTS

The motivating discussions and scientific support of B. Petrovski and continued support of the work by K.-H.Schwalbe and R. Wagner are gratefully acknowledged. Thanks are also due to J.Granacher of T.U.Darmstadt for carrying out creep tests.

REFERENCES

1. Int. Conf. Life Assessment of Hot Section Gas Turbine Components, Organised by IOM, James Watt Centre, Edinburgh, UK, Oct.5–7, 1999.
2. W. Smarsly and L. Singheiser, *Materials for Advanced Power Engineering, Part II*, D. Contouradis et al. eds., Kluwer Acad. Publications, 1994, 1731.
3. M.V. Nathal and S.R. Levine, in *Superalloys 1992*, S.D. Antolovich et al. eds., TMS, 1992, 329–.
4. D.M. Dimiduk, D.B. Miracle, Y.W. Kim and M.G. Mendiratta, *ISIJ Int.*, 1991, **31**, 1223–.
5. T. Khan, S. Naka, P. Veyssiere and P. Costa, *High Temperature Materials for Power Engineering 1990, Part II*, E. Bachelet, R. Brunetand et. al. eds., Kluwer Acad. Publications, 1990, 153–.
6. Y.-W. Kim, 'High Temperature Ordered Intermetallic Alloys IV'. *MRS Symp. Proc.*, 1991, **213**, L.A. Johnson, D.P. Pope and J.O. Stiegler eds., 777–794.
7. R. Wagner, F. Appel, B. Dogan, P.J. Ennis, U. Lorenz, J. Müllauer, H.P. Nicolai, W. Quadakars, L. Singheiser, W. Smarsly, W. Vaidya and K. Wurzwallner, *Gamma Titanium Aluminides*, Y.-W. Kim, R. Wagner and M. Yamaguchi eds., TMS, 1995, 387–404.
8. Y.-W. Kim, *Proc. Symp. Gamma Titanium Aluminides*, Y.-W. Kim, R. Wagner and M. Yamaguchi eds., TMS, 1995, 637–654.
9. F. Appel and R. Wagner, *Materials Sci. and Engg.*, 1998, **R22** (5), 187–268.
10. M. Yamaguchi, H. Inui and K. Ito, *Acta Mater.*, 2000, **48**, 307–322.
11. Y.-W. Kim, *JOM*, 1995, 39–41.
12. C.M. Austin, T.J. Kelly,, K.G. McAllister and J.C. Chesnutt, in *Structural Intermetallics 1997*, M.V. Natal, R. Darolia, C.T. Lui, P.L. Martin, D.B. Miracle, R. Wagner and M. Yamaguchi eds., TMS, 1997, 413–.
13. T. Tetsui, *Current Opinion in Solid State & Materials Science*, 1999, 4, 243.
14. W. Vaidya, K.-H. Schwalbe and R. Wagner, *Proc. Symp. Gamma Titanium Aluminides*, Y.-W. Kim, R. Wagner and M. Yamaguchi eds., TMS, 1995, 867–874.
15. B. Dogan, P.A. Beaven and R. Wagner, *Proc. Int. Conf. IRC 92*, M.H. Loretto and C.J. Beevers eds., MCE Publications Ltd, 1992, 429–434.
16. B. Dogan, D. Schöneich, K.-H. Schwalbe and R. Wagner, *Intermetallics*, 1996, 4 (1), 61–69
17. B. Dogan and K.-H. Schwalbe, *Engineering Fracture Mechanics*, 1997, **56** (2), 155–165.
18. K.S. Chan and Y.-W. Kim, *Met. Trans. A*, 1993, **24A**, 113–125.
19. A. Fuji, M. Tabuchi, A.T. Yokobori Jr., and T. Yokobori, *Engineering Fracture Mechanics*, 1999, **62** (1), 23–32.
20. VAMAS (*Varsailles Project on Advanced Materials and Standards*) TWA19: High Temperature Fracture of Brittle Materials, 1993–1998.
21. ASTM E1457–92, *Standard Test Method for Measurement of Creep Crack Growth Rates in Metals*, Annual Book of ASTM Standards, 1992, **03.01.**, 1031–1034.

22 K.T. Venkateswara Rao, Y.-W. Kim, C.L. Muhlstein and R.O. Ritchie, *Matls Sci. and Engg. A*, 1995, **192/193**, 474–482.

23 B. Dogan, H. Martens, K.-H. Blom and K.-H. Schwalbe, *Proc. IITT Int.Conf. Laser 5*, 10–11 April 1989, 1989, 188–194.

24 ASTM E1737–96, *Standard Test Method for J-Integral Characterisation of Fracture Toughnes*, Annual Book of ASTM Standards, 1996, **03.01.**, 994–1017.

25 ASTM E399–90, *Standard Test Method for Plane-Strain Fracture Toughness of Metallic Materials*, Annual Book of ASTM Standards, **03.01.**, 485–515, April 1991.

26 A. Saxena, D.E. Hall and D.L. McDowell, *Engineering Fracture Mechanics*, 1999, **62**, 111–122.

27 O. Kwon, K.M. Nikbin, G.A. Webster and K.V. Jata, *Engineering Fracture Mechanics*, 1999, **62**, 155–165.

Application of Thermal Barrier Coatings on NiAl Intermetallic Material

E. LUGSCHEIDER, K. BOBZIN and A. ETZKORN

Materials Science Institute, Aachen University of Technology, Templergraben 55, 52056 Aachen, Germany
http://www.rwth-aachen.de/ww

INTRODUCTION

Reaching higher efficiency and longer life times of gas turbines and gas turbine components are some of the main objectives of today's research work in this area. To reach the target of higher efficiency one possible way to go is to increase the gas turbine inlet temperature. In doing so the life time of gas turbine parts will be reduced. Consequently the aim of research work has to be to introduce better materials and thermal protection systems.[1, 2]

Because of their good oxidation behaviour and their high melting point, NiAl-alloys are of great interest as a replacing material for superalloys. To use the whole capacity of this material also thermal barrier coatings should be used.

In the present work the coatability of the high temperature material NiAl-intermetallic was examined. The advantages of NiAl are low density, high melting point of the stoichiometric phase, high thermal conductivity, low thermal expansion and good oxidation behaviour. Because of the oxidation resistance of NiAl alloys, it is not necessary to use MCrAlY-coatings as bond coats. Also the mismatch of thermal expansion is lower between NiAl and zirconia than it is between zirconia and MCrAlY. The thermal barrier coatings were deposited by the EB-PVD technique. The deposition process was varied in the pre-treatment of the substrate surface and the temperature state of the substrate itself. During the pre-treatment a rf-plasma etching was used to clean and to activate the substrate surface. By SEM, the microstructure of the coatings was examined. The coatings were also analysed by EDX and XRD.

EXPERIMENTAL PROCEDURE

The EB-PVD facility contains two electron beams, one for the vaporisation of the coating material and the other one as substrate heating. The vapour can be produced from bulk material or powder which makes a change of the materials composition very easy. By measuring the heating electron current flowing through the substrate, heating power can be determined and related to the substrate temperature and the resulting microstructure.[3, 4] The coating equipment allows a rf-plasma etching of the substrate surface using different gases. The pressure of the deposition atmosphere is kept constant. The principle of the EB-PVD facility which was used for the deposition of the zirconia coatings is shown in Fig. 1.

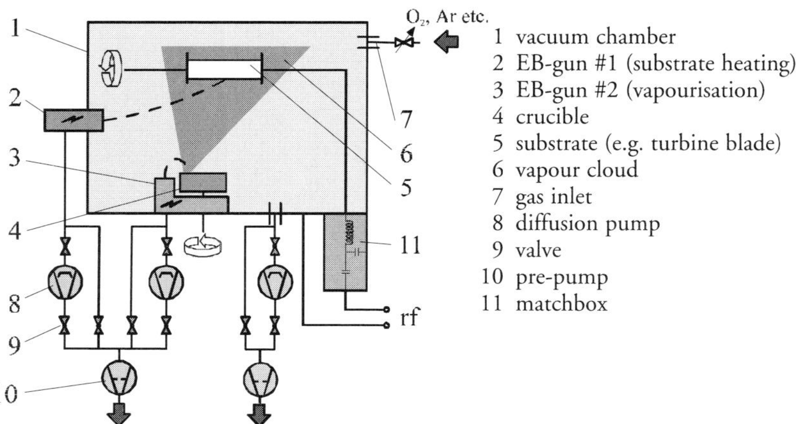

Fig. 1   Schematic of the EB-PVD facility.

All samples were cleaned by rf-plasma etching at 400 W plasma power and argon as plasma gas. Afterwards a 60 minute heat treatment at approx. 1100°C and an oxygen partial pressure of 1 Pa was performed to get a thermally grown oxide. As coating material a powder with a zirconia/yttria ratio of 92.98/7.02 was used. The deposition experiments were performed at both different oxygen partial pressures and substrate temperatures. The variation of the oxygen partial pressure started up with the deposition of zirconia without additional oxygen went on with 0.5 Pa and 1 Pa oxygen partial pressure and ended at 1.5 Pa. The variation of the heating electron current varied in 5 mA steps from 15 mA to 35 mA which is about 600°C to 1100°C corresponds to a homologous temperature of 0.18–0.36.

RESULTS AND DISCUSSION

The coatings microstructures as a function of the substrate current variation are shown in Fig. 2, those related to the oxygen partial pressure variation in Fig. 3.

It can be seen that the microstructure becomes less homogeneous with an increase of substrate temperature, the diameter of the columns is growing. For the coating, which was deposited at 15 mA the columns are hardly perceptible. The variation of the oxygen partial pressure shows very little influence on the microstructure. With the presence of additional oxygen the microstructure becomes clearer. Compared to the microstructures which would be achieved during the deposition of zirconia on MCrAlY under the same deposition conditions, there is no difference recognisable. The coatings can also be compared with the structure zone models by using the homologous temperature.[5] At low temperatures the approaching particles condense immediately, so a regular structure is not possible. With an increase in temperature crystallisation, the surface mobility of the adatoms becomes more important. Reaching a certain energy state by a further increase of substrate temperature big grains grow faster by overgrowing smaller ones.

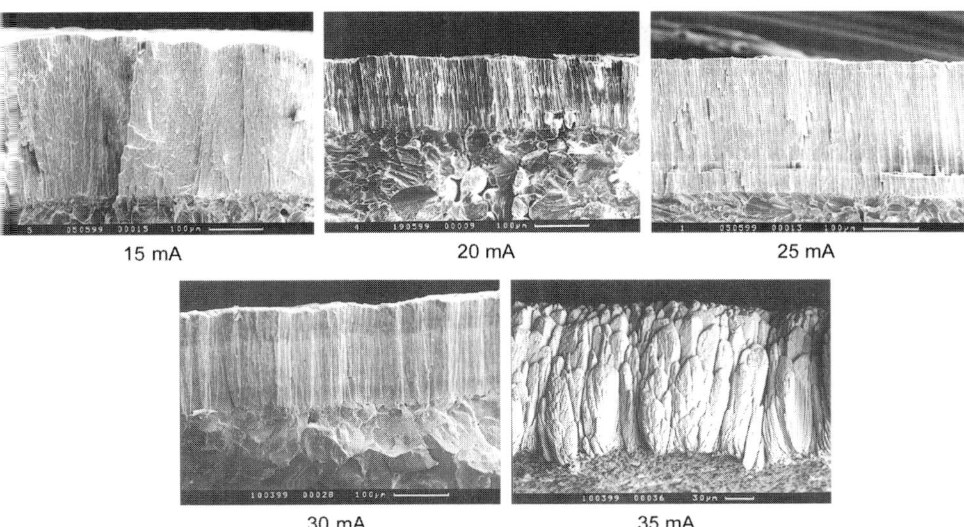

Fig. 2 Microstructure of the different coatings for the variation of substrate current by SEM (oxygen partial pressure = 1 Pa).

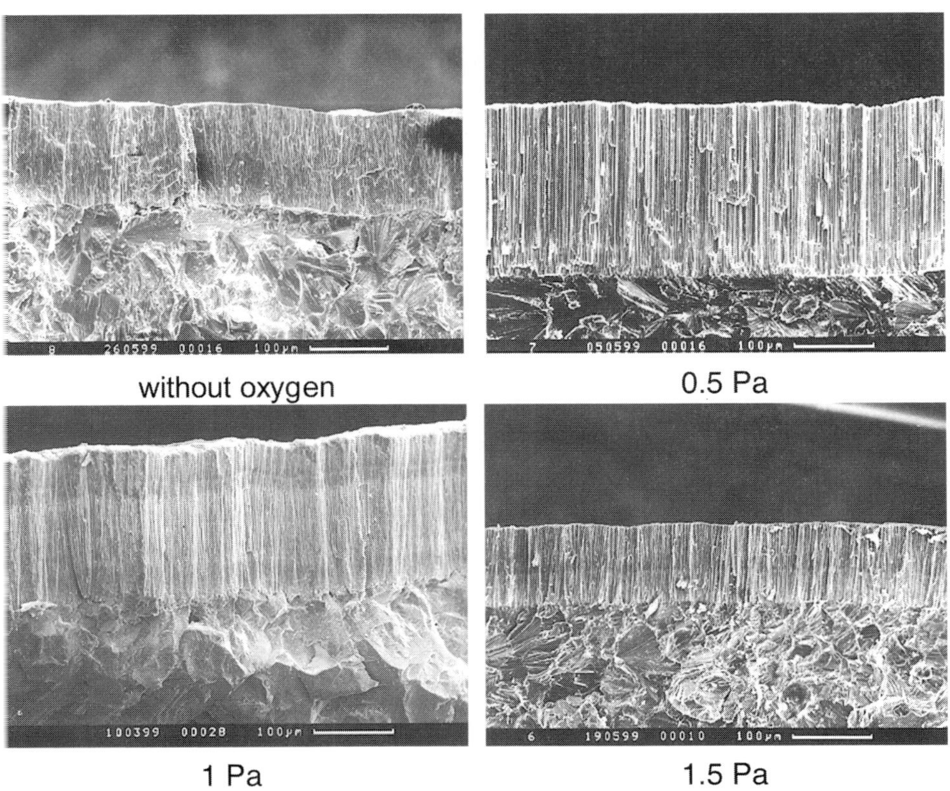

Fig. 3 Microstructure of the different coatings for the variation of oxygen partial pressure by SEM (substrate current = 30 mA).

The relation of the deposition rate on the oxygen partial pressure or on the substrate current is shown in Fig. 4. Without any additional oxygen the vapour cloud becomes wide because there is just a very low atmosphere which stabilises the vapour cone. This causes a lower density of the vapour in the cone which leads to a lower deposition rate. The deposition rate decreases from 0.5 Pa with an increase of oxygen partial pressure. When the deposition atmosphere becomes denser, the mean free path becomes shorter and less vapour particles reach the substrate. With an increase of substrate current which means an increase of substrate temperature, the deposition rate decreases because the energy state of the vapour becomes higher and with this the mobility of the coating material. Vapour particles which reach the substrate can be re-evaporated.

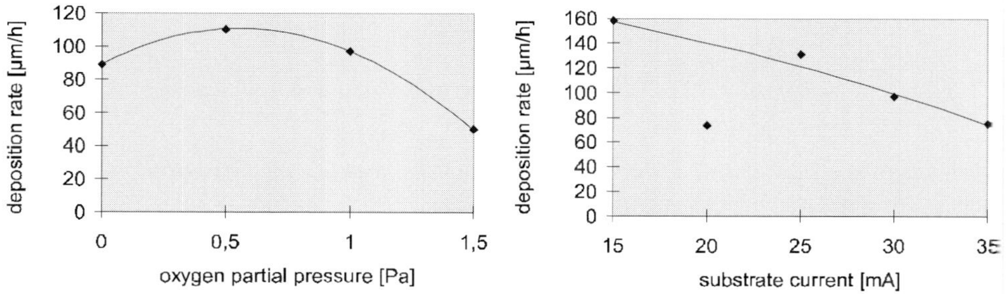

Fig. 4 Deposition rates due to oxygen partial pressure (left) and substrate current (right).

The EDX-analysis show a dependence of the coatings zirconium content on either the oxygen partial pressure or the substrate current (Fig. 5). The vapour pressures of zirconia and yttria are very similar[6] so this can't be the reason for differences in the zirconium content. This result led to the need to analyse the phases present. Therefore X-ray diffraction measurements were performed.

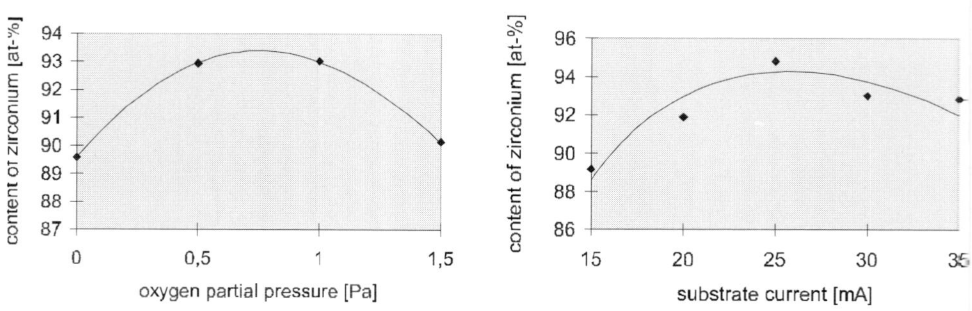

Fig. 5 Content of zirconium due to oxygen partial pressure (left) and substrate current (right).

The aim of the XRD-analysis was to check if there are phase transformations caused by different compositions of zirconia and yttria at different deposition conditions. The measurements were performed with a Seifert X3000 applying Cobalt radiation and a Fe-filter. In the system zirconia/yttria it is very difficult to distinguish between the cubic and tetragonal phase especially because the characteristic peaks are close together. Furthermore, the results measured on the solid coatings are compared with measurements which were performed on powders. The main result during X-ray diffraction measurement confirmed that the high zirconium content samples show a different behaviour compared to the lower ones.

All peaks were found in all samples but with different intensities. The result for the substrate current variation is shown in Fig. 6. The coatings consist of a solid solution of cubic and tetragonal phase with a low content of monoclinic phase. During the substrate current variation the 1 1 1 orientation is dominant. Scan 3 which is the 25 mA coating shows a higher intensity at the 2 0 0 orientation. Scan 5 was performed on a coating sample which wasn't in the same orientation to the vapour source during the deposition process as the other samples and is not comparable.

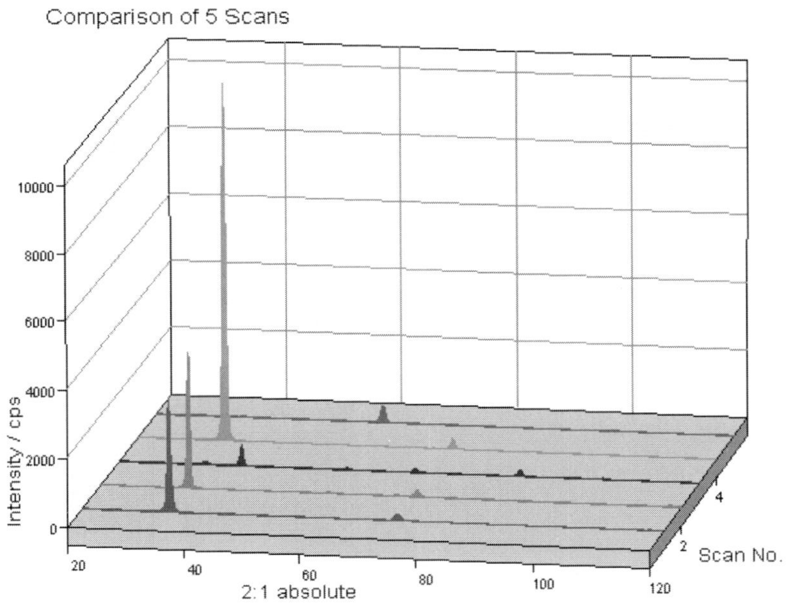

Fig. 6 XRD-analysis of samples at different substrate currents (1–5: 15 mA – 35 mA).

The variation of the oxygen partial pressure shows similar results as the variation of the substrate current. The X-ray results could be seen in Fig. 7. With additional gas pressure the coatings consist of a solid solution of cubic and tetragonal phase with a low content of monoclinic phase. The main orientation is 1 1 1. A high amount of additional oxygen results in the 2 0 0 orientation. The X-ray measurement of the sample which was coated without additional oxygen shows a content of non-stochiometric phases, which was expected because of the colour of the coatings. All coatings showed the typical white of stochiometric zirconia except the sample which was coated without additional oxygen. The colour of this coating was black.

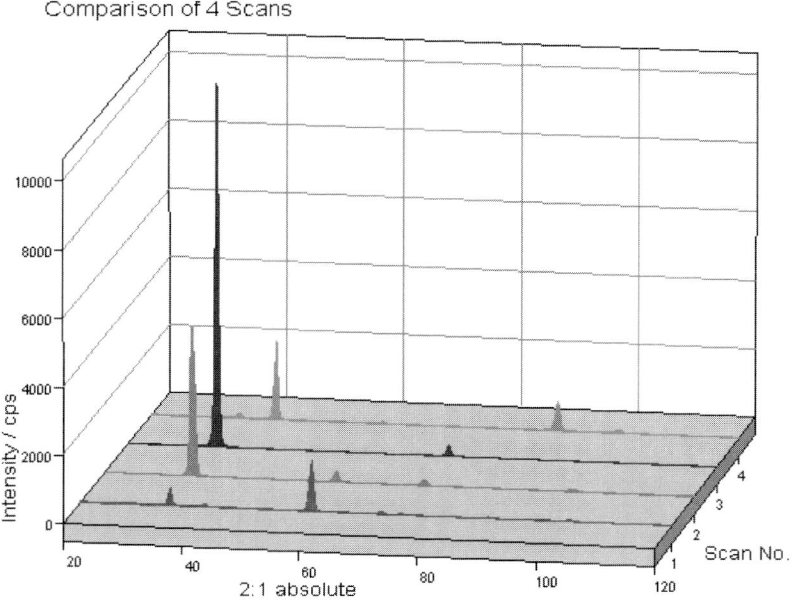

Fig. 7 XRD-analysis of samples at different oxygen partial pressures (1–4: without oxygen – 1.5 Pa).

CONCLUSIONS

Introducing new materials and materials compositions in the gas turbine technology seems to be a sensible way to reach the desired lives. NiAl shows a high potential as a replacing material for superalloys because of its low density, high melting point of the stochiometric phase, high thermal conductivity, low thermal expansion and good oxidation behaviour. Using the whole capacity of NiAl means using protection systems like thermal barrier coatings.

The experiments showed that in principle the deposition of zirconia thermal barrier coatings on NiAl without a bond coat is possible. Our investigations determined the dependency of deposition rate, zirconium content and the phase composition to the oxygen partial pressure and the substrate current (a measure of substrate temperature). There was no difference in the microstructure recognisable if the TBC was deposited on NiAl or MCrAlY.

The change of the deposition properties like oxygen partial pressure or substrate current led to different yttria/zirconia compositions over a small range. This change in the stabilisers content didn't cause a change in the phase composition which is worth mentioning.

To underline the results adhesion tests like thermal cycling or burner rig have to be performed. Also a static oxidation should be done. During these tests the growth of the interface oxide layer has to be examined. By a variation of the interface pre-treatment advances in adhesion could be reached.

ACKNOWLEDGEMENTS

The authors gratefully acknowledge the financial support of the Deutsche Forschungsgemeinschaft (DFG) within the Collaborative Research Center (SFB) 561 'Thermally highly loaded, porous and cooled multilayer-systems for combined cycle power plants'.

REFERENCES

1. E. Lugscheider, G. Doepper, H.G. Mayer, A. Seidel and W. Dreier, 'EB-PVD Zirconia Thermal Barrier Coatings for Experiments in Space', *High Temperature Surface Engineering*, J. Nicholls and D. Rickerby eds, IOM Communication, 2000, 77–84.
2. E. Lugscheider, C. Barimani and G. Doepper, 'Ceramic Thermal Barrier Coatings Deposited with the Electron Beam-Physical Vapour Deposition Technique', *Surface and Coatings Technology*, 1998, **98**, 1221–1227.
3. E. Lugscheider, C. Barimani and G. Doepper, 'Possibilities and Limits of Temperature Control in EB-PVD Equipment Used for Deposition of Zirconia Thermal Barrier Coatings', *High Temperature Surface Engineering*, J. Nicholls and D. Rickerby eds, IOM Communications, 2000, 85–94.
4. E. Lugscheider, C. Barimani, C. Wolff, S. Guerreiro and G. Doepper, 'Comparison of the structure of PVD-Thin films deposited with different deposition energies', *Surface and Coatings Technology*, 1996.
5. J.A. Thornton, 'The microstructure of sputter-deposited coatings', *J. Vac. Sci. Tech.*, 1986, A4(6).
6. G.V. Samsonov, *The Oxide Handbook*, IFI/Plenum, 1973.

Residual Life Assessment: Approaches, Techniques and Questions

M.I. WOOD

ERA Technology, Leatherhead, Surrey, UK

INTRODUCTION

Hot section components are high value items, and the costs associated with the repair or replacement of these components forms a significant part of the maintenance costs of gas turbines. As such, it is not unreasonable that the owners of the units would wish to maximise the safe economic engineering life of these components. How such a decision is arrived at is dependent upon a large number of factors, not all of which are technical.

The first distinguishing factor is whether one is considering aero engines or those used for ground based industrial purposes (power generation, combined heat and power etc). In the former case the lifing of critical components has to take place within some form of regulatory framework, e.g. the Civil Aviation Authority. The safety considerations for flight engines severely limits the scope of action that such operators have for life extension. In contrast, ground based units are not constrained by such tight frameworks and consequently have more scope for independent action. Whilst the amount of publicly available information on life assessment, i.e. residual life assessment, of hot section parts is limited, it is predominantly concerned with industrial units.

On the technical front, whilst the idea of a residual life assessment may seem relatively straightforward, the methodologies, techniques and appropriate material databases are not as well developed as is the case for the high temperature engineering steels used in conventional power generation plant (steam turbines, boilers etc). In that sense, one is dealing with an immature technology where there are a variety of approaches in use.

In addition, the assessment of acceptable risk can differ between an original equipment manufacturer (OEM) and an operator. The manufacturer can be more interested in the determination of a recommended or design life for their machine based on fleet averages. The operator is ultimately only interested in the specific components in his units, with their own specific operating history.

The purpose of this paper is to briefly review the general topic and to highlight a number of areas where uncertainties exist.

APPROACHES

There would seem to be three basic approaches, although these differ more in the detail than in some of the general objectives. The first is essentially a design based approach, within an overall framework of:

(i)   Calculate life consumed
(ii) Non destructively inspect the component at an appropriate time to determine their condition.
(iii) Destructively examine a sample of components
(iv) Assess the information

As an example of an OEM, ABB has set out its approach clearly in several papers (e.g. Refs 1, 2). Its starting point is a sound design procedure and the resultant design life. This life is then refined through feed back from the information generated by the examination of service run components (Fig. 1). In addition to sound design procedures, they also emphasise the need for:

- accurate service data (filtered and validated)
- correlations of operating history with component degradation, including component trending and age-degradation relationships
- relationships between microstructures and residual properties to the degradation conditions
- statistical aspects (Weibull/Bayesian)

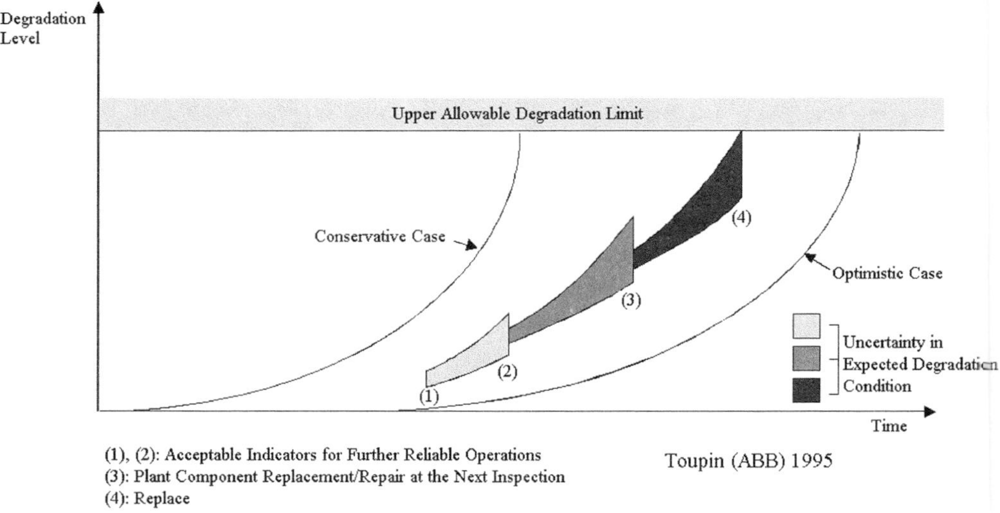

Fig. 1   ABB component trend monitoring.

This general approach has also been followed by the Electric Power Research Institute (EPRI).[3, 4] To assist in the first step, they have developed finite element packages for a number of blade designs as well as a number of virgin material property databases. These have then been combined into their REMLIF PC based life assessment programmes for a number of units, e.g. GE Frame 7E and 6B. A similar route has also been followed by some other operators, e.g. Powergen for the Siemens V94.2 unit, although in this case they have also incorporated less conventional damage accumulation models into the models.[5] As well addressing the main structural member of the blade, the superalloy, this modelling based approach is also now being extended to address the oxidation behaviour of the coating, and its associated durability.

In both cases, OEM and non OEM, it must be remembered that the quality of the output depends not only on the standard of the analysis but in the appropriateness of the assumptions made for the various parts of the engine as a whole. This would include factors such as:

- temperatures, both absolute and their spacial distribution (circumferential, radial and axial).
- allowances for manufacturing tolerances (e.g. cooling passage sizes and position variability)
- engine build standard

Whilst the non OEM has the possible advantage of more information from service units to assist them on certain engine types, in other cases they would need to have analysed the units more thoroughly than the OEM for their analysis to realistically address the shortcoming of the original design and life expectancy.

In the absence of design information, it is sometimes appropriate to follow what seems to be reversal of the above approach:

(i)   Non destructively inspect the component (in service, or during an overhaul)
(ii)   Destructively examine components taken form service
(iii) Carry out a more detailed calculation of loading etc if necessary
(iv)   Assess the information

This approach has been used very successfully in the military aerospace sector (e.g. Ref. 6). For this approach to be used in a more (rather than less) quantitative manner to assess the potential for life extension, relationships have been established between observable microstructural features, the results of destructive examinations and parameters describing the operational profile of specific engines (Fig. 2). The nature and importance of these are discussed a little later in the paper.

A third approach can be considered as a quantification of historical information, in essence a sophisticated version of experience. Since calculational approaches inevitably involve assumptions, there is an important role to be filled by detailed feedback from service experience. At one extreme, if there is a large enough fleet of similar units, then this information can be used in its own right for assessment purposes.

This approach has been implemented in the case of nozzle guide vane cracking in GE Frame 7E/E units. A detailed statistical analysis was carried out relating detailed inspection results on the size, location and frequency of cracking to the operational records for the units.[7] This then allowed the probability of a given level of damage to be present in the future to be determined purely from a knowledge of the operational pattern and history of the unit (Fig. 3).

EXAMINATION TECHNIQUES

All the assessment approaches rely on information acquired from an examination of the component. Whilst the routine non destructive inspections (visual, fluorescent penetrant

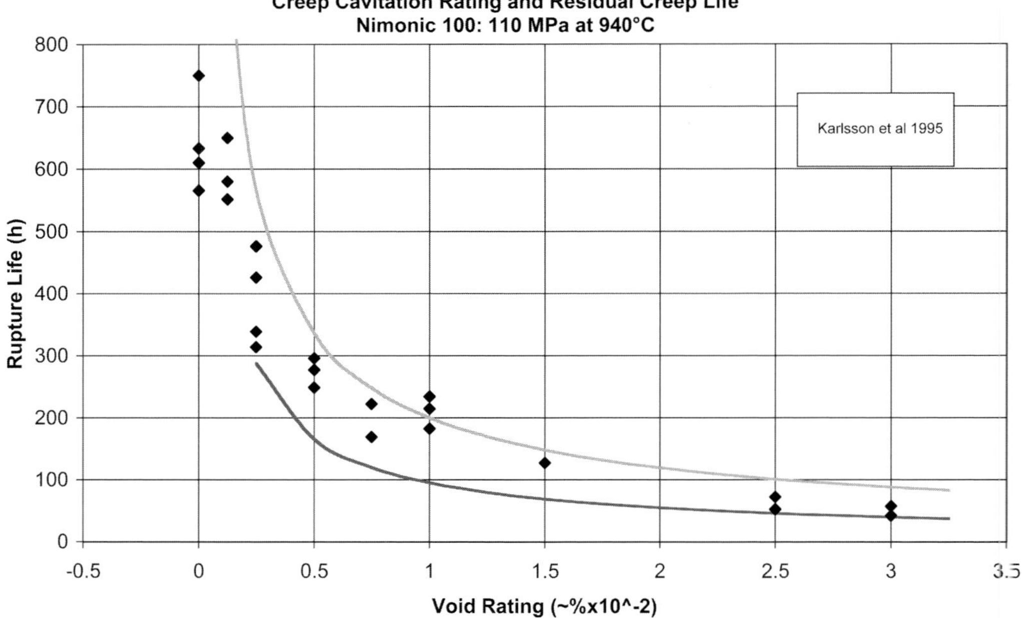

Fig. 2 Creep cavitation rating and residual life for Nimonic 100.

etc) carried out at overhaul, or in a repair shop, are necessary, they are essentially pass-fail inspections, revealing little about the actual damage state of the component (except that it has not proceeded beyond a certain level). Indeed, it is a presumption of such an inspection regime that critical, life limiting features are evident at the surface of the component. This is not always the case, an example being certain past designs of the GE Frame 6B row 1 blade which have suffered failures initiated from the internal cooling passages.[8]

There have been recent efforts to investigate a number of unconventional examination (nde) techniques. These have included electrochemical techniques (relating polarisation behaviour and changes in the γ' structure.[9]. X-ray line broadening[10] and rocking curve analysis,[11] small angle neutron diffraction[12] and electron backscattering patterns.[13] Whilst some show some promises, albeit in limited areas, none have been developed to the stage where they are generally used in component assessments. The only other non destructive inspection (NDE) technique which is used from time to time is surface replication/surface metallography. Under certain circumstances this technique can be extremely informative. It has been applied very successfully to older (wrought) superalloy blades used in military engines to monitor the development of creep voiding in service.[6]. Its success in this case relies on their being a clear relationship between measurable creep cavitation and life consumption (see below for further comment).

Because of the serious limitations of the current NDE techniques, the most reliable way of determining the condition of the component is to carry out a destructive examination. The types of features which are usually considered include:

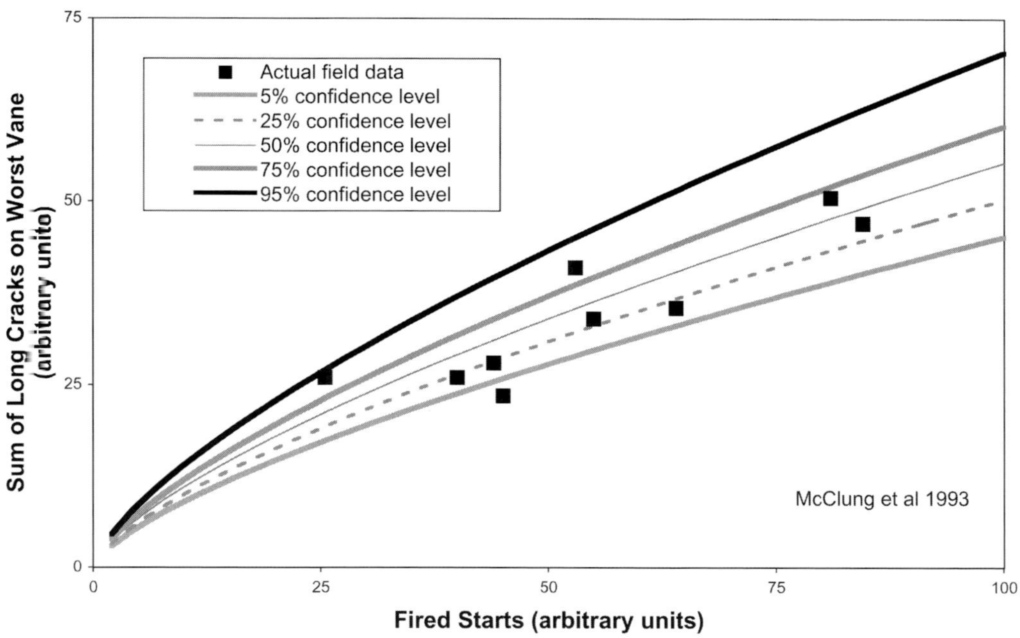

Fig. 3   Probability of the extent of nozzle cracking with time in service.

- microstructural condition
- associated 'damage'/'damage accumulation'
- surface and coating condition
- estimate of the metal temperatures through quantified microstructural structural changes

The information which is generated is often only semiquantitative There are very few quantitative relationships available between what can be observed microstructurally and the level of damage which has been sustained (or conversely, the residual properties) – an exception has already been cited above for wrought alloys between creep cavitation and life consumption.[6] Additionally, it would seem that the relationship between the commonly observed microstructural features in degraded superalloys and some of the (appropriately determined) residual mechanical properties (Fig. 4) is rather weak.[14] Despite this, the microstructural appearance is widely used to pass judgement on the serviceability of components. It is interesting to note that one of the points raised by ABB (cited above) was the need to better determine the relationship between degraded microstructures, residual properties and degradation conditions.

Metal temperatures (from steady state operation) can be estimated by quantifying microstructural features such as γ' coarsening, coating-substrate interdiffusion, and depths of oxidation. Whilst some of these techniques are widely used, in particular γ' coarsening, very little information is available in the open literature about their real accuracy, and their sensitivity to the precise metallographic and quantification procedures. Almost no data allowing for a detailed intercomparison is available

Fig. 4   (a) Isostress rupture for Nimonic 115 for virgin and ex-service conditions at 100 PMa; (b) virgin Nimonic 115; (c) ex-service Nimonic 115 assessed above.

Some residual mechanical properties can be assessed by machining small test pieces from components. Leaving aside the problem of identifying the critical locations within a cooled component, there is no commonly accepted approach to determining the relevant residual properties. Indeed, in some cases it is even not clear which properties are actually relevant in controlling the component's engineering life. Tests vary from stress rupture tests using the original material pass-off test conditions, stress or temperature accelerated rupture or creep tests,[14] through to stress relaxation.[15] In addition, in common with many high temperature materials, the apparent residual properties can depend on the test conditions used to evaluate them (Fig. 5). Only when the tests are conducted at the appropriate 'service stress' levels do the results give a realistic indication of the residual properties relevant to the component: Stress accelerated tests give a conservative indication of the residual properties.

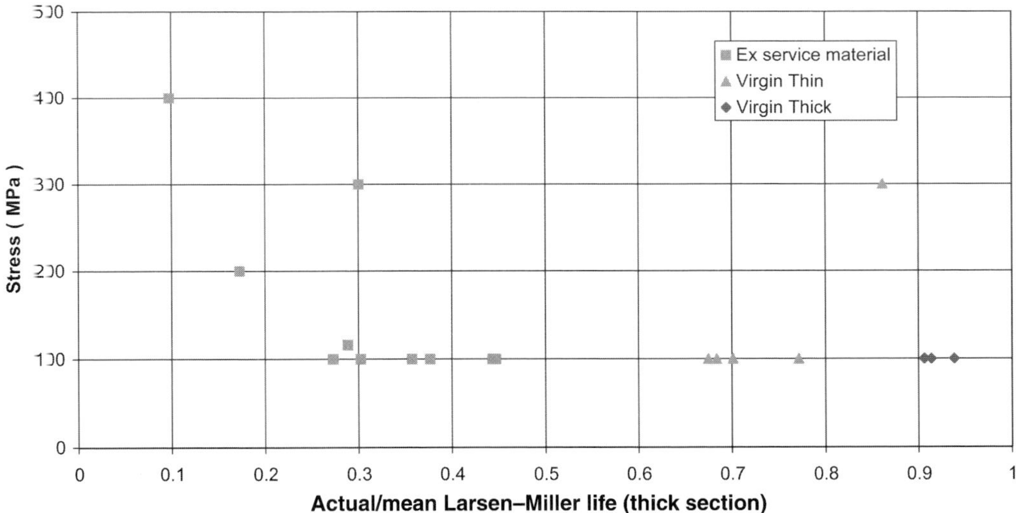

Fig. 5   Effect of assessment stress on the apparent residual life of degraded Nimonic 115.

Surface effects in thin section components, be they from coating interdiffusion or oxidation, are hard to determine quantitatively through an examination of ex-service components. Test pieces taken from components do not usually include the original surfaces.

Microstructural changes in service will also affect properties such as impact toughness and ductility. Although in some alloys it is commonly linked to the precipitation of $M_{23}C_6$ carbides on grain boundaries, the exact behaviour is alloy dependent and requires a specific database to characterise the alloy behaviour. For some current alloys (such as GE's GTD111) little information is available in the public domain. Consequently there are some R&D programmes lead by user organisations or advisors which are effectively re-evaluating the alloy to generate such information.

FURTHER QUESTIONS

In addition to the various shortcomings and difficulties mentioned in the preceding sections, there are a number of other issues which have not yet been touched on.

METALLURGICAL

Components are routinely refurbished at part life. At its simplest, after stripping the coating, the component is recoated and re-heat treated. This reverses, or is supposed to reverse, the microstructural changes which took place in service. Such re-heat treatment can be applied a number of times. It is unclear whether there is a real upper limit to the number of times which this can be done (for metallurgical reasons). Additionally, on many components, the inner surfaces are uncoated. The reheat treatment, or occasionally Hot Isostatic Pressing (HIP) and reheat treatment, will diffuse the oxygen and nitrogen in the surface of the cooling passages deeper into the superalloy. How much does the local embrittlement so caused compromise the engineering integrity of the component?

DEFINITION OF END OF LIFE

Most new components come with some form of recommended life. As one looks towards extending the life of the component, then one can be sure that further running will take it closer to its real end of life that was initially the case. Under such circumstances, rather than assessing the components deterministically, there are steps towards the more extensive use of probabilistic types of life assessment. In such an approach, the risk of failure (howsoever failure is defined) is quantitatively determined as a function of time in service, both past and future. The amount of life extension achievable is then dependent on the level of risk which is acceptable. An example of such a cumulative probability plot is shown in Fig. 6 for a set of stator vanes, where the objective was to assess the risk of extending operation beyond 100 000h[16] under different assumptions, e.g. amount of temperature inhomogeneity in the inlet.

It should be appreciated at this point that the amount of life extension which is achievable in part depends on the degree of conservatism inherent in the original design. In general terms, this is higher in the older technology machines. The latest technology machines are, to all intents and purposes, designed to operate closer to their inherent capabilities than their predecessors.

LIMITED BASE OF EXPERIENCE AND SAMPLING

All the techniques and methodologies set out above are necessary for engineering decisions to be taken in a more informed manner. However, the feedback of service experience is vital, since it allows the assumptions made at the various stages of the design or assessment to be reviewed. Components, engines and operators are variable entities, yet the number of service run components which are examined in detail in a methodical manner is very

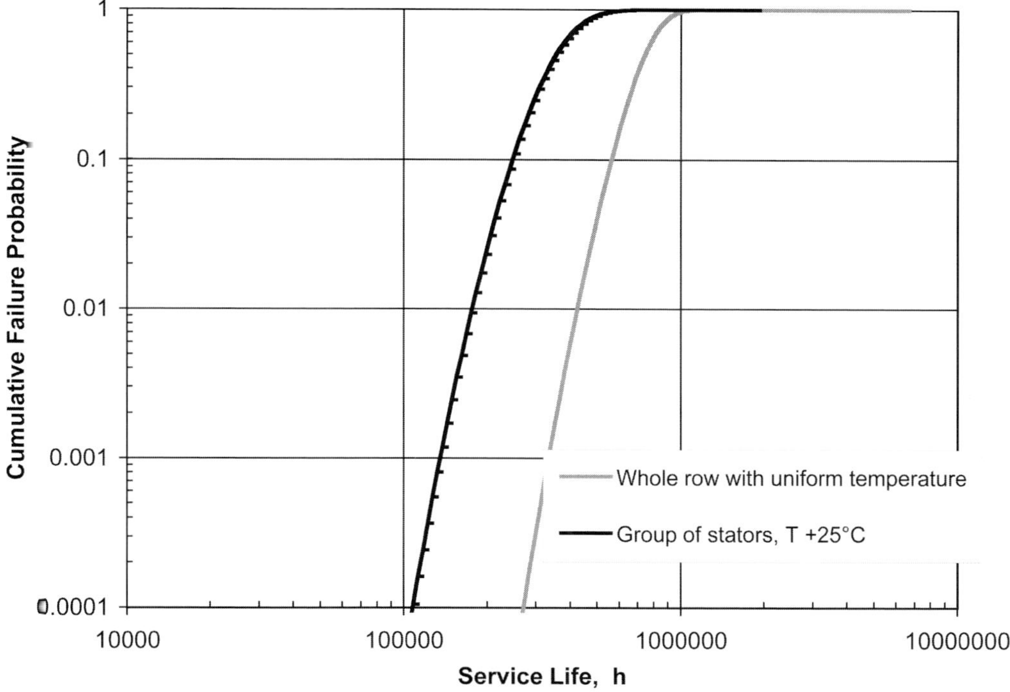

Fig. 6   Cumulative probability of failure with time in service for a set of stator blades.

limited. Additionally, since one of the purpose of such examinations is to assess how the component has performed, so that it's future potential can be assessed, it requires that detailed operational information is available. This is often not the case. A few operators or operator organisations are taking steps on selected machines to address this through installation of better data capture, archiving and analysis, coupled with installation of additional sensors such as pyrometers for blade metal temperature determination. This is, though, a minority position.

This limited base of information is also inherent in the very basis of many assessment, the destructive examination of a component. In most cases this is chosen at random from the engine set since there is no way of distinguishing the operating conditions of one blade as opposed to another. One therefore does not know whether it is the hottest or coolest in the set. Only through additional monitoring (such as pyrometry) is it possible to reduce this uncertainty.

THERMAL FATIGUE

One of the prime damage mechanisms in units which are subjected to a large number of cycles is thermal fatigue cracking of the hot section components. For nozzle guide vanes, which have a high degree of tolerance to this form of damage, one can treat the situation probabilistically, as already mentioned above for GE Frame 7B/E units.[7] For rotating blades, this a somewhat more problematic subject from a residual life point of view blades are less tolerant to cracks (in certain locations). Most work which has been done concentrates on characterising and modelling the behaviour of coatings under thermomechanical fatigue (e.g. Refs 17, 18), since the response of the coating can have a considerable role in governing the thermomechanical fatigue life of a coated system. Whilst this is a necessary step, it does not allow residual life to be assessed except through the calculation-design approach: i.e. the residual life is the expected calculated cyclic life less the number of cycles already imposed. It does not address the issue of safe crack sizes in blades, for which one needs a more detailed structural analysis with appropriate crack growth data. It also interacts with the issue of repair, since certain aerofoil cracks are repaired through welding or brazing to extend component life.

RATE OF CHANGE OF TECHNOLOGY

In the industrial sector the last 10 years has seen a marked acceleration in the introduction of new(er) technology. Some machines have been introduced and superseded before they have operated out to their design, or even their initial major overhaul, life. This affects the level of confidence one can have in the inherent behaviour and capability of the actual design, and the reliability of assessment made on the component.

CONCLUDING COMMENTS

Whilst some of the technology is available for assessing the condition and potential useful safe life of engine run hot section components, there are a number of areas which require further attention. Some of these are generic subjects, independent of specific units. These can be summarised as:

- metallurgical damage quantification techniques
- metallurgical damage accumulation models
- improved engine sensor data (as an input into the historical operational conditions for the unit)
- the consequence of uncertainties in data and methodologies on residual life estimation

Other topics are rather more unit or material specific

- the behaviour of specific alloys and their response to refurbishment
- component operating conditions (stresses, temperatures, temperature gradients and thermally induced strain ranges)
- service experience and evaluation of specific hot section components
- acceptable levels of damage within components

Many of these are active areas of research and application. Some of the work is being done by the turbine manufacturers, but significant efforts are also being made by many utilities and industry linked associations. It will continue to be a challenging area because of the currently continuing introduction of 'new' technology into the industrial units (such as thermal barriers, single crystals, steam cooling and so on). Even in the technically more stable field of high temperature steels, life assessment techniques and procedures have been worked on for the last 20 years, and work is still continuing.

ACKNOWLEDGEMENTS

The author is grateful to the Directors of ERA Technology for permission to publish this article.

REFERENCES

1. W. Hoffelner, *Lifetime Analysis of Highly Loaded Hot Section Components in Industrial Gas Turbines, Life Assessment and Repair: Technology for Combustion Turbines Hot Section Components*, EPRI GS-7031, ASM International, 1990, 35–39.

2. T. Toupin, 'GT field data management: a prerequisite for reliability centred maintenance implementation and reliability-availability-maintainability optimisation', *PowerGen '95 Asia*, Penwell, Vol. 2, 1995, 335–354.

3. R. Viswanathan and A.C. Dolbec, 'Life assessment technology for combustion turbine blades', *Journal of Engineering for Gas Turbines and Power*, Jan. 1987, **109**, 115–123.

4. R. Viswanathan, 'Damage mechanisms and life assessment of high temperature components', ASM International, 1990, 415–479.

5. J. Wilson, G. Touchton and F. van Zeveren, 'Siemens V84.2/V94.2 combustion turbine blade life management system', *Quadrennial Int. Conf on Power Station*, Assocation des Ingenieurs de Montefiore, 1997, 335–338.

6. S.-A. Karlsson, C. Persson and P.-O. Persson, 'Metallographic approach to turbine blade life time prediction', *Materials and Manufacturing Processes*, **10**(5), 1995, 939–953.

7. R.C. McClung, H.L. Bernstein, J.P. Buckingham, J.M. Allen and G.L. Touchton, 'Probabilistic analysis of industrial gas turbine durability', ASME Conference pre-print 93-GT-427, American Society of Mechanical Engineers, NY.

8. M.I. Wood, 'Internal damage accumulation and imminent failure of an industrial gas turbine blade: Interpretation and implications', ASME Conference preprint 96-GT-510 American Society of Mechanical Engineers, NY.

9. T. Kubo, K. Saito, I. Komura, K. Kimura, 'Non destructive evaluation of material degradation for gas turbine hot gas path components', in *Proc ASME PVP Conference*, 1994, vol.276/nde vol.12, 137–143.

10. M.I. Wood and D. Raynor, 'Condition assessment techniques for degraded gas turbine superalloy materials', *Int. J. Pres. Ves. and Piping*, **66**, 1996, 341–350.

11. R. Dupke and W. Reimers, 'X-ray diffraction investigations on individual grains of polycrystalline IN939 during cyclic loading: part 1: X-ray rocking curve broadening', *Z. Metallkunde*, 1995, **86**, 371–377.

12. P. Krautwasser, M. Widera, H. Schuster and H. Nickel, 'Determination of creep strain in IN100 by small angle neutron diffraction scattering', in *Superalloys 1992*, The Mineral, Metals and Materials Society, 665–674.

13. P.N. Quested, P.J. Henderson and M. McLean, 'Observations of deformation and fracture heterogeneities in a nickel based superalloy using electron beam backscattering patterns', *Acta Met*, 1988, **36**, 2743–2752.

14. M.I. Wood, 'The assessment of service induced degradation of nickel based superalloy gas turbine blades', *Mat. and Manufacturing Processes*, 1995, **10**, 903–924.

15. D.A. Woodford, 'Test methods for accelerated development, design and life assessment of high temperature alloys', *Mat. and Design*, 1993, **14**, 231–242.
16. M.I. Wood, J.M. Brear and R.M. Cotgrove, 'Life extension and risk assessment of gas turbine hot section components', *Power-Gen Europe 99*, Penwell Publishing.
17. M.I. Wood, D. Raynor and R.M. Cotgrove, 'Thermomechanical fatigue of coated superalloys', in this volume.

Life Management System for Combustion Turbine Blades and Vanes*

R. VISWANATHAN, J. SCHIEBEL and D.W. GANDY

EPRI, 3412 Hillview Avenue, Palo Alto, California 94304-1395, USA

ABSTRACT

Recently, there has been considerable growth in the selection of natural gas-fired advanced gas turbine combined cycle systems by electric utilities for new generation capacity. Increasingly higher rotor inlet temperatures (RIT) in these machines have necessitated the use of advanced and expensive materials and coatings for hot section components. The durability of these materials and coatings, is of great concern to the users of the equipment. EPRI is developing a life management system that attempts to integrate life assessment of blade materials and coatings, non-destructive inspection (NDE), and repair and coating life extension into a single framework. It is anticipated that successful completion of this effort will result in optimised blade designs, inspection/refurbishment intervals, repair guidelines and operating guidelines. This paper will present an overview of these activities.

1   INTRODUCTION

In recent years, gas turbines (GTs) have become the equipment of choice for power generation by both electric utilities and independent power producers. Some interesting statistics reported by Zink are worth noting.[1] According to his article, 11% of the total US power generation is now in the hands of independent power producers who typically opt for GT power plants. It is also anticipated that over the next 10 years, utilities will shut down nearly 25 GW of nuclear capacity reaching its end of life and that most of this will be replaced by GTs. Worldwide, GTs now comprise 40% of all new capacity additions. This is twice the percentage contributions of GTs to total capacity 20 years ago.

The reasons for the popularity of GTs are not difficult to understand. Among these are: availability of abundant low priced gas fuel, short lead times and ease of installation of GT plants, modular nature of plant and flexibility to increase plant size incrementally in the face of uncertain load growth, lower CO_2 emissions and cost of environmental compliance, substantial fuel cost savings due to increased efficiency in combined cycle configuration and availability of GT plants at power ratings as high as 280 MW. For base load applications, combined cycle plants approaching 60% thermal efficiency can be installed at a cost as low as $400 per kW.[1] The cycle ability of simple cycle GTs also makes them an attractive part of the power generation mix.

Continuing advances in design concepts and in structural materials and coatings for hot-section components have increased compressor turbine (CT) rotor inlet temperatures to nearly 1316°C (2400°F) and 1427°C (2600°F) is likely to be reached in the near future.

---

* The words 'blades' and 'buckets' and 'vanes' and 'nozzles' are used in equivalent sense by different manufacturers. For consistency, 'blades' and 'vanes' will be used throughout this paper.

For rotating blades, prolonged exposure to high temperatures and stresses requires structural materials to exhibit excellent oxidation and hot corrosion resistance, high thermal and low-cycle fatigue resistance, long-term microstructural stability, creep resistance, and high-cycle fatigue resistance to withstand various sorts of vibratory loading. In today's highly cooled airfoils, blades also need to tolerate large, highly localised stresses adjacent to cooling passages. Other considerations include good castability, good coatability and relatively good weldability for ease of manufacture and repair. Table 1 lists the materials and coatings used for first row blades and vanes in various designs.[2]

Table 1 Combustion turbine alloys and coatings for vanes and blades.

OEM	Model	Vanes	Blades	Coatings
ABB	11N2	IN939	IN738LC	NiCrAlY+Si
	GT24/GT26	DS CM247LC	DS CM247LC	TBD (N)/NiCrAlY+Si (B)
GE	7/9EA	FSX-414	EAGTD111	RT22→GT29In+ (B)
	7/9FA	FSX-414	DS GTD111	GT33 In+ (B)
	7H	SC Rene 5	SC Rene N5	TBD (B&N)
Siemens	V84/94.2	IN939	IN738LC	CoNiCrAlY+Si
	V84/94.3A	SC PWA1483	SC PWA1483	TBC (EB-PVD) (B)
W/MHI	501D/701D	ECY-768	U-520	TBC (N)/MCrAlY (B)
	501F/701F	ECY-768	IN738LC	TBC (N + B)
	501G/701G	IN939	DS MM002 More recently DS CM247	TBC (EB-PVD) (N+B)

B – Blades; N – Vanes; DS – Directionally solidified; SC – single crystal; EA – Equiaxed
Rene N5: 7.5Co, 7Cr, 1.5Mo, 5W, 3Re, 6.5Ta, 6.2Al, 0.05C, 0.2B, 0.01Y.
SC PWA 1483: 12%Cr Alloy similar to INCO792
GT29: CoCrAlY; GT29+: CoCrAlY + Diffusion Aluminide Top Coat
GT33: MCrAlY with Improved Oxidation Resistance.
GT33+: MCrAlY + Diffusion Aluminide Topcoat

The move from strong polycrystalline materials to directionally solidified alloys to single crystal alloys for the blades (buckets) and vanes (nozzles) is readily seen. Similarly, coating technology is moving from simple aluminide coatings to more complex dual layer coatings and thermal barrier coatings.

The rapid progress and deployment of advanced blade and vane materials have outpaced coatings, life assessment and repair capabilities. Degradation of protective blade coatings represents a major profitability challenge for combustion turbine (CT) owners. Coating life usually dictates blade refurbishment intervals – which typically are shorter than desired for baseload units – and downtime for coating inspection and replenishment requires dispatch of less-efficient generating equipment or purchase of replacement power. Given present life prediction and condition assessment capabilities, however, conservatism is warranted: Coating failure can lead to rapid, severe damage to the superalloy substrate. Effective, long-lasting repair methods are not yet available, and replacement of a single conventionally cast alloy blade can cost $10,000 and of an entire row up to $1 million. Single-crystal super-alloy blades may cost three times as much, while directionally-solidified superalloy blades fall somewhere in between. Thus, there is considerable incentive to develop life manage-ment systems to achieve maximum possible life from the components. EPRI is directing a coating and repair technology program for CT blades. The program focuses on the industry's needs for extending maintenance intervals, increasing coating lifetime and re-pairing damaged components. Elements include:

- Remaining life assessment tools for blades and vanes.
- Predictive models and nondestructive evaluation methods for in-service assessment of coating degradation and remaining life.
- 'Diffusion barriers' for delaying coating degradation and
- Laser welding methods for reliable, cost-effective repairs. This paper will provide a broad overview of the activities.

2 REMAINING LIFE ASSESSMENT TOOLS FOR BLADES AND VANES

The purpose of life assessment for combustion turbine blades and vanes can vary from one context to another. Knowing how much useful life is left in the component is helpful in avoiding forced outages and planning replacement schedules. A more common need is to determine the proper intervals for inspection, repair, and rejuvenation. In addition, assess-ment techniques also enable the operator to optimise the operating conditions so as to get maximum life from the components. Turbine blades are high stressed components whose failure can lead to more consequential damage than vane failures. The tolerance of blades to cracks and defects is, therefore, much less than that of vanes. Hence, the failure criterion for blades is in terms of crack-initiation events. Vanes, on the other hand, can usually tolerate large cracks, and hence life-prediction techniques for vanes have emphasised crack-growth-based approaches. The objective of the life-assessment exercise and the failure criterion to be used must be kept clearly in focus in selecting the appropriate technique.

In general, a three-phased approach is utilised for life assessment of components. The first phase involves calculation of the life consumed based on operating history, component geometry and material properties. This is then followed by actual inspections using non-destructive evaluation (NDE) methods. The last phase is obtaining component specific data using destructive metallography, accelerated mechanical testing, etc. on sacrificial samples.

NDE procedures for determining base metal damage in CT blades are unavailable. Destructive procedures involve sacrificial blades. Often, the damage in the blade can be localised and may not be typical of the damage in other blades in the same row. Suitable accelerated tests to reproduce the long term thermal fatigue and creep are also unavail-able. Use of microstructural information as a way of estimating the creep life consumed has been used to some extent[3] and much of this information has been reviewed elsewhere.[4] Calculational methods are the most non-invasive and therefore, the least expensive and most commonly used.

The primary life limiting modes for the first stage components of the advanced engines are creep and thermal mechanical fatigue of the first stage blades and crack growth of the first stage vanes. Additionally, the protective coating on the first stage blades is consumed during gas turbine operation and provides another criterion dictating the service intervals of the first stage components. Material properties determined in the laboratory are combined with analytically determined boundary conditions to model the thermal mechanical fatigue and creep of the first stage blades. Similarly, a coating degradation model based on the consumption of the coating is also based on material properties and first principles. The vane cracking algorithm, however, is generally based on empirical approaches due to the complex boundary conditions present at the first stage vane.

Figure 1 illustrates the general framework for a life management system for blades. The technical scope of effort is usually divided into four logical steps. The first involves obtaining details of the external and internal geometry from sample blades as required, to prepare the respective computer models, and to collect operating data by which the analytical results can be made meaningful in terms of their relevancy to an operator. The second step is to perform the aerothermal analysis by which the blade airfoil temperatures are calculated for steady state and transient operating conditions. Step three is to apply the thermal loads calculated for the different operating conditions in order to profile the

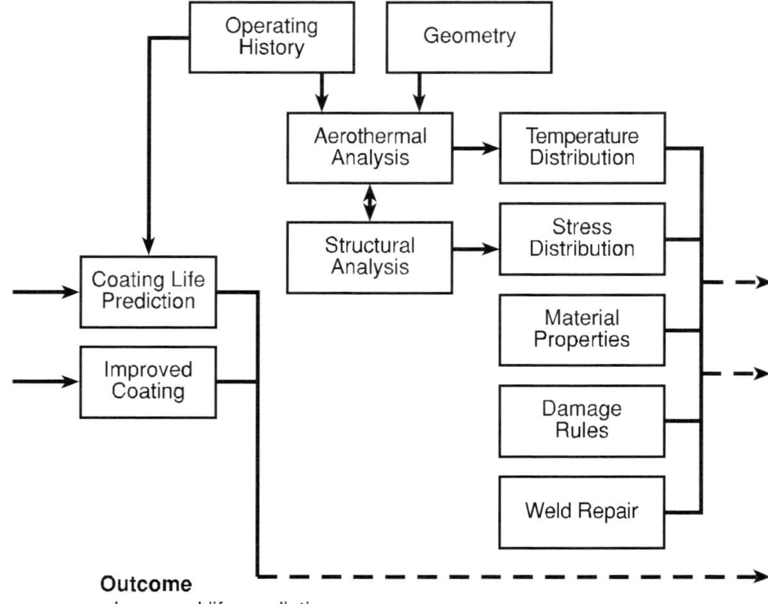

Fig. 1   Typical steps in a life management program and benefit derived.

magnitude and distribution of stresses that occur at critical sites within the blade. The final step relates blade stresses with material properties and operating parameters to the fatigue and creep damage that is accumulated for a given period and profile of unit operation.

The earliest attempts by EPRI to develop a life management resulted in a software package, REMLIFE, which predicts the remaining life of first stage components for GE MS7001 B, E and EA engines. REMLIFE, available since December of 1993, is an expert system designed to assist in maximising the useful service lives of critical first stage components. REMLIFE addresses the primary concerns of cracking of the first stage vane as well as thermal mechanical fatigue and creep of the first stage blades.[5] With the exception of coating degradation, these are the life limiting modes of degradation for the first stage components, which play a pivotal role in determining maintenance intervals.

REMLIFE addresses vane cracking through empirically based algorithms based on the results of 13 inspections for the Frame 7E engines and 9 inspections for the Frame 7B engines. The algorithms are based on the sum of long cracks present on the worst vane of the entire vane, where a long crack is considered any crack longer than 2.54 cm (1 inch) in length. It is important to note that the cracking of the two vanes occurs at different locations due to the varied design between the 7E and 7B vanes. Namely, the 7E vane segments have two vanes per segment and crack primarily on the pressure side of the vanes while the 7B vanes have four vanes per segment and crack primarily on the suction side of the vanes. Although the design of the first stage vanes differs between the Frame 7E and Frame 7B algorithms, the shape of the curve of cracking as a function of starts remains the same, as shown in Fig. 2.[6]

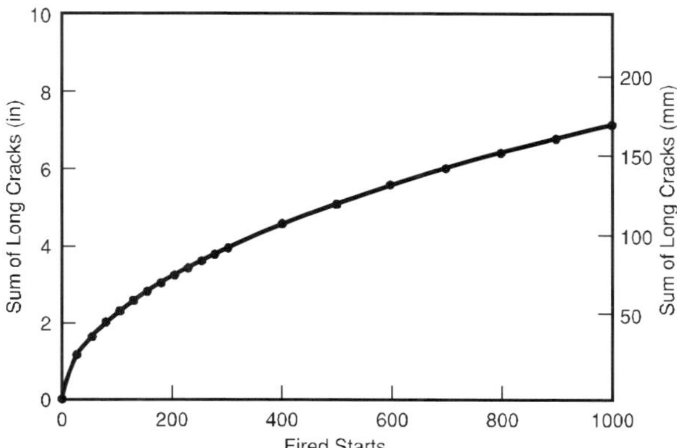

Fig. 2 Shape of predicted nozzle cracking curve, based on field experience[5,6] with GE frame 7B machines.

The thermal mechanical fatigue model for the first stage blades combines laboratory test results with the analytical results of thermal and mechanical stress analyses. Thermal mechanical fatigue test results on IN-738 specimens with a GT-29 coatings provided the fatigue life characterisation necessary for modeling and predicting the thermal mechanical fatigue behaviour of the blade material system.[7] It is important to note that the material behaviour is of the base material and the coating, as the cracks initiate in the coating and then propagate into the base metal. The material characterisation for the Frame 7 TMF model evaluated the effects of strain range, dwell time, and strain ratio on the fatigue life. The material property algorithm was determined to be:

$$N_f = C_0(\Delta\varepsilon)^{C_1} (t_h)^{C_2} exp \left(\frac{C_3}{A}\right)$$

(1)

Where constants, C_0 through C_3 represent the material system properties. The number of cycles to failure, N_f is predicted based upon the strain range ($\Delta\varepsilon$), dwell time (t_h), and the strain ratio (A) of the application. A later modification to the algorithm incorporated metal temperature as an additional variable by allowing C_0 to be a function of the metal temperature. This algorithm correlated very well with both field experience and results of independent efforts available in the literature as shown in Figs 3 and 4. Fig. 3 shows the predicted life consumed for a dozen blades which were metallurgically evaluated. The limit line divides the cracked and the uncracked regimes. These analytical predictions were entirely confirmed by metallographic examination. In addition, the farther the point resides from the critical line, the more severe the cracking found in the metallurgical inspection.[8] As mentioned, Fig. 4 shows the predictive capabilities of the TMF model when applied to data found in the open literature.[9–12] The model predicts the data, within a factor of two, the results of numerous investigators including Kuwabara *et al.*,[9] Gabrielli, *et al.*,[10] Nazmy,[11] and Ostergren.[12] Detailed test results and analysis of the TMF data is also reported in Ref. 4.

A basic creep algorithm is also built into REMLIFE, although analysis as well field experience have indicated that creep is not a problem in the Frame 7 design.

Since the original development of the REMLIFE code, the methodology has been extended to first stage GTD 111 equiaxed (EA) blades of GE Frame 6B.[13] Currently work

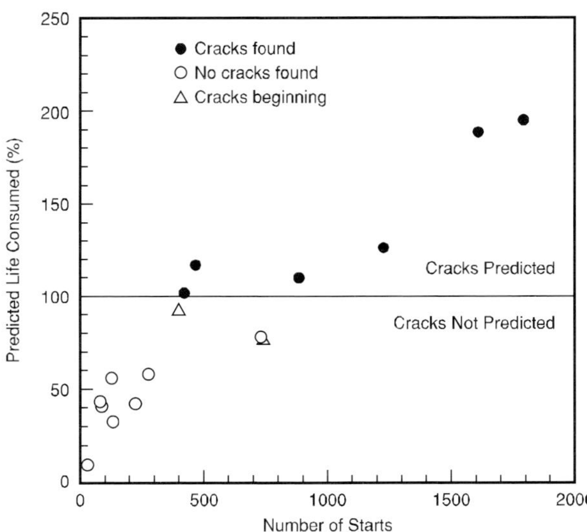

Fig. 3   Comparison of predicted bucket TMF life consumed with metallurgical observations.[8]

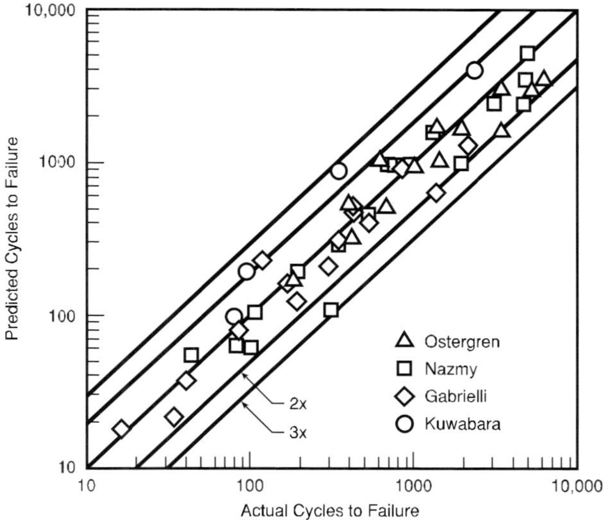

Fig 4   Comparison of TMF model to literature data.[8]

is in progress on first stage GTD 111 directionally solidified (DS) alloy blades in GE 7FA/9FA machines. In this project, the material characterisations include the determination of both physical and mechanical properties of GTD 111-DS material. The physical properties measured are the mean coefficient of thermal expansion, thermal conductivity, thermal diffusivity, density and specific heat as well as the Poisson's ratio and modulus of elasticity. The mechanical properties determined include tensile and creep properties. These material properties will serve as the inputs for the numerical analyses to be conducted. For the TMF testing planned, recommendations are awaited regarding test specimen design (hollow or solid), minimum hold time required for realistic results, measurement of coating cycles to failure or crack initiation of the substrate, and the number and type of different cycles to test. With regard to the number of cycles, it would seem that at least two cases must be considered: (1) normal start/normal shutdown cycle and (2) normal start/full load trip shutdown cycle. With regard to the cycle type, it is likely that it will turn out to be the 180 degree out-of-phase type, which is the most damaging type, because the maximum compressive strain will probably occur at steady state and maximum temperature rather than during the start-up transient. Measurements of blade and vane dimensions and metal temperatures by optical pyrometry have been completed on 9FA blades. Inspections and metallurgical examinations of a blade after 14 000 hours service, have also been completed.

In the metallurgical evaluation, extensive cracking of the aluminide topcoat was observed on both suction and pressure surfaces (Fig. 5). The cracks were radial in direction on the pressure surface and varied in direction on the suction side with the horizontal direction being the most dominant. These cracks run perpendicular to the direction of maximum principal stress and are similar to what has been seen many times in cooled blades. Metallographic view of the cracks is shown in Fig. 6. In addition, cracks were seen on the inside walls, which is not surprising in view of the high tensile stresses that occur on the inside walls due to the cooling and brittle aluminide coating used to prevent oxidation on the inside walls.

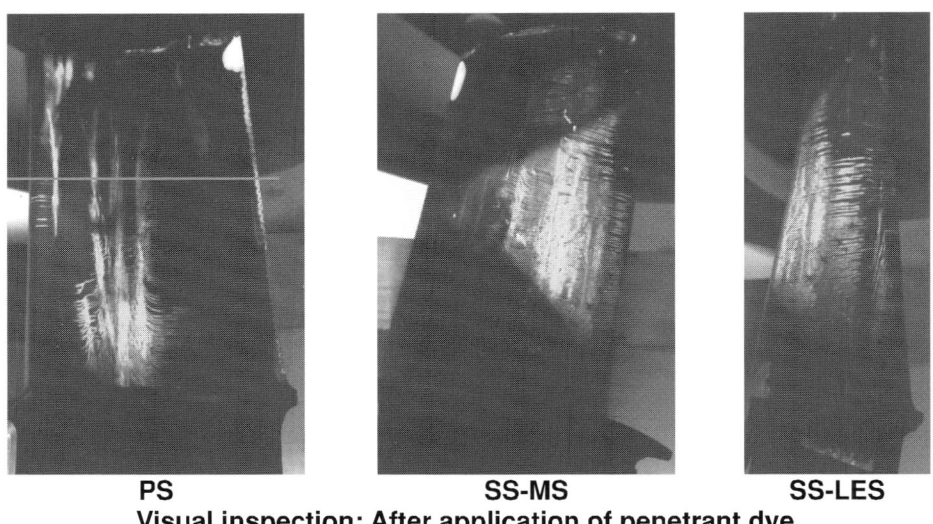

Fig. 5 Oxide cracks in pressure side and suction side.

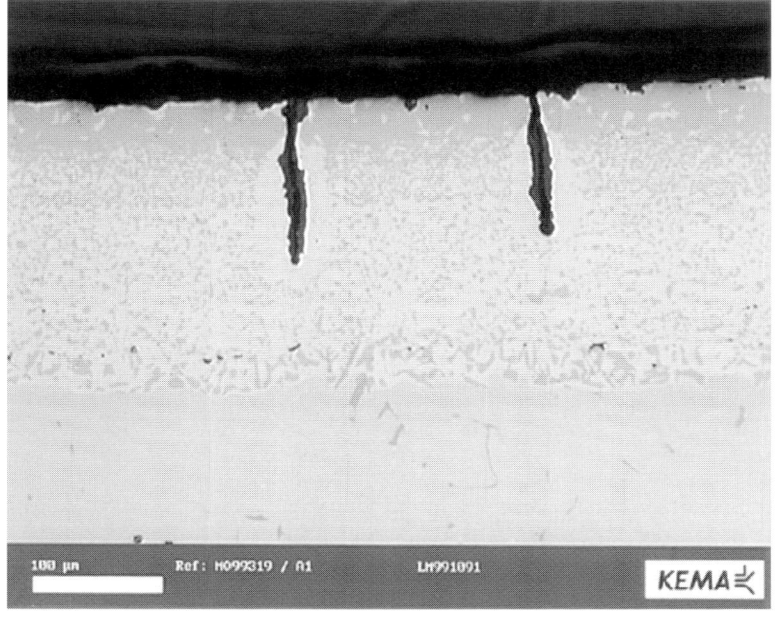

Fig. 6 Metallography of cracks found on the coating surface.

Cuts were also made along the pyrometer trace lines on the blade and these sections as well as horizontal sections at mid-height and higher will be submitted for metallurgical examination to estimate temperature based coating interdiffusion growth and % beta phase (NiAl) depletion. The heavily oxidised regions on the pressure surface at mid-chord and trailing edge area of the tip section of the blade, will also be examined. Limited data available shows a clear correlation between temperature – time of exposure and such microstructural features as beta phase depletion in coating and width of interdiffusion zone (Fig. 7).* This type of information can be used as an additional tool for estimating local temperatures. They will be especially valuable if the microstructural changes in the coating could be measured non-destructively by eddy current or other techniques.

The FEA model has been completed and the steady state aerothermal modeling done based on best estimates of gas temperature. Because the gas temperature distribution is not actually known and because motion of the blade past the stator and its effects on turbulence and secondary flow cannot be included in the computational fluid dynamics (CFD) model, extensive temperature measurements are necessary to calibrate and validate the model. In addition to the optical pyrometer data, which currently only covers the suction side of the blade below mid-height, additional data are needed on the pressure surface and above mid-height. To obtain such data, the aforementioned metallurgical correlations with temperature will be made and data from optical pyrometer measurements on the pressure side of Frame7FA blades will be used for calibration.

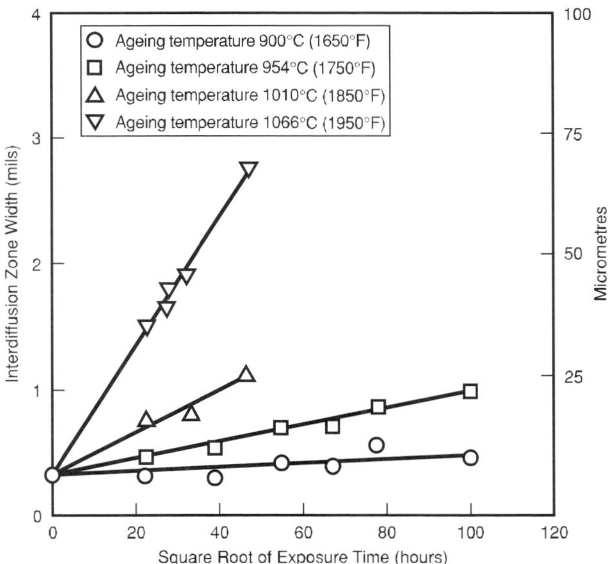

Fig. 7   Influence of aging temperature on interdiffusion zone width in substrate. (NiCoCrAlY on IN-738).

---

* Beta phase is an intermetallic CoAl or NiAl phase in the coating that serves as a reservoir for aluminium. The microstructural features described here are illustrated in Figure 7.

Results of temperature distributions from the aerothermal analysis agree reasonably well with the leading edge suction side pyrometer trace (Fig. 8). It is clear however that further refinements were needed in the aerothermal analysis. Interestingly, the high temperature locations identified by analysis were precisely those where the majority of cracks and spallation of coating on the blade have occurred (Figs 9 and 10). These results are encouraging and reinforce our confidence in predicting critical damage locations based on analysis.

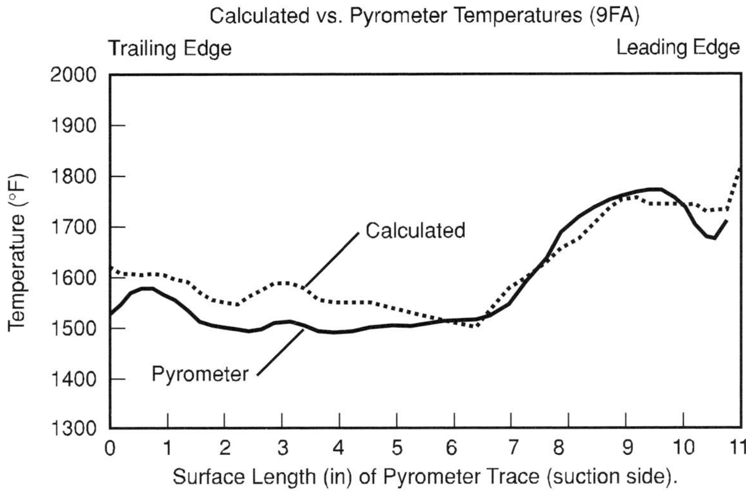

Fig. 8   Correlation between predicted temperatures and measured temperatures (using pyrometry).

To ensure that the advanced computer analyses do not become just an exercise in generating vast quantities of data, the program features a series of checks and balances at critical points in the sequence of proceeding from model generation to damage assessment. Engine operating flows, temperatures and pressures are to be characterised from actual DAS operating data collected from the host plant. Measurements of airflow through the internal passages of the blades have been conducted to check and refine the CFD modelling to be performed for the internal gas flow through the cooling passages. A check of the heat transfer analysis has been performed by comparing predicted metal temperatures against optical pyrometer data previously collected from operating ~ stage blades as discussed earlier. The emphasis on these multiple checks of the computer results is to establish a foundation of techniques and procedures – a sound, consistent methodology – by which other hot section components can be systematically evaluated.

3   PREDICTING COATING LIFE

The main life-limiting degradation modes for CT blades are oxidation and cracking. Oxidation of the coating and spallation of the protective Al_2O_3 scale that forms on the surface and interdiffusion of the coating and substrate reduces the amount of Al available

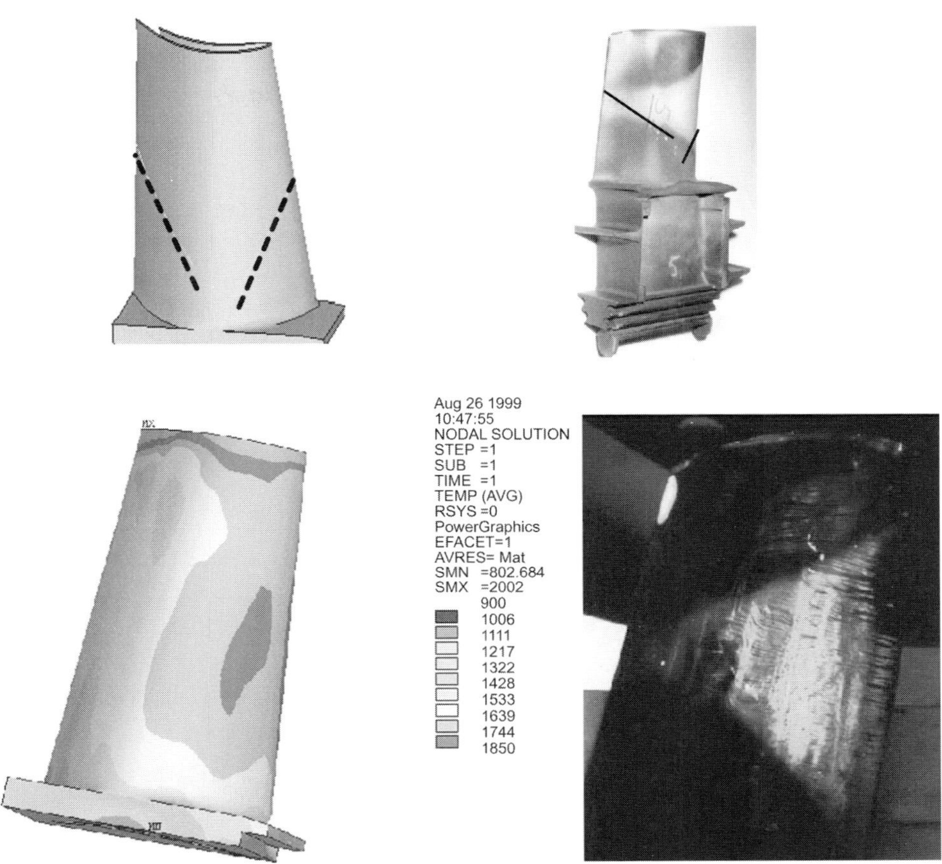

Fig. 9   Comparison of predicted temperature distribution with cracking found on suction side.

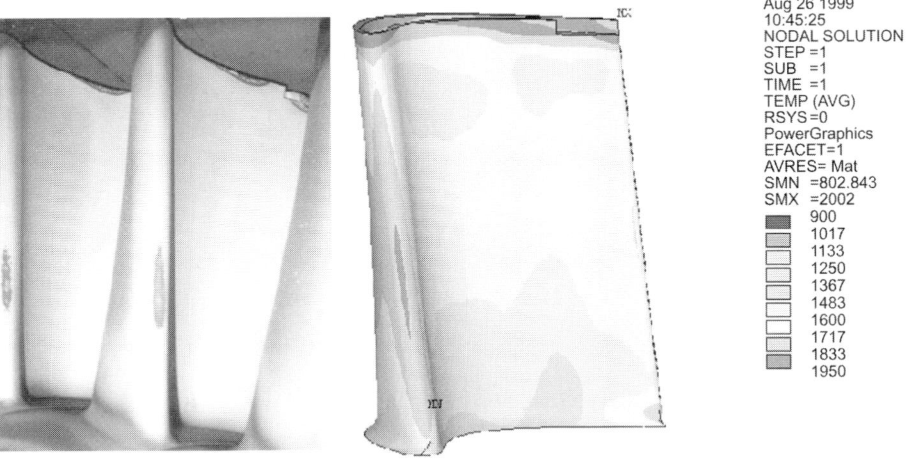

Fig. 10   Comparison of predicted temperatures with spalling of coating at leading edge.

to form the protective oxide scale. An additional failure mode, especially in peaking machines is the cracking of the oxide by TMF and penetration of the crack into the base metal.

3.1   COATLIFE MODEL

COATLIFE, a new mechanistic model for predicting the remaining useful life of in-service coatings on a cycle-by-cycle basis,[14,15] is based on these physical degradation modes (with the exception of TMF). This model incorporates oxidation kinetics, oxide fracture and spallation, material differences, overall kinetics of cyclic oxidation and depletion of Al, and a life prediction scheme based on the Al concentration required to form a protective oxide layer. Critical to its success is use of a fracture mechanics approach to derive explicit relationships between the weight of oxide spalled and Al lost as a function of the number of thermal cycles, and the oxide's critical physical and mechanical properties. Earlier models have used empirical correlations, limiting accuracy. The coatings being investigated include aluminide, platinum-aluminide, CoCrAlY (similar to GT29), GT29+, GT33+ and PWA 286.

COATLIFE's constants were obtained by testing coated coupons from GTD-111 blades in a cyclic oxidation facility in which specimens are typically heated for 55 minutes at various temperatures of 954°C, 1010°C and 1066°C (1750°F, 1850°F and 1950°F) and cooled to room temperature for 5 minutes. These constants were validated successfully by comparing predicted degradation levels with actual levels in GTD-111 blades from an operating GE Frame 6B engine, for which blade temperatures were monitored with an optical pyrometer. The model leads to an estimate of the decreasing Al concentration in the coating as a function of number of cycles for various cycle time as shown in Fig. 11. When the Al level decreases to a critical threshold level, failure of the coating is deemed to have occurred.

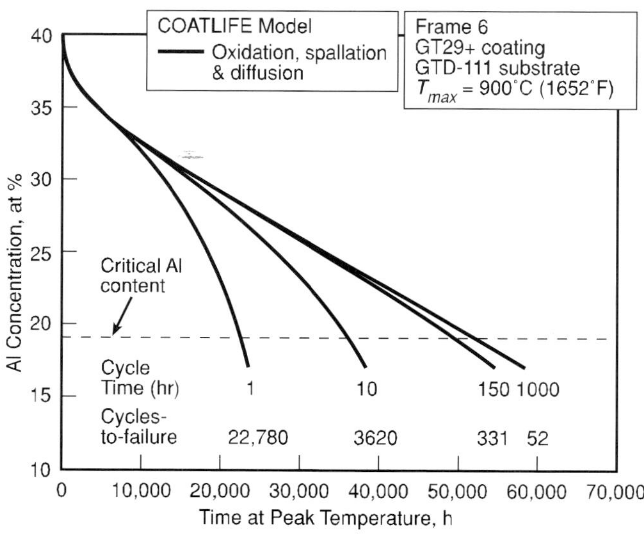

Fig. 11   Estimated decreases in aluminium concentration in coating 9 as a function of cycles to failure.

Based on the COATLIFE model and supporting data, a novel life diagram (Fig. 12) has been developed to provide a simple and rigorous means for forecasting the remaining life of an in-service coating and for scheduling blade refurbishment before excessive oxidation of the base metal occurs. This diagram can be used to predict time to failure for any combination of cycle time and number of starts, including switches from peaking to baseload service or vice versa. As shown in the figure, the controlling damage mechanism along the coating failure boundary varies with cycle time and temperature. In general, oxidation and spallation dominate at short cycle times, oxidation and inward diffusion dominate at long cycle times, and all three mechanisms are involved at intermediate cycle times.

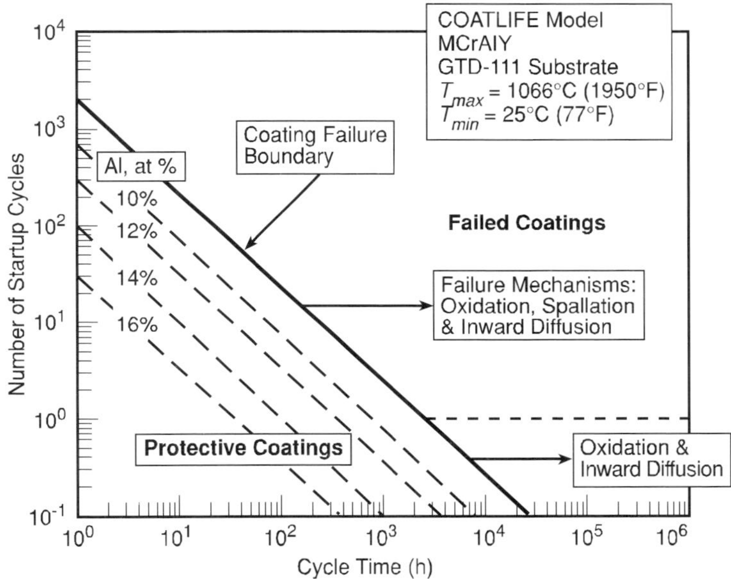

Fig. 12 Coatlife prediction of protective and p aided regions.

3.2 TMF TESTS ON COATED PARTS

To incorporate coating cracking by TMF, TMF tests will be conducted using out of phase strain cycles. A limited amount of data have already been gathered in a related project at EEA on similar systems.[16] In this effort, six different alloy/coating systems were subjected to a 180° out-of-phase cycle with maximum temperature of 900°C (1650°F) or 1050°C (1950°F). The alloy/coating systems included:

(1) uncoated IN 738 at 900°C (1650°F).
(2) IN 738 with an overaluminised CoCrAlY coating akin to GT29+ at 900°C. Since GT29+ is a GE proprietary coating, and not commercially available, a coating with a similar composition was substituted.
(3) IN 738 with a Pt–Al coating at 900°C (1650°F). This is a commercially available coating used on heavy industrial units.

(4) IN 738 with a CoNiCrAlY coating at 900°C (1650°F). This is the type of MCrAlY coating used by Siemens–Westinghouse. Two different thicknesses were studied (3 mils (0.07 mm) and 6 mils (0.15 mm)).

(5) Uncoated CMSX 4 at 1060°C (1922°F).

(6) CMSX 4 with Pt–Al coating at 1050°C (1922°F). This was an aeroengine version of the Pt–Al coating.

IN-738 was the only equiaxed (EA) cast material tested and CMSX 4 the only DS/SC material. The 270°F difference in test temperature is larger than the allowable difference for the two alloys based on creep considerations, which is about 200°F. The TMF life plotted in Fig. 13 is the number of cycles to the onset of sustained tensile load drop.

RT-122 is the coating used by Siemens–Westinghouse. It's composition is Co32Ni21Cr8Al.7Y, and is similar to GE's GT33 (Co32Ni22Cr10Al.3Y). It was applied at an average thickness of 5.9 mils (0.15 mm).

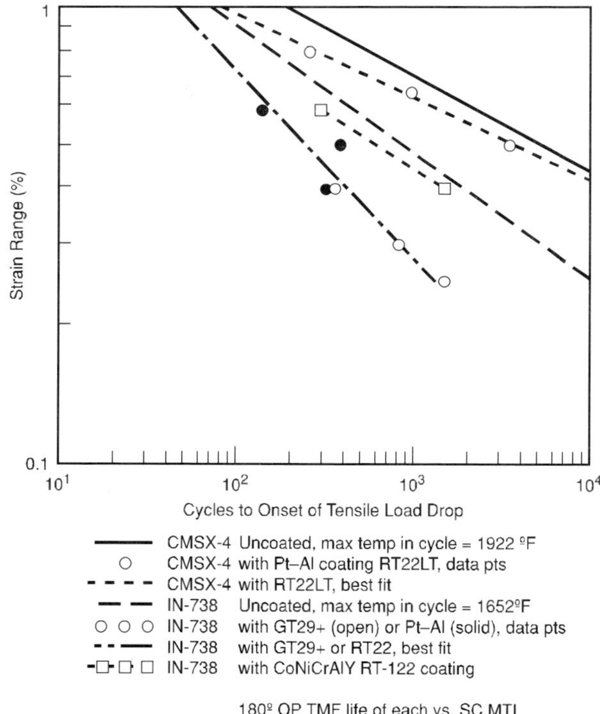

Fig. 13　Results of 180° out-of-phase cycle TMF test uncoated CMS × 4 and INCO 738 alloys[16]

Since the coating GT29 is GE proprietary and not commercially available, a coating with similar composition was substituted. This substitute coating had the composition Co3OCr8AlY. This compares to Co29Cr6AlY for GT29. A standard aluminide top coat was applied to create a GT29+ 'look alike'. Total thickness of the coating averaged 5.9 mils (0.15 mm). The thickness of the outer CoAl layer was about 1.2 mils (0.03 mm).

RT-22A is a Pt–Al coating applied to IN-738. Average thickness of the coating was 3.5 mils (0.09 mm)with a platinum content of about 50%. RT-22LT is a proprietary Pt–Al coating supplied to Rolls–Royce for their CMSX-4 hardware. It had a thickness of about 2 mils (0.051 mm)and a maximum platinum content of 40%.

It is readily seen from Fig. 13 that the TMF life N_f and strain range $\Delta\varepsilon$ (in %)* are related by the standard relationship

$$N_f = C\ \Delta\varepsilon^n \qquad\qquad (2)$$

It can be seen in Fig. 13 that the TMF life of uncoated CMSX-4 at a representative strain range of 0.4% is about 8 times that of IN-738. With regard to coating effects, the GT33 type coating does not have a large effect, but an effect of coating thickness was found. TMF life of specimens with 3 mil (0.076 mm)thick coatings generally exceeded that of specimens with 6 mil (0.15 mm) thick coatings. Possibly this is because cracks in the substrate grow faster due to higher stress intensity factors associated with deeper cracks in the thicker coating. Another possible explanation is that load drop is more influenced by failure of the thicker coating in that the wall thickness of the test specimen was only 30 mils (0.76 mm).

The effect of the overaluminised CoCrAlY (similar to GT29+) and Pt–Al on IN-738 is much more significant. At a normalised strain range of 0.35%, it reduces TMF life by a factor of 4.5 relative to plain IN-738 and 3.5 relative to RT-122 coated IN-738. This is because the aluminide fails almost immediately due to tensile overload below the DBTT (ductile to brittle transition temperature) in the low temperature high tensile stress part of the cycle. At lower cyclic strains where brittle failure of the aluminide does not occur (below about 0.35 and 0.17 normalised strain for the Pt–Al and overaluminised CoCrAlY coatings respectively), the TMF life of the aluminide coating is as good as that of the substrate, but these thresholds are too low to be avoided in service because occasional full load trips cannot be avoided. The lower threshold to cracking for the overaluminised CoCrAlY is due in part to the fact that CoAl has a lower strain to fracture than NiAl (Ref. 17). See Fig. 14.

Pt–Al on CMSX-4 did not have much effect even though it developed cracks early in the test at strain ranges above 0.6 (normalized). Two possible reasons that coating cracks have less effect in this case are: (1) the reduced thickness of the RT22LT (about 44%), and (2) one-third less stress due to the lower modulus of elasticity in the direction of stressing. Both of these factors reduce stress intensity factor and hence crack growth rates. Below 0.6 strain (normalized), TMF cracks would be expected to initiate earlier in the Pt–Al since it is an EA material with lower TMF properties than DS and SC alloys.

3.3 NDE Methods for Coating Degradation

Another condition assessment method shows promise for nondestructive field inspection of coatings. In recent work proof of concept has been established for the use of 'sweptfrequency eccy-current' (SFEC) testing to determine coating integrity. SFEC techniques measure resistivity changes in a coating at specific blade locations: these changes can be

---

* The strain has been normalised by dividing the actual strain by a factor.

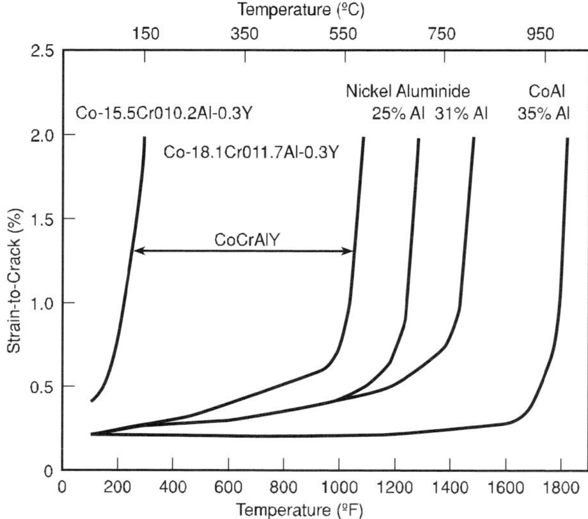

Fig. 14   Ductibility of coatings as a function of temperature[17].

empirically correlated to the equivalent thickness of the coating's β-NiAl layer, which indicates the amount of protective aluminide phase remaining, as well as to the width of its interdiffusion layer. The relationships between formation of the β-NiAl depletion zone, Al depletion, and coating deterioration are so strong that the depleted zone widths can be used as an indicator of coating condition and to estimate the operational temperature of service-exposed blades.

The SFEC method has successfully determined the condition of a GT-29PLUS coating on a GTD-111 substrate. In ongoing research, the technology is being used to assess the remaining life of other common coatings with and without Ni-Re diffusion barriers applied to GTD- 111 first row FA buckets. Software is also being developed for integration with the SFEC field inspection system to present graphic representations of key coating/ base metal microstructures with depth.

4   EXTENDING COATING LIFE

By inhibiting degradation of CT blade coatings, the life of blades can be extended and the frequency of coating refurbishment can be reduced. Suppressing Al diffusion into the superalloy substrate represents one approach for stretching the life of a coating's protective Al_2O_3 layer. Preliminary results indicate that deposition of a Ni-Re barrier between MCrAlY coatings and superalloy blades can substantially reduce Al diffusion rates, and a patent has been issued for this novel development.[18,19]

Researchers began looking at Re to provide a diffusion barrier following previous work indicating that this element significantly reduces the creep rate of nickel-base superalloys. The exact mechanisms by which Re enhances creep resistance are unknown, but because Re is a very large atom and creep is known to be controlled by diffusion, it was hypothesised that Re might significantly reduce diffusion rates.

To show the beneficial effects of Re, it is necessary to illustrate the changes occurring in a coated component with exposure. These changes include reducing the proportion of beta aluminide phase in the coating and its eventual disappearance, growth of the interdiffusion zone, increase in the amount of the dark aluminum phase in this zone and formation and growth of an aluminide depleted zone at the surface. These changes are shown in Fig. 15. In initial testing of this hypothesis, thin (0.5 μm) interlayers of Ni–Re were deposited onto coupons of the blade superalloy IN-738LC prior to the application of MCrAlY coatings. Results indicate that the Ni–Re interlayer can significantly decrease the growth rate of the inner β-NiAl depletion zone under oxidising conditions, suggesting that it provides an important impediment to the inward diffusion of Al into the substrate. Subsequent tests have demonstrated that this technique, which promises to be both simple and cost effective in commercial application, can delay coating degradation on other common blade superalloys, including GTD-111 and CM247 (Fig. 16). A patent is pending for a practical method of applying the Ni–Re layer.

EPRI is assessing the potential for thicker Ni–Re interlayers to further prolong coating lifetime and is seeking a site for field testing of this novel life extension method.

5   WELD REPAIR OF BLADES

CT blades that have been damaged due to foreign objects passing through the turbine, operational stresses, or inherent defects in the high-stress location of the airfoil are difficult or impossible to repair using traditional welding methods, as well as costly to replace. Certain precipitate-forming elements (mainly aluminum and titanium) added to blade superalloys to achieve unmatched high-temperature strength are primarily responsible for the poor welding record. Weldability is limited principally by the behaviour of these

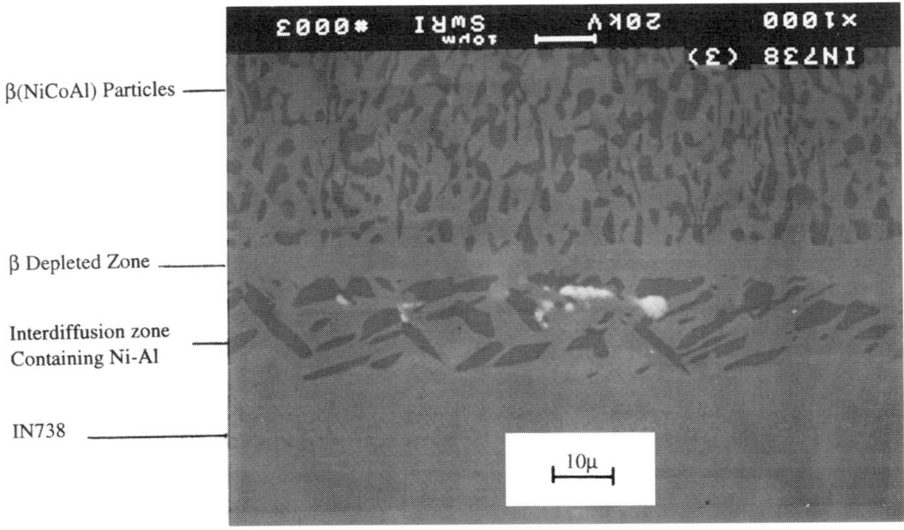

Fig. 15   Features of coating degradation at high temperature NiCoCrAlY on INCO 738.

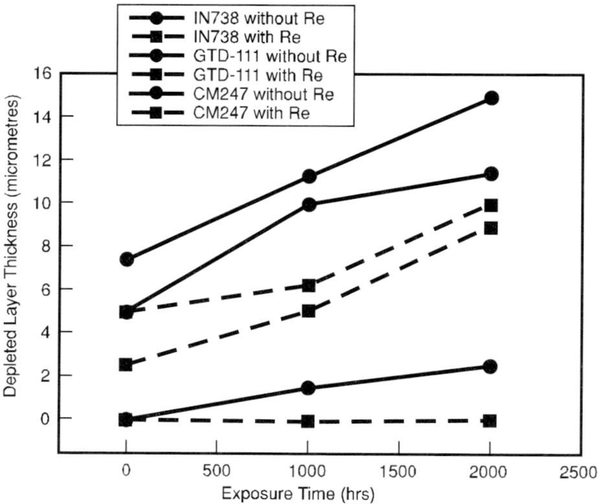

Fig. 16   Depletion zone thickness vs. exposure time.

elements under high-temperature stress, which contributes to hot cracking during the welding process and strain age cracking after welding, particularly during postweld heat treatment.

Weld filler materials identified as most effective in the repair of precipitation-strengthened blade superalloys are simpler, solid-solution strengthened alloys. The lower strength of these filler materials significantly limits applicability. Current industry practice permits welding only in areas of very low stress; some 80 to 90 percent of blade surfaces are considered non-repairable. Controlled, low-energy welding processes were determined to offer the highest potential for extending repairs to more highly stressed blade regions because their precise heating can reduce cracking.[20,21]

EPRI's Repair and Replacement Applications Center and several CT repair vendors are undertaking a collaborative program to develop and test laser welding technology, the most promising controlled-energy process. A Nd:YAG laser system has been successfully applied in other complex power industry applications such as steam generator tube repair, is being used in welding trials on 1N738 test coupons. In recent months, Nd:YAG laser weld repair procedures and precipitation-strengthened filler materials have been identified and demonstrated that can be readily used to repair 1N738 blade materials. Figure 17 provides an example of a crack-free laser weld deposit using the new precipitation strengthened alloy. Results of rupture tests are shown in Fig. 18.

In combination with proprietary material conditioning methods, these advances offer the potential for repaired areas to achieve 75 to 80 percent of the original blade material's stress rupture characteristics. This will enable repair of more highly stressed regions of the airfoil, dramatically extending the limits over conventionally used solid-solution-strengthened alloys.

6 SUMMARY

Life prediction models for first row buckets used in several designs of CTs have been developed and validated against limited field experience. Aerothermal analysis and structural analysis have been shown to successfully characterise the steady state temperature distributions and identify the critical damage locations. Microstructural techniques have been developed that can be used to estimate local temperatures in the bucket. These techniques combined with aerothermal analysis and optical pyrometry provide a powerful set of tools in identifying the distribution of one of the most critical parameters affecting component life, i.e., temperature. Models have been developed to predict coating remaining life based on operating conditions. NDE methods are also being investigated to

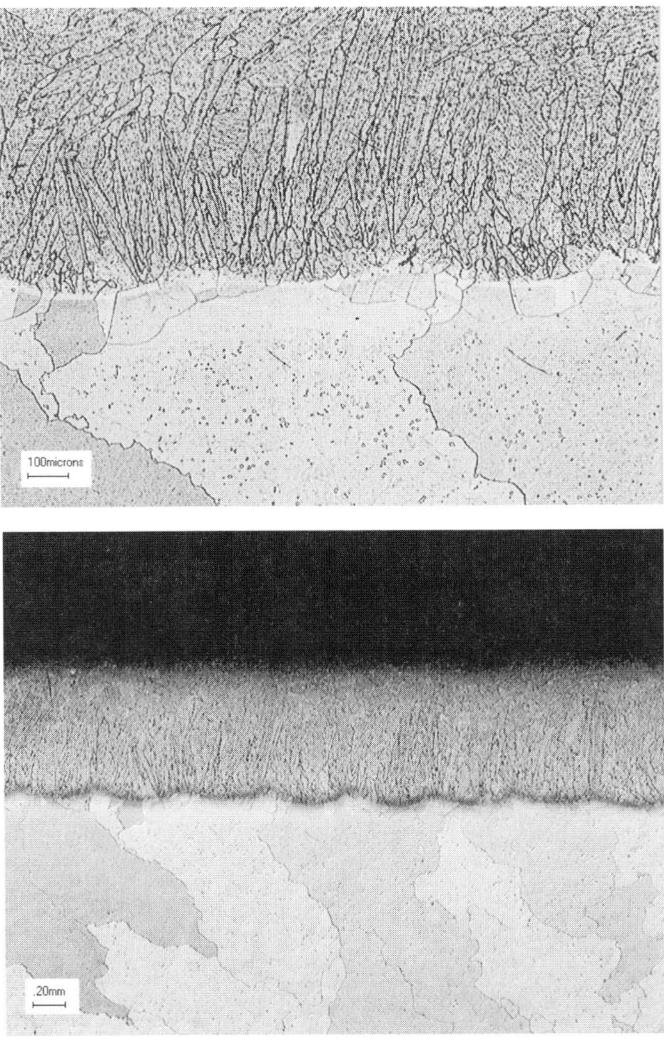

Fig 17   Crack free weld deposit on INCO 738 by LASER welding.

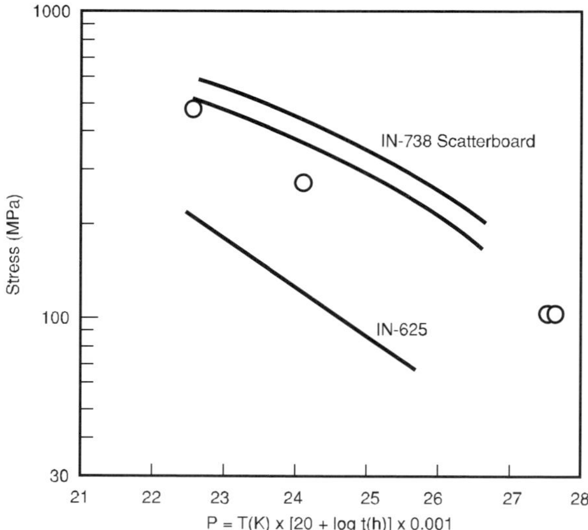

Fig. 18   Rupture lives of repair welds compared with IN-738 and IN-625.

estimate coating remaining life. The feasibility of extending coating life by Re implementation to the substrate has been demonstrated. The feasibility of performing sound weld repairs at high stress locations has also been demonstrated. It is hoped that these results will help utilities substantially in reducing the lifetime cost of components, by reducing the operating cost and in extending the life of blades and vanes.

ACKNOWLEDGEMENTS

This paper is based on technical work carried out by a number of organisations under contract to EPRI. The contributions of K. Chan, S. Cheruvu and G. Leverant at the Southwest Research Institute, H. Huishoff and G. Koopmans and T. Biejers of KEMA (Netherland), E. Wan and R. Dewey of TTI, C. Wells and A. Mucciardi of Structural Integrity Associates are gratefully acknowledged. Their assistance as well as that of J.M. Allen in reviewing this paper and his contribution to the TMF data analysis are also acknowledged.

REFERENCES

1. J.C. Zink, 'Combustion Turbines Move Markets', *Power Engineering,* March 1998, 20–22.
2. R. Viswanathan, S. Schetrer and J. Stringer, *Materials Advancements in Land-Based Gas Turbines,* PowerGen International, Orlando, December 1998, Pennwell Publicatons.
3. J.A. Daleo and J.R. Wilson, 'GTD 111 Alloy Material Study', International Gas Turbine and Aero Engine Conference, Birmingham, U.K., June 10–13, 1996, Paper 96-GT-520.
4. R. Viswanathan, *Damage Mechanisms and Life Assessment of High Temperature Components,* ASM International.

5 T.S. Grant, H.L. Bernstein and D.T. Lincoln, *EPRI Life Management System for Gas Turbine Engines,* Manual for REMLIFE™, Version 1.0, 1 December 1993, Electric Power Research Institute.

6 H.L. Bernstein, R.C. McClung, T. Sharron and J.M. Allen, 'Analysis of General Electric Model 7001 First Stage Nozzle Cracking', International Gas Turbine and Aeroengine Congress and Exposition, ASME No. 92-GT-311, Cologne, Germany, 1–4 June 1992.

7 E.S. Russell, 'Practical Life Prediction Methods for Thermal-Mechanical Fatigue of Gas Turbine Buckets', *Proceedings: Conference on Life Prediction for High-Temperature Gas Turbine Materials,* V. Weiss and W.T. Bakker eds, EPRI AP-447, Electric Power Research Institute, 1986, 3–1–3–39.

8 H.L. Bernstein, T.S. Grant, R.C. McClung and J.M. Allen, 'Prediction of Thermal-Mechanical Fatigue Life for Gas Turbine Blades in Electric Power Generations', *Thermomechanical Fatigue Behaviour of Materials,* ASTM STP 1186, H. Sehitoglu ed., ASM, 1993, 212–238.

9 K. Kuwabara, A. Nitta, and T. Kitamura, 'Thermal-Mechanical Fatigue Life Prediction in High-Temperature Component Materials for Power Plant', *Proceedings of the ASME International Conference on Advances in Life Prediction Methods,* Albany, NY, April 1983, 131–141.

10 F. Gabrielli, M. Marchioinni and G. Onofrio, 'Time Dependent Effects on High Temperature Low Cycle Fatigue and Fatigue Crack Propagation of Nickel Base Superalloys', *Advances in Fracture Research,* Proceedings of the 7th International Conference on Fracture (ICF7), Vol. 2, Houston, TX, March 1989, 1149–1163.

11 M.Y. Nazmy, 'High Temperature Low Cycle Fatigue of IN738 and Application of Strain Range Partitioning', *Metallurgical Transactions A,* 1983, **14A**, 449–461.

12 W.J. Ostergren, 'A Damage Function and Associated Failure Equations for Predicting Hold Time and Frequency Effects in Elevated Temperature, Low Cycle Fatigue', *journal of Testing and Evaluation,* 1976, 4, 327–339.

13. Blade Life Management System for GE Frame 6B Gas Turbines. Report TR-109196, Vols. 1 and 2, EPRI, 1998.

14. K.S. Chan, N.S. Cheruvu and G.R. Leverant, 'Coating Life Prediction Under Cyclic Oxidation Conditions', *Journal of Engineering for Gas Turbines and Power,* 1998, 609–614.

15. K.S. Chan, N.S. Cheruvu and G.R. Leverant, 'Coating Life Prediction for Combustion Turbine Blades', ASME Paper 98-GT-478, presented at the International Gas Turbine and Aeroengine Congress, Stockholm, Sweden, 1998.

16. M.I. Wood, D. Raynor and R.M. Cotgrove, 'Thermo Mechanical Fatigue of Coated Superalloys For Gas Turbine Blading', ERA Report 96–1088, Final Report, ERA Technology, May 1999.

17. J.A. Daleo and D. Boone, 'Failure Mechanisms of Coating Systems Applied to Advanced Turbine Components', ASME Paper 97-GT-486, ASME, 1997.

18. G.R. Leverant, 'Diffusion Barrier for Protective Coatings', U.S. Patent 5,556,713.

19. R.A. Page and G.R. Leverant, 'Inhibition of Interdiffusion from MCrAlY Overlay Coatings by application of a Ni–Re Interlayer', ASME Paper 98-GT-375, presented at the International Gas Turbine and Aeroengine Congress, Stockholm, Sweden, 1998.

20 EPRI TR-108272, Weld Repair for Nickel-based Superalloy Blading, Final Report, Electric Power Research Institute, April 1998.

21 EPRI TR-110355, LASER Welding Survey for Power Generation Industry, Final Report, Electric Power Research Institute, April 1998.

Predicting the Remnant Life of Corrosion Resistant Coatings

N.J. SIMMS, J.E. OAKEY and J.R. NICHOLLS

Power Generation Technology Centre, Cranfield University, Cranfield, Bedfordshire, MK43 0AL, UK

ABSTRACT

The hot gas environments found in gas turbines can have the potential to cause both erosion and corrosion damage to hot gas path components, depending on the fuel and air qualities. These materials degradation modes can be life limiting for some components, for example first stage vanes and blades. Of the several different potential materials degradation modes, this paper develops methodologies to predict oxidation/corrosion damage to metallic coatings and thus evaluate the coating remnant life under hot corrosion conditions.

To allow the statistically valid assessment of materials performance an approach has been developed for the collection of compatible quantitative data on materials degradation. The approach requires a high measurement precision to allow lives for the best materials out to 25,000h to be realistically predicted. This approach has been applied to burner rig exposures and realistic laboratory-simulated environments for gas turbine coatings and bulk alloys validated against short term power plant exposures. Accurate measurement techniques based on pre-exposure contact metrology and post-exposure optical microscopy/image analysis have been developed. The method is valid for both the low level of damage required for practical systems and the localised nature of hot corrosion damage. Variations of the general measurement methodology are necessary to obtain data from different shaped components, and to take account of the changing localised environments around plant components.

The data produced have been used to derive and test quantitative models for the prediction of the performance of candidate materials for use in systems burning dirty fuel gases. These models take into account such process variables as component temperature, gas composition and alkali deposition fluxes, and permit the prediction of corrosion rates based on the application of extreme value statistics to the data set. The models have been validated against data from plant exposures, with predictions of a measure of 'maximum' damage to metallic coatings within $\pm 7\mu m$ $1000h^{-1}$.

1   INTRODUCTION

Gas turbines can be fired on a wide variety of fuels, ranging from natural gas to sour gases and heavy fuel oils, with growing interest in the potential to use fuel gases derived from coal, biomass and waste products.

The environments found within the hot gas paths of gas turbines depend on the contaminants present in the fuel and air entering the turbine, as well as the turbine operating conditions. For example, fuel gases can be produced from solid fuels via a variety of processes including pressurised gasification and/or combustion systems. The fuel gases from gasification systems need to be burnt in a special combustor (for low CV gases) prior to entry into the gas turbine. The combusted gases entering the turbine hot gas path have the potential to cause deposition, corrosion and erosion damage depending on the balance of contaminants within them. The presence of particles may cause erosion, abrasion or deposition, and gaseous species (e.g. SO_x and HCl, alkalis and other trace metal species) may cause deposition and/or enhanced corrosion. Considering trace metals, for example, in coal fired systems, combustion derived gases tend to have significant levels of alkali metals, whereas gasifier derived fuels gases (when used with hot gas cleaning processes) have higher levels of heavy metals, e.g. lead and zinc.[1,2] Co-firing with biomasses or waste fuels changes the levels of all the contaminants, e.g., for trace metals, wheat straw has higher potassium levels and sewage sludges have higher zinc levels.[1] The effects of different fuels on vapour deposition fluxes and deposit compositions (e.g. melting points) requires careful consideration for each process and fuel.

There are standards for fuel and air qualities entering gas turbines (e.g. Refs 3 and 4). However, these standards were developed for gas turbines that were expected to be fired on natural gases or relatively clean fuel oils, with hot gas path components lasting for at least three years. The desire to move towards using dirtier fuels, with a different mix of contaminants to the traditional fuels, means that such standards need to be critically reviewed, both in terms of allowable contaminants and mixes, as well as targets lives.

Materials degradation data is required in a form that can be readily incorporated into plant design and maintenance processes. Both erosion and corrosion damage are often quoted as averages, whereas in practice the regions of deepest damage will lead to failures: for example by providing a site for fatigue crack initiation (e.g. a pit), or by reducing the section thickness to such an extent that creep rupture failure is possible. Most studies report erosion and corrosion damage as mass change, rather than dimensional metal loss. It is possible to convert mass change data to metal loss data, but this involves several assumptions. Mass change data have the effect of producing average measures of the damage that are inappropriate for evaluating many forms of corrosion damage (e.g. localised pitting or internal corrosion) or when deposits form on components. In practice both average and 'maximum' materials damage rates are desirable, with the different forms of damage identified and measured as a thickness loss/depth of attack.[5–7]

This paper reports the use of a metrology methodology for accurately measuring materials degradation by corrosion (and/or erosion) in gas turbine environments. (There are, of course, other potential degradation modes, e.g. thermal shock and thermal fatigue, which are beyond the scope of this paper). The method is based on the accurate measurement of sample dimensions before and after exposure. The methodology was developed specifically to generate compatible sets of materials degradation data from laboratory, burner rig and short term plant runs. Such data can subsequently be used for modelling materials performance in such environments and validating the models against plant experiences.

2   DATA REQUIREMENTS

Three issues need to be addressed to define the materials performance data requirements:

- The data collection methods must allow all relevant forms of materials damage to be assessed. Destructive methods of monitoring metal damage are required which provide measurements of metal loss, depths of internal corrosion, pit depths, etc. The calculation of metal losses requires dimensional measurements to be made before and after exposure.

- The accuracy of the data must reflect its potential use. More accurate data is needed if trial exposure periods are short compared to required component life. Different target lives and corrosion allowances will result in different levels of acceptable damage within a test period. The overall metrology system must be capable of greater accuracy than the target acceptable damage levels.
- Data must allow the assessment of both average and a measure of 'maximum' damage. If sufficient data are produced for statistical analysis, it becomes possible to define and evaluate statistical parameters that describe the damage.[8–10] The minimum number of data points needed to assess the behaviour of a material in a single environment, e.g. in a laboratory test, has been found to be ~24.[5,11] However, components in plants tend to be exposed to a number of different physical/chemical environments around their surfaces and so require far more data points to be measured so that different zones of behaviour can be assessed independently of each other.

3   SUMMARY OF METROLOGY METHOD

The basic experimental approach developed for the assessment of materials performance consists of five main stages:

- Component manufacture
  To minimise the difficulties that may be encountered in later stages of the materials assessment procedures, careful specification and/or monitoring of the dimensions and surface finishes of the manufactured components are needed: machined surfaces are usually desirable, but other surface finishes can be acceptable, e.g. investment cast. Two features of the samples prior to exposure are particularly important:
  – surface roughness
  – the rate of change in sample dimensions perpendicular to the measuring plane, since this determines how close to the original measurement plane the polished section is needed for post-exposure measurement.
- Pre-exposure metrology
  Contact metrology is used to determine the dimensions of components or samples prior to exposure. The components need to be measured on planes that relate to reference points on the samples that will still be present after the exposure has been completed (Fig. 1) and will allow the matching of the pre- and post-exposure data sets. For solid cylindrical laboratory corrosion test samples, only sample diameters need to be measured. For simple cylindrical samples that will not experience any bore damage during their exposure, only the tube wall thicknesses need to be measured, which can be carried out conveniently with purpose-built jigs. For more flexibility, or complex component shapes such as gas turbine blading, 3-D co-ordinate metrology machines need to be used, which can have repeatabilities of < 2 µm. Hundreds of dimensional measurements may be required for a given component or sample, but this is easily achieved using automated computer controlled measurement systems.

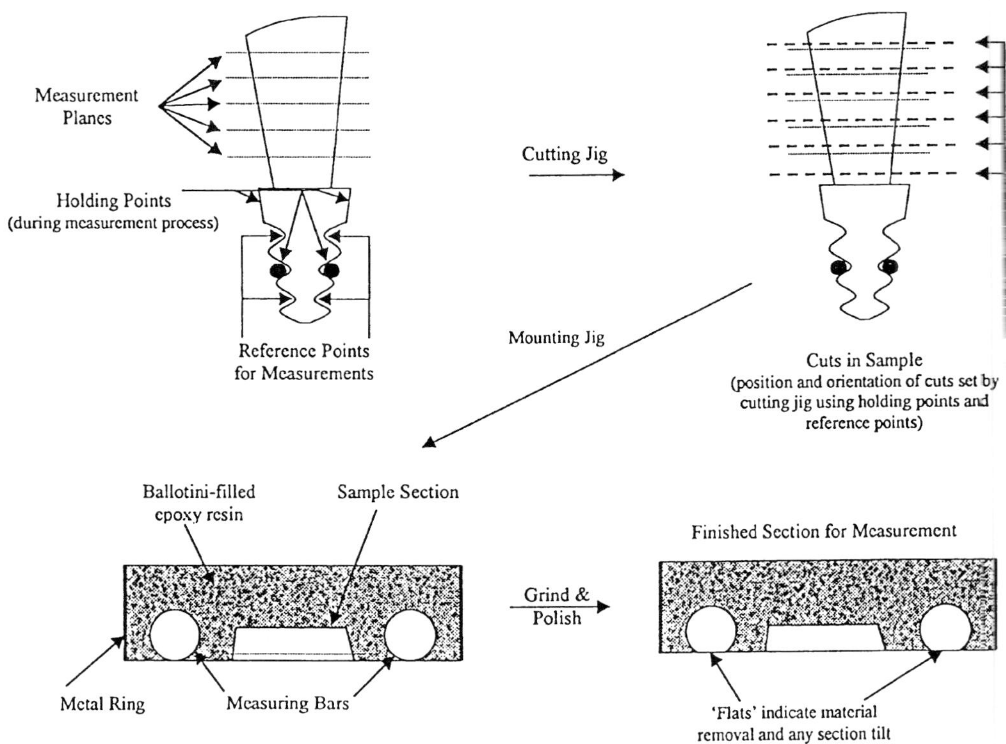

Fig. 1   Sample preparation route for turbine aerofoils.

- Sample preparation

  Post-exposure cross-section preparation is aimed at producing polished cross-sections through the components or samples which are both parallel to and close to the pre-exposure measuring planes.

- Post-exposure metrology

  This is carried out to accurately determine the final dimensions of the components/ samples and to characterise the extent of any corrosion damage or other required parameter (e.g. deposit thicknesses). The cross-sections are measured using reflected light microscopes with very accurately calibrated x-y stages. For laboratory samples with only 24 measurement points, a manually controlled stage on a projection microscope can be used. For larger and more complex sample/component shapes, it is necessary to use an image analyser with a motorised stage to automatically move the cross-sections and allow the measurement of the required features at several hundred reference points around each cross-section.

- Data analysis

  For each sample/component, comparison between the sets of pre-exposure and post-exposure dimensional measurements results in a set of materials loss measurements. Additional post-exposure parameters (e.g. depths of internal penetration, deposit thicknesses) at known locations around components/samples are evaluated purely from the post-exposure measurement data. Various levels of analysis of these data are possible, including statistical analyses.[7–11] These can be used to generate, for example, the probability of materials damage exceeding a particular value. As all measurements are made at known positions around a cross-section's surface, it is possible to select only data from a particular region of a component for data analysis (e.g. if it is believed that different exposure conditions have been experienced by different parts of the surface).

4 APPLICATION OF METROLOGY METHODOLOGY

4.1 LABORATORY HOT CORROSION TESTS AND MODEL DEVELOPMENT

The tests carried out form part of a systematic investigation of the effect of realistic levels of exposure parameters on the corrosion of candidate gas turbine materials for use in solid fuel fired systems. The data produced from these tests is being used to produce models of hot corrosion damage as functions of exposure conditions for gas turbine materials in 'dirty' gas conditions. Test work is continuing to extend the range of conditions covered by the models.

Laboratory studies of hot corrosion processes have been carried out using the general metrology method throughout its development (e.g. Refs 2,5,10,11). Many of these tests have been targeted at evaluating the performance of materials in gas turbine environments produced by combusting solid fuel derived fuel gases. These tests have used samples in the form of solid cylinders. A variety of base alloys and coatings have been tested, but all tests have included a standard gas turbine material (IN738LC) both uncoated and with a RT22 coating. All samples were measured prior to exposure with an accurate micrometer and rotating jig. Samples have been exposed at 600, 650, 700, 800 and 900°C for periods of up to 1000 hours, with cycling (every 20 or 100 hours) to replenish surface deposits (i.e. discontinuous cyclic corrosion testing). In different tests, various deposit compositions containing sodium/potassium/lead/zinc sulphates/chlorides have been sprayed on to the sample surfaces to simulate deposition fluxes and compositions spanning the ranges that are anticipated for gas turbine hot gas paths using solid fuel derived fuel gases. The gaseous environments used in these tests have included realistic ranges of SOx and HCl partial pressures. Post exposure metrology was carried out by optical microscopy/image analysis on polished cross-sections.

Figure 2 illustrates the results obtained from IN738LC at 700°C, with different deposit compositions, with the 'maximum metal damage' (evaluated at the 4% probability of excedence) used to characterise the corrosion damage. For all materials, the dependence of corrosion rate on deposition flux was found to be approximately sigmoidal with three distinct behaviour regimes:

(a) for low deposition fluxes there were low corrosion rates;
(b) at intermediate deposition fluxes, much faster corrosion rates were found with a dependence on deposition flux close to linear;
(c) at high deposition fluxes, a thick scale/deposit layer was found which effectively creates 'a buried in ash' scenario and slightly reducing corrosion rate with further increases in flux.

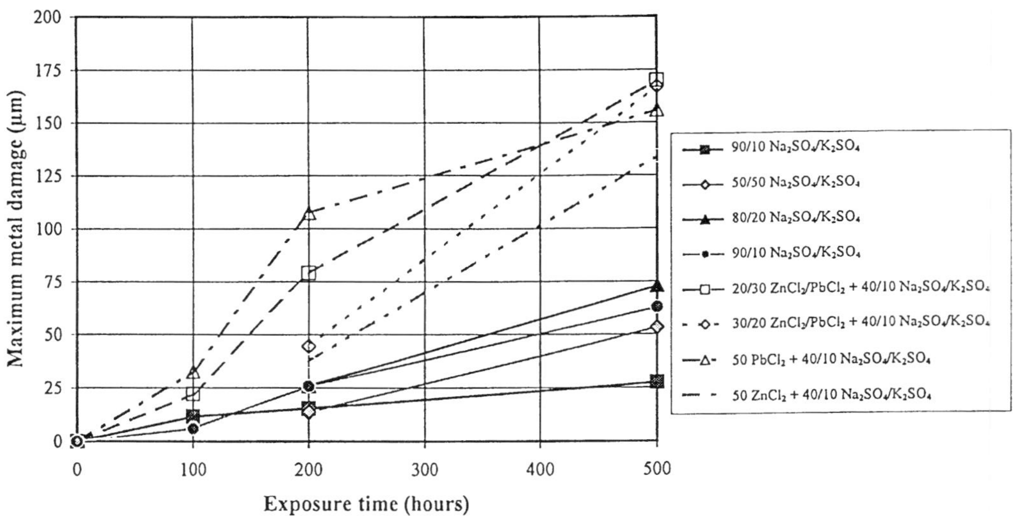

Fig. 2   Performance of IN738LC at 700°C with different deposit compositions.

Under these corrosive conditions for metallic coatings, it has been found that the deposition flux and deposit composition dominate the corrosion behaviour of the materials, with no discernible effect of the number or frequency of the thermal cycles that are an intrinsic part of this type of corrosion testing.

From the growing set of materials performance results, models have been produced from statistical analysis of the metal loss data sets to indicate the sensitivity of corrosion damage to the various exposure parameters. The approach used for corrosion modelling is based on a statistical analysis of the data produced to identify the extreme metal loss corrosion data and then to use a Gumbel type I model of maxima to analyse these data. The characteristic parameters obtained from this analysis are used to generate the overall corrosion model by curve fitting with respect to exposure parameters. The data sets gathered also allow average (median) values of corrosion damage to be found and modelled for comparison with the extreme value data.

Mathematical models for coating/base alloy performances were deduced by fitting equations to the data generated at each of the different test temperatures. For type II corrosion behaviour in regimes (a) and (b) with alkali sulphate deposits at a set temperature and HCl partial pressure, the equations were found to have the following form:

$$Corrosion\ Rate = (a + bx + cx^2) + (d + ex + fx^2) \log (pSO_2) \tag{1}$$

where a–f are constants, x is the alkali sulphate deposition flux and pSO_2 is the partial pressure of SO_2. The values of these constants are temperature sensitive due to the nature of hot corrosion, so the corrosion test temperatures correspond to the temperatures of the corrosion models. For the temperatures appropriate for type II hot corrosion, the effects of coating/base alloy interdiffusion on the corrosion rate with increased exposure time are very limited, especially compared with the effects of deposition flux, deposit compositions and local gas partial pressures.

4.2 BURNER RIG

A burner rig facility has been developed to expose materials in more realistic continuous deposition environment.[2] The rig consists of a natural gas fired burner capable of producing combusted gas temperatures controllable in the range 600–1500°C. With an air flow rate of 0.2 – 0.6 kg s⁻¹, it is possible to test materials in gas streams of ~10 – ~350 m s⁻¹ using different configuration sample exposure sections. For corrosion tests, a low velocity exposure section is used, which was designed to operate with hot internal wall faces and allow up to 24 air-cooled probes to be exposed simultaneously. Each cooled probe has a separate air supply which allows its temperature to be individually controlled. Appropriate contaminants are fed into the rig just upstream of the outer edge of the burner flame. The gaseous contaminants (SO_2 and HCl), are supplied from individual drums of liquefied gases via pressure let-down and gas mass flow control systems. The vapour phase contaminants are supplied as a solution in water, which is fed into the test rig as a spray via an atomising nozzle.

Corrosion tests have been carried out in this test rig at gas temperatures of 1100–1200°C with different levels and mixes of sodium/potassium/lead/zinc vapour phase contaminants and SO_x/HCl gaseous contaminants. Sample surface metal temperatures have been maintained at different temperatures in the range 650–900°C. The dimensions of all samples used in this rig have been accurately pre-measured. After exposure, cross-sections of all samples have been prepared, measured on the image analysis system and the data processed to generate materials performance results. Corrosion tests have included uncoated IN738LC and both RT22 and GT29+ coated IN738LC. Test durations have been up to 2500 hours.

Metal loss results from these exposures compare well with those obtained from the laboratory tests in terms of damage rates. The corrosion morphologies observed are also similar for equivalent exposure conditions.

4.3 PLANT EXPOSURES AND MODEL TESTING

An extensive metrology programme, using the methods described in Section 3, was carried out on a gas turbine, a static aerofoil cascade and arrays of probes in a topped coal-fired PFBC system at Grimethorpe, UK.[5,6] The probe exposure locations were selected in the hot gas path of this plant to allow gas turbine materials to be exposed in realistic gas conditions and also in related environments, i.e. with only limited changes in exposure conditions (e.g. high/low dust loading) to provide data for the materials performance models. The materials were exposed as a rainbow in the turbine and cascade and on a series of specially designed cooled and uncooled probes to allow a range of gas-metal temperature differences to be studied. During the course of these exposures, the following exposure parameters were monitored: gas composition (including SO_2, SO_3 and HCl); gas temperature and pressure;

dust loading and particle size distribution; deposition rates (fluxes) and deposit compostions; alkali compound dewpoints; and metal temperatures.

Models produced from laboratory tests using compatible metrology (Section 4.1) have been used to predict the damage observed in the plant. Figure 3 illustrates this for two of the materials tested (IN738LC uncoated and with a RT22 coating) for the selection of the plant environments causing type II hot corrosion. The scatter observed is mainly a result of uncertainties in determining the pilot plant exposure conditions, but limitations in the model and measurement errors (which increase in significance at low damage levels) will also be factors. It is clear from Figure 3 that the model provides a good fit to plant data. Use of the models at the lower level deposition fluxes expected in service allows predictions with an accuracy of $\pm 7\mu$m 1000 h^{-1}. The target accuracy of the plant data was better than $\pm 5\mu$m 1000 h^{-1}. Extrapolation of the models allows the environmental conditions (deposition flux, contaminant gas partial pressures, etc) to be identified that would be required to achieve a 25 000 hour component life.

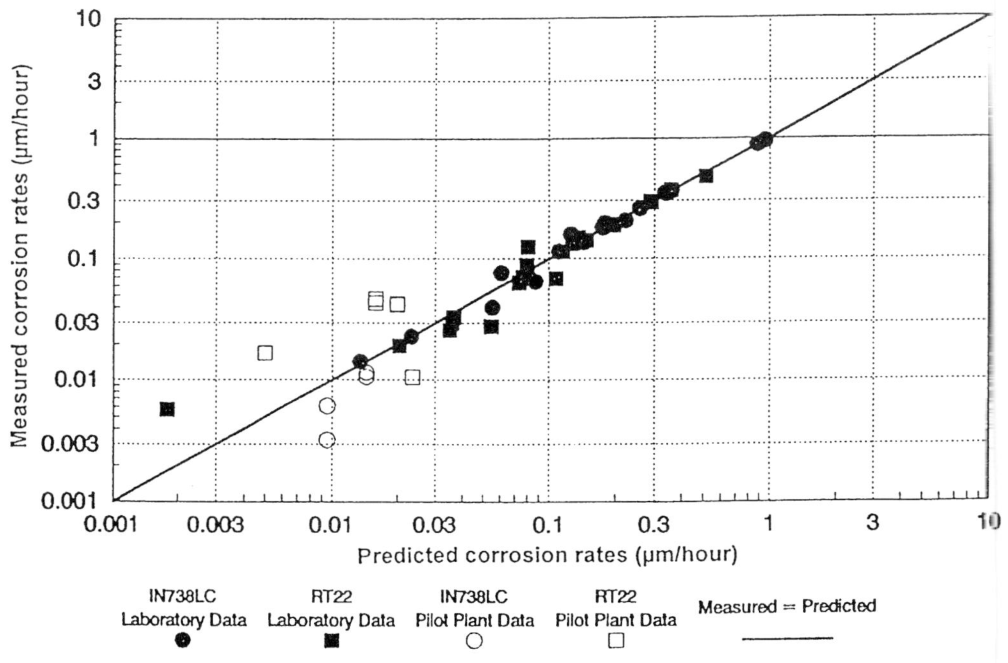

Fig. 3 Comparison of model predictions and plant results for type II hot corrosion.

5 CONCLUSIONS

Models of the hot corrosion behaviour of gas turbine materials have been developed from laboratory studies for dirty fuel fired gas turbines. These models have been successfully tested against corrosion data generated from a newly developed burner rig facility and a coal-fired topped PFBC plant.

Underlying the development and testing of these models is a general methodology for quantitatively determining the level of damage to components and samples exposed in plant, burner rig and laboratory test environments. This method of evaluating materials performance allows statistical analysis of the materials degradation data to generate models of materials performance for particular degradation mechanisms.

REFERENCES

1 *Co-gasification of Coal/Biomass and Coal/Waste Mixtures*, Final Report EC APAS Contract COAL-CT92–0001, University of Suttgart (Germany), 1995.
2 *Effects of Contaminants on Materials Performance in Industrial Gas Turbines for Advanced Combined Cycle Power Plants*, Coal R&D Report COAL R129, ETSU for UK Department of Trade and Industry, 1997.
3 *Standard Specification for Gas Turbine Fuel Oils*, ASTM D2880–90a, 1990.
4 M. Decorso, D. Anson, R. Newby, R. Wenglarz and I. Wright, ASME Paper 96-GT-76, ASME, Atlanta (U.S.A.), 1996.
5 *Erosion/Corrosion of Advanced Materials for Coal-fired Combined Cycle Power Generation*, JOUF-0022, EC JOULE Project Final & Summary Reports, 1993.
6 N.J. Simms, J.E. Oakey, J.R. Nicholls, P.J. Smith and D.J. Stephenson, in *Erosion by Liquid and Solid Impact*, J.A. Little and I.M. Hutchings eds, *Wear*, 1994, **186–187**, 247.
7 J.E. Oakey and N.J. Simms, *ASME Turbo Expo '96*, Paper No. 96-GT-396, ASME, 1996.
8 J.R. Nicholls and P. Hancock, in *High Temperature Corrosion*, R.A. Rapp ed., NACE, 1983, 198.
9 M. Kowaka ed., *Introduction to Life Prediction of Industrial Plant Materials: Application of the Extreme Value Statistical Method for Corrosion Analysis*, Allerton Press, 1994.
10 J.R. Nicholls and P. Hancock, in *'Plant Corrosion: Prediction of Materials Performance'*, J. E. Strutt and J. R. Nicholls eds, Ellis Horwood, 1987, 257.
11 J.R. Nicholls, P.J. Smith and J.E. Oakey, *Materials for Advanced Power Engineering*, D. Coutsouradis et al. eds, Kluwer, 1994, 1273.

A Modelling Approach to Gas Turbine Blade Life Management

J.K. HEPWORTH, J.E. FACKRELL, L.W. PINDER and J.D. WILSON

Power Technology Centre, Ratcliffe-on-Soar, Nottingham, NG11 0EE, UK
e-mail: john.hepworth@powertech.co.uk

1   INTRODUCTION

As a result of the drive towards higher efficiencies, the first stage blading of modern, large power generation gas turbines has become increasingly sophisticated and correspondingly more and more expensive. As the life of the blading is limited, this has a significant impact on maintenance costs, making it important to obtain the maximum safe life from a set of blades. In order to be able to do this, it is necessary to have a detailed understanding of the damage mechanisms active within the blade, to predict growth rates for cracks and other damage mechanisms. The paper outlines an approach to achieving a quantitative understanding of damage growth, which is illustrated diagrammatically in Fig. 1. The procedure starts with an aerothermal model of the turbine to provide boundary conditions for a thermal and stress analysis of the blade, the results of which are used with material damage models to estimate damage rates and hence life.

It is of course vital that any blade life extension is safe as the costs associated with a failure or even an unplanned outage would exceed any hoped for savings. It is therefore

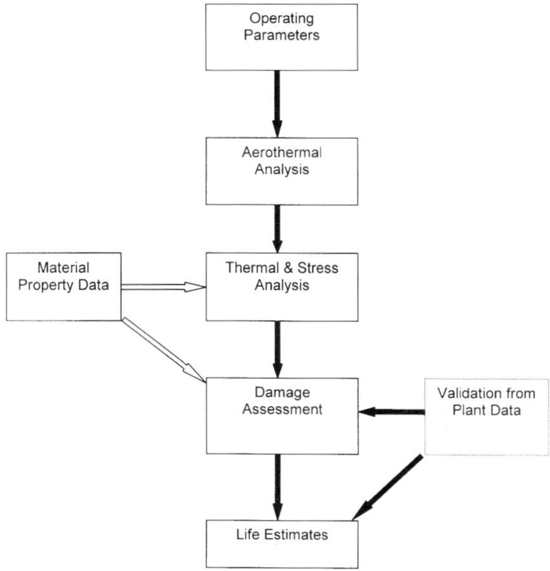

Fig. 1   Overall methodology.

important that the predictions of the modelling work are compared with observations made on blades removed from service after different periods of operation.

The primary focus of this paper is on the material damage aspects of blade life estimation, accordingly the aerothermal, thermal and stress analyses will only be described in outline.

2   AEROTHERMAL ANALYSIS

Gas flow and heat transfer calculations have been performed for both Siemens V84.2 and V94.2 gas turbines and for the GE 9FA machine. The information on turbine geometry and operating conditions was input to a 3D turbine flow analysis, see for example Denton,[1] which provided values of external flow variables, such as gas velocities, pressures and temperatures. The code solves for both the fixed and moving blade rows at the same time and has allowances for the effect of injected coolant flows and for blade losses. External heat transfer coefficients were estimated using a separate boundary layer flow code. The internal coolant flow and heat transfers were calculated using additional codes and an iterative procedure was employed to allow for variations in the inner wall temperatures. A typical gas temperature distribution is illustrated in Fig. 2.

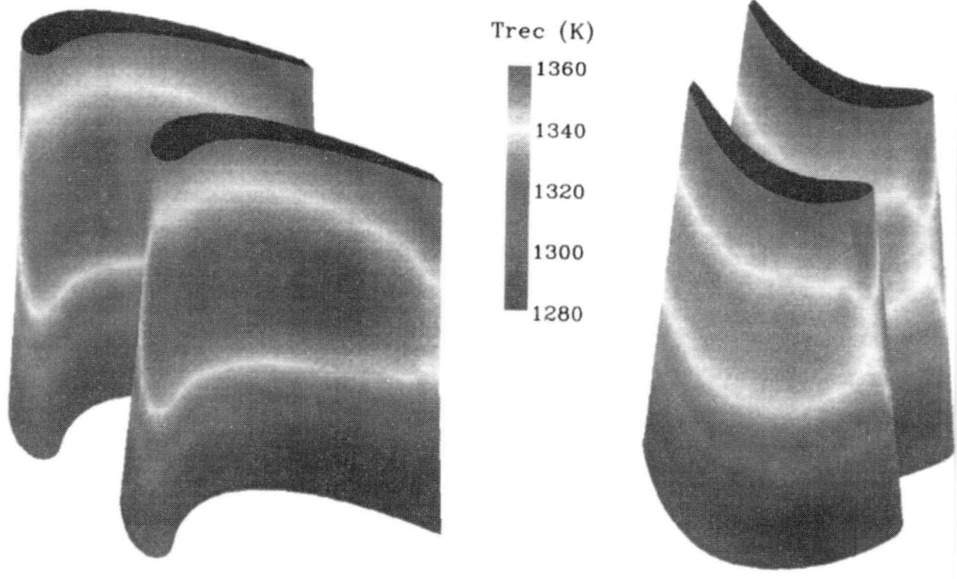

Fig. 2   Gas temperature distribution on blade surface.

3   BLADE THERMAL ANALYSIS

This was conducted using the finite element program, Abaqus,[a] using gas temperatures and heat transfer coefficients from the aerothermal analysis as boundary conditions. Values for

---

[a] Abaqus is a registered trademark of Hibbitt, Karlsson & Sorensen Inc.

the inner wall metal temperatures were used to repeat the coolant analysis and the process re-iterated until the temperatures converged. For the boundary conditions at the root end of the airfoil, it was assumed that heat transfer to the root section could be neglected. Checks on the temperature gradient near the root confirmed this to be a reasonable assumption. Figure 3 shows the temperature distribution across the mid-height section of a cooled first stage blade of a GE9FA machine.

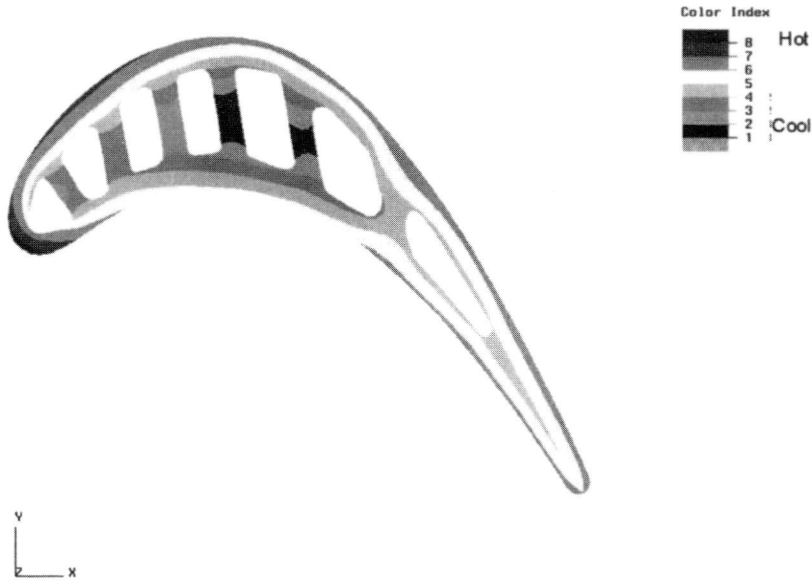

Fig 3 Temperature distribution on mid-height section of Stage 1 rotating blade.

4   STRESS ANALYSIS

The temperature distributions calculated are used as the loading conditions for the stress analysis of the blade. An initial elastic calculation is carried out, but in most cases this needs to be followed up by a creep analysis, as the effects of relaxation of the thermal stress under creep needs to be taken into account and the simplified methods for doing this are not really sufficiently accurate. In many modern machines, the blades in the hottest stages of the turbine are made of directionally solidified or even single crystal material. In such cases the stress analysis must take account of the anisotropy of the elastic and creep properties. The studies carried out so far have involved conventionally cast and directionally solidified material. For the latter, measurements were made of the elastic and creep properties in the longitudinal direction. In the transverse direction, properties were inferred from data in the literature for comparable materials in conjunction with the measured longitudinal data. The anisotropy has a fairly significant effect as Young's Modulus is lower, particularly in the longitudinal direction, which reduces the thermal stress from the through wall temperature gradient. The anisotropic creep properties were modelled using the Hill potential, as implemented in Abaqus.

Examples of the results of the stress analysis are shown for the elastic condition in Fig. 4 and the creep results in Fig. 5. These are for a directionally solidified blade and show the radial component of stress on the airfoil mid height plane. The stresses display the

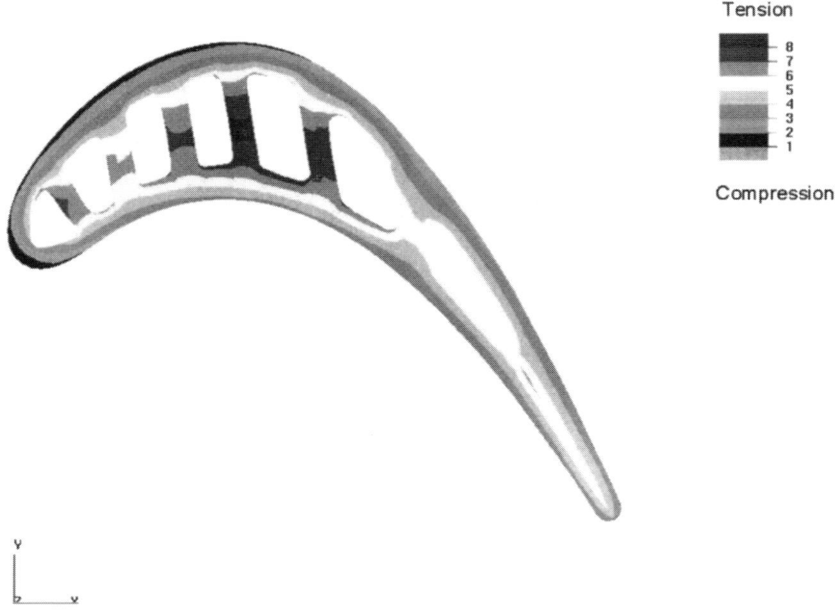

Fig. 4 Elastic stresses on mid-height section of rotating blade.

Fig. 5 Stress on mid-height section of rotating blade after 20 000 fired hours.

expected patterns, with generally compressive stresses on the outer surface and tensile stresses on the inner surface, due to the temperature gradient in the walls of the cooled blade. The creep results show relaxation of the thermal stress, though of course the centrifugal stresses cannot relax. There is evidence of more rapid relaxation on the hotter outer surface.

The effect of transients resulting from starts, shut downs or trips must also be considered. It has been shown that, in general, such transients do not produce excessive stresses in the blade walls. This is presumed to be the case because the blade walls are relatively thin and have a fairly large through wall temperature gradient during steady state operation. The most rapid change in external gas temperature occurs during a trip and then the external chilling serves to reduce the existing through wall gradient.

5 DAMAGE MODEL

The purpose of the modelling work described above is to allow estimates to be made of damage accumulation within the blade material and hence to provide life estimates. The damage analysis considers three mechanisms: namely oxidation, fatigue and creep and also considers interactions between the mechanisms.

5.1 OXIDATION DAMAGE

The outer surface of a blade is coated with an oxidation and corrosion resistant layer to prevent damage to the blade. Earlier coatings with aluminium, platinum aluminide or chromium, but more recent coatings are usually MCrAlY, where M is nickel and/or cobalt and the other components are represented by their chemical symbols. Sometimes the internal passages may be aluminised to prevent oxidation damage, though different manufacturers have different policies on whether to coat internally or not. In this work it is assumed that no oxidation damage in a coated surface occurs in the base metal until the coating fails. Oxidation damage is considered in two ways. Firstly as a general loss of material from the surface and secondly as oxide penetration of grain boundaries.

5.1.1 *Loss of Material by Oxidation*

It is assumed that the oxidation of an exposed surface of the superalloy material of the blade proceeds according to the conventional parabolic law. This has then to be modified to allow for the fact that the oxide will spall from the surface, which increases the oxidation rate. Spallation is assumed to be caused by the differential thermal expansion between oxide and metal, when the blade temperature is cooled at shut down. The implementation of this mechanism has been described in the literature, see for example Ref. 2. One consequence of this is that the rate of loss of metal due to oxidation is *strongly* influenced by the number of operational cycles during a period of service. This is illustrated in Fig. 6, which compares metal loss in a 10 000 hour period with 10 hour cycles with the loss over the same period with 100 hour cycles.

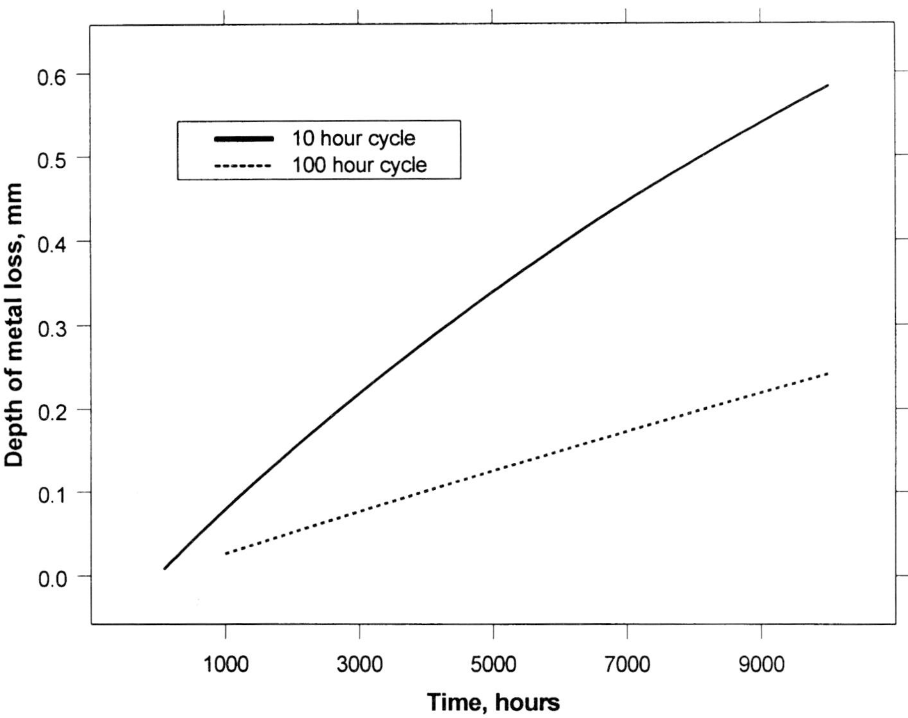

Fig. 6   Comparison of metal loss by oxidation for different cycle times.

5.1.2   *Oxide Penetration of Grain Boundaries*

There is evidence, see for example Ref. 3, that superalloy grain boundaries are preferentially attacked during oxidation, leading to oxides penetrating the grain boundaries. The significance of these oxide spikes is that they can form stress concentrating features which enhance fatigue and creep damage rates. According to Remy,[3] grain boundary oxidation is described by a $t^{1/4}$ law. It turns out that crack initiation is dominated by this oxidation process, which implies that crack initiation is strongly temperature dependent.

Once a crack has initiated and is growing, an oxidation growth mechanism is still incorporated, which now combines a grain boundary mechanism with a contribution from matrix oxidation. Generally, this only gives a modest increase to crack growth resulting from other mechanisms.

5.2   FATIGUE

For uncoated material, the initiation phase of fatigue should be taken into account, but where a coating is present, it can reasonably be assumed that when the coating is breached, there will be a sufficient number of initiation sites produced to assume that cracking has initiated. This assumption should be conservative, but in any case it is likely that any initiation time will be very short. When it is required to estimate initiation times, material test results are used in the form of a relationship between cycles to initiation and total strain range. The strain range derived from the finite element results is enhanced by a strain concentration factor derived from the depth of oxide penetration along a grain boundary.

Fatigue crack growth is calculated on the basis of experimental data which relates crack growth rate to the stress intensity factor, K, for deep cracks, greater than 2mm. For shorter cracks the growth rate depends on the applied strain range and the crack depth. The stress intensity factor and the strain range are obtained from the finite element results. It is known that fatigue crack growth rates are influenced by the presence of an oxide affected zone around the crack tip. In this work it has been assumed that this effect is included in the experimental data used.

5.3 Creep

Where it is necessary to calculate initiation times, creep damage is estimated using a strain fraction rule, as recommended in the R5 high temperature assessment route.[4] Again, account is taken of the stress enhancing feature provided by oxide penetration of grain boundaries.

Creep crack growth rates are estimated using experimentally determined relationships between C^* and growth rate. The energy integral characterising creep conditions at the crack tip, C^*, is estimated using a reference stress method due to Ainsworth.[5] The reference stress is also used to estimate creep continuum damage in the remaining ligament. In the event of the crack arresting, it would be creep damage in the remaining ligament that would lead to failure at that position.

6   COMBINATION OF DAMAGE MECHANISMS

In general, all three damage mechanisms will be operating, although their relative importance will vary from case to case. It is assumed that the mechanisms are essentially additive, apart from the stress enhancing effect of oxide penetration as described above. Thus, for initiation the damage fractions for fatigue, creep and oxidation are added and crack initiation deemed to occur when the total damage fraction is equal to unity. The damage fraction for oxidation is taken as the ratio of oxide penetration depth to an assumed initiation depth, usually 0.1 mm.

In the case of crack growth, the growth increments from oxidation, fatigue and creep are added to give the total growth increment for the cycle under consideration. The contributions of the various mechanisms to crack growth are illustrated in Fig. 7, a sample growth plot for an inner wall crack on a v94.2 first stage blade. The influence of oxygen on the properties of the crack tip zone is assumed to be included in the experimental data. In terms of oxidation, there is an issue concerning the effective depth of the crack, which will depend on whether the oxide filling the crack is ruptured during a thermal cycle. It is the effects of crack oxide fracture that may well be a factor in some of the results of thermo-mechanical fatigue testing. The model contains a simple criterion for oxide cracking based on a critical value of stress intensity factor.

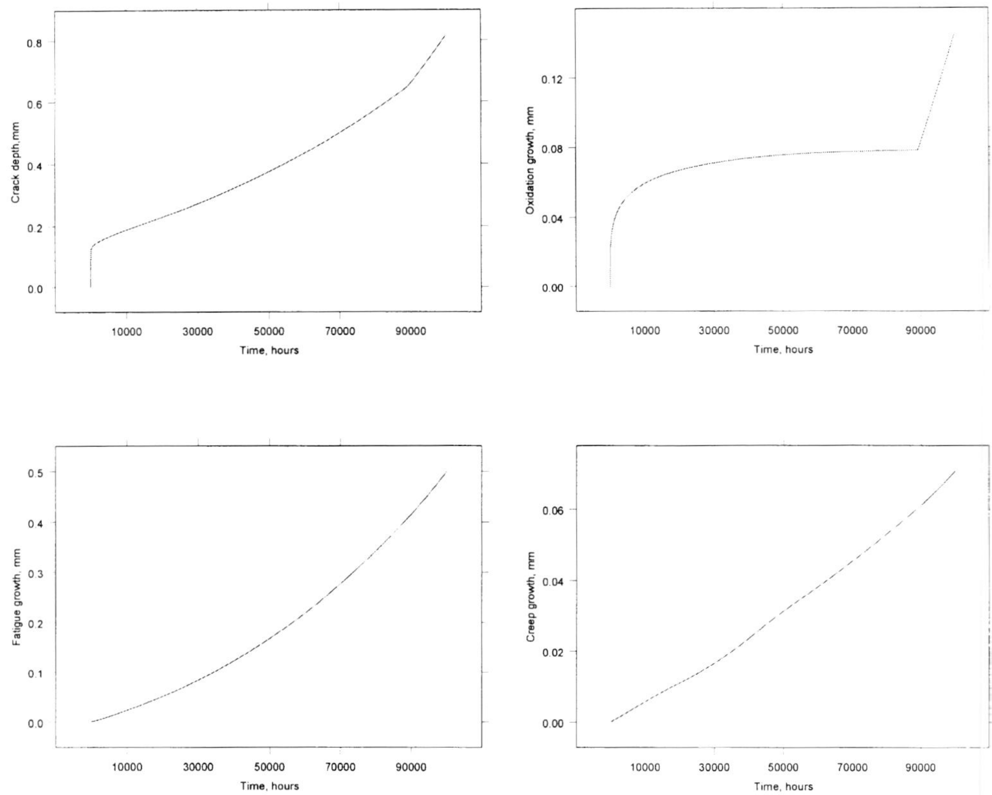

Fig. 7   Examples of growth predictions from inner wall of blade.

7   COMPARISON WITH PLANT DATA

It is vitally important that any calculated life and damage estimates are compared with data derived from plant in order to validate and in some respects calibrate the model. The main source of data comes from blades removed from service and examined metallographically for evidence of damage. Examples of sections of blades taken from service are shown in Figs 8 and 9. It can be seen that there is a significant spread in the observed damage, which in these blades is at the uncoated inner surfaces of the cooling ducts. The predictions of the model show that cracking should occur in the places where it is observed. After the service that these blades had seen, the predicted crack depth is about 0.4 mm, which compares reasonably with the measured crack depths, bearing in mind that, for the samples examined, the deepest cracking, shown in Fig. 8, appears to be associated with a region of material affected by casting porosity.

Comparisons have also been made with blade metal temperature measurements from plant. These have been derived from two different sources. There are pyrometers installed in some of the PowerGen gas turbines and data from these has been compared with calculated metal surface temperatures, showing good agreement as illustrated in Fig. 10.

288

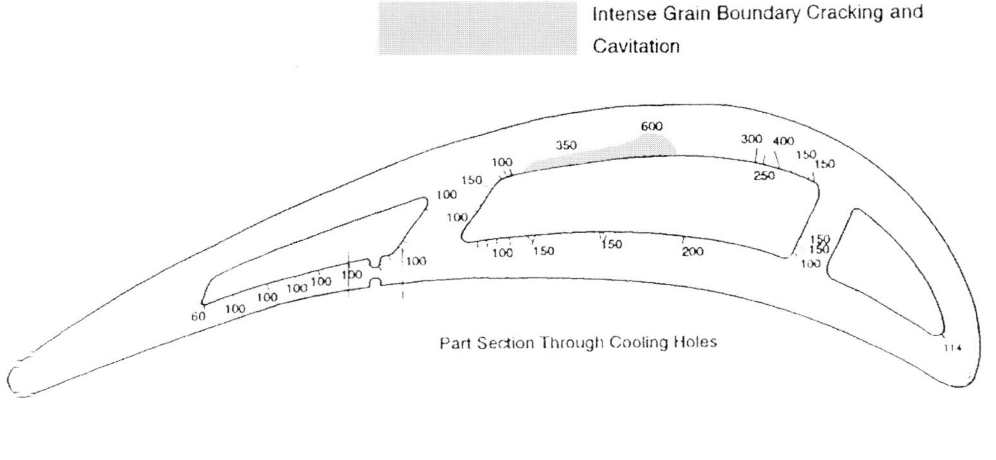

Fig. 8   Example of cracking in first stage rotor blade.

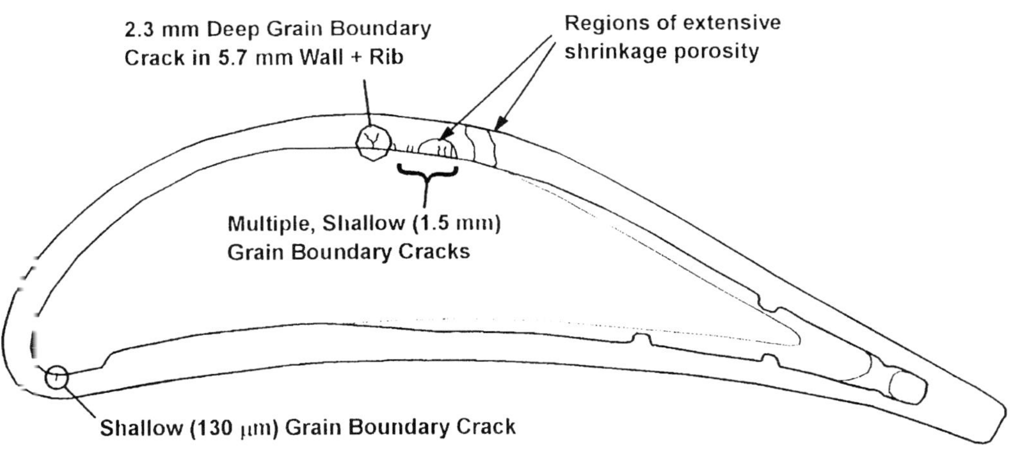

Fig. 9   Example of cracking in stator vane.

Temperatures across a section through the blade have been obtained from observation of gamma prime coarsening and again good agreement is obtained.

8   APPLICATIONS

There are three main areas where the life estimation model is being applied. Firstly, it enables the effects of changes in operating mode on blade life and hence costs to be assessed. To facilitate this application, the model is being incorporated into computer

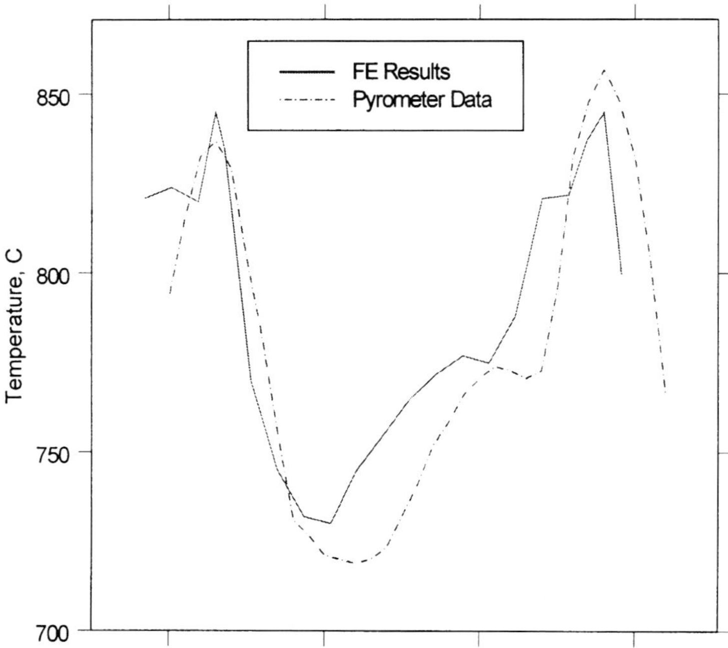

Fig. 10   Comparison of calculated blade temperatures with pyrometer data.

software packages, which can be used by power station engineers. The second concerns the problem of dealing with unexpected blade damage that may be observed during an inspection. The understanding of blade behaviour provided by the model allows an optimum strategy to be developed for dealing with the problem. The third application is concerned directly with life extension issues. This would include assessment of some blading which is life expired according to its design parameters, but which may well have significant remaining life in practise. Another possibility may be extending the outage intervals, which would involve running the blading for longer than originally recommended by the OEM.

Clearly, any decisions about blade life management involve both technical and economic factors, which may interact. Thus, when estimating the life of a blade, one may be concerned with the maximum life for the blade to be capable of refurbishment or the maximum safe life, beyond which there would be an unacceptable risk of blade failure.

9   CONCLUSION

A model has been developed for estimating the life of gas turbine hot section components, particularly the blading. The model relies on a computational procedure to establish the conditions seen by the blade metal and then applies the results of materials testing to estimate the resulting damage to the blade material. The model is being validated by comparison with damage observed in blades removed from service and sectioned for metallurgical examination. The resulting estimates of blade life can then be used to optimise blade replacement and utilisation.

REFERENCES

1. J.D. Denton, 'The Calculation of Three-Dimensional Viscous Flow through Multistage Turbomachines', *ASME Journal of Turbomachinery*, 1992, **114**, 18–26.
2. K.S. Chan, 'A Mechanics-Based Approach to Cyclic Oxidation', *Met. Mat. Transactions A*, 1997, **28A**, 411.
3. L. Remy, 'Fatigue Damage and Lifetime Prediction in Alloys for Gas Turbine Components submitted to Thermal Transients', *Behaviour of Defects at High Temperatures, ESIS 15*, MEP, London, 1993, 167–187.
4. *R5 – An Assessment Procedure for the High temperature Response of Structures*, Nuclear Electric, Gloucester, 1998.
5. G.A.Webster and R.A. Ainsworth, *High Temperature Component Life Assessment*, Chapman & Hall, London, 1994, 106.

Stress Relaxation as a Basis for Blade Creep Life Assessment

DAVID A. WOODFORD[a] and JOSEPH A. DALEO[b]

[a]*Materials Performance analysis, Inc., 1707 Garden Street, Santa Barbara, CA 93101, USA*
[b]*BWD Turbines Ltd, 1-601 Tradewing Drive (SS1), Ancaster, Ontario, L9G 4V5, Canada*

ABSTRACT

An integrated approach to life management incorporating a new methodology for evaluating the creep strength and fracture resistance of serviced blades is described. Creep strength is determined from a stress relaxation test (SRT) which is capable of generating data covering five decades in creep rate in less than one day. The SRT test may be augmented with a separate constant displacement rate (CDR) test to evaluate the current fracture resistance.

There are five basic steps involved in the blade life assessment. Metallographic evaluations identify the effects of engine specific operating conditions on coating life, structural stability and the state of the components cooling system. Finite element analysis allows a three dimensional centrifugal stress distribution to be established. Next, temperature profiles are estimated using heat transfer analysis in conjunction with microstructural based temperature estimations. Allowable stresses may then be estimated as a function of temperature using conventional design procedures and compared with the computed values. Finally, the properties of the material using the SRT (and CDR) data allow for condition assessment or life prediction. This paper will outline the testing methodology and present results relative to current CC, DS and MX blade alloys.

INTRODUCTION

Specific operating conditions can control the life of gas turbine hot section blades. An understanding of the original design parameters and their effect on component life is essential to understanding degradation modes, especially those resulting from 'off-design' operation. Today's advanced alloy and coating systems are expected to withstand very high temperatures and stresses for periods up to 48 000 hours. Steady state centrifugal stresses range from a few MPa to 250 MPa, and the thermal stress in the airfoil as a result of temperature gradients can range from −350 MPa to 500 MPa. Typical steady state metal temperatures range from 750°C to 1000°C. Specific conditions affect the system balance and degradation modes, and hence the life. The material degrades at different rates and in different modes in different areas of the same component.

The authors have established a procedure for remaining life assessment, which involves an analysis of the stress and temperature profiles coupled with a calculation of allowable stresses based on conventional design procedures. The first step is to perform a detailed metallographic analysis, which includes a study of the precipitate size distribution and coating diffusion layer. From these observations, and a heat transfer analysis, temperature profiles are established. A three dimensional centrifugal stress distribution is next computed from finite element analysis. This is then compared with the calculated allowable stresses.

The final important step in the analysis is the measurement of the current creep strength and fracture resistance. This information must be obtained in a short time, requiring the use of an accelerated test procedure. One approach is to run standard creep rupture tests at a higher temperature and/or higher stress and apply normal extrapolation procedures such as time/temperature parameter analysis in conjunction with linear cumulative damage. However, the information obtained from a few short creep rupture tests is limited and restrictive. An alternative approach has been developed[2] which provides extensive data and addresses some of the flaws in the current approach. This methodology may be summarised as follows:

- creep strength and fracture resistance are decoupled and measured in separate tests or as distinctly different properties
- both properties are measured as current values so that time dependent changes during the test are minimal
- creep strength is evaluated in terms of creep rate rather than time
- fracture resistance is evaluated in terms of crack extension or ductility rather than time to failure.
- state of damage is assessed in terms of current values of the critical properties.

Specifically, for creep strength a high precision stress relaxation test (SRT) is used. The specimen is deformed to a fixed strain which is held constant during a typical one-day test. The stress vs. time response is converted to a stress vs. creep rate response by differentiating and dividing by the modulus measured on loading.[3] This is, in effect, a self-programmed variable stress creep test. Typically, a test lasting less than one day may cover up to five decades in creep rate. The accumulated inelastic strain is usually less than 0.1%, so that several relaxation runs at different stresses may be made on a single specimen with minimal change in the mechanical state. Thus, an enormous amount of creep data can be generated in a short time on a single specimen. This test is meant to be performance based, i.e. it provides a measure of the current mechanical state in terms of creep strength, since time-dependent microstructural changes are deliberately minimised. Thus, the test allows a determination of the consequences of microstructural evolution and damage development. It does not attempt to incorporate these changes in the actual test as in the traditional creep rupture approach.

Many failures in engineering alloys occur at intermediate temperatures where there is a ductility minimum.[4] This reduced ductility may be accentuated for notched tensile tests because of the triaxial stress state at the notch root. A constant displacement rate (CDR) notch tensile test was originally proposed as a means to accelerate the development of notch sensitivity in long time notch rupture tests.[5] It has since been used in tests at Materials Performance analysis as a basis for fracture resistance evaluation in short time tests at a temperature and strain rate where the alloy is most vulnerable to fracture. The constant displacement rate ensures that once a crack initiates it will grow under control until a critical crack length for brittle fracture is exceeded. Thus, the displacement at failure and the extent of unloading at failure provide measures of the fracture resistance of the alloy. This test is also meant to be performance based rather than prediction based. It thus provides a comparative measure of the current fracture resistance. It is particularly suitable for assessing embrittlement in superalloys which may occur in service due, for example, to intergranular oxygen or sulfur penetration,[6] segregation of embrittling species to interfaces from grain interiors,[7] and precipitation of brittle phases.[8]

The SRT test is currently used as the primary evaluation tool. Although the test data do not capture time dependent changes during the test they can, nevertheless, be presented in the form of projected time to conform to current design practice. Since modern equipment is frequently operating in the cyclic mode it may be argued in any case that the long time test is inappropriate. Thus, we may retain a time-based design without the need to run long time tests.

This paper summarises the steps taken in the formal approach and then evaluates test results on several superalloys to illustrate the general utility of the test methodology.

MICROSTRUCTURAL BASED TEMPERATURE ESTIMATION MODELS

Temperature profiles are estimated using two distinct approaches. The first is based on the kinetics of γ' coarsening. Figure 1 for DSGTD111 shows typical $t^{1/3}$ kinetics. The rate constant was found to be independent of volume fraction and could be plotted as a simple inverse function of temperature. For a known service time the extent of coarsening provides a measure of the rate constant which leads directly to a temperature estimate. Taking into account the calculated constants used in the various formulas and the possible errors in measurements and estimates of the starting and ending gamma prime sizes, the

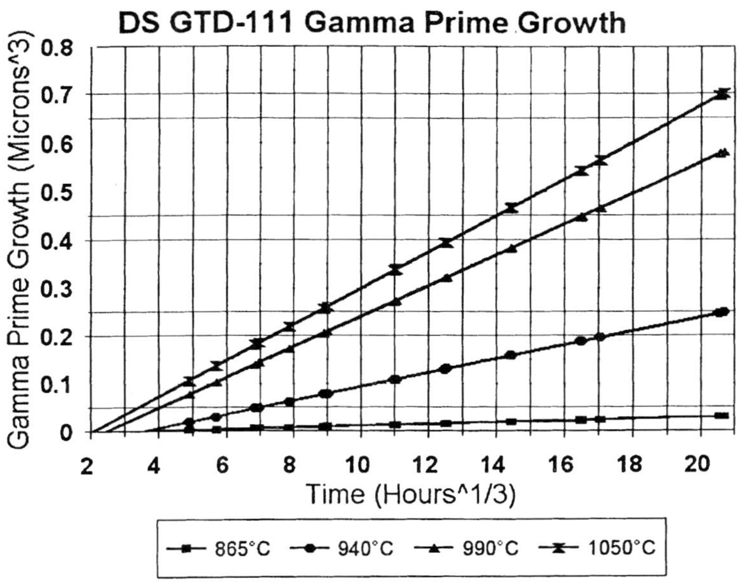

Fig. 1 Gamma prime growth as a function of temperature.

temperature estimates are believed to be accurate to approximately ± 30°C. This assumes that the aging rate is independent of gamma prime volume fraction and that material is exposed at relatively constant temperature. The technique has been used successfully to analyze components manufactured from DS GTD-111, IN-738LC, Udimet 520 and Udimet 710/720 alloy systems.

The second approach uses a MCrAlY coating interdiffusion growth rate method. A temperature estimation model based on inter-diffusion reactions between an aluminised CoCrAlY overlay coating (GT-29ªPlus) and a directionally solidified nickel base superalloy substrate (GTD111) was recently developed to analyze first stage buckets removed from a General Electric MS7001F industrial gas turbine.[9] The inter-diffusion kinetics between the GT-29 In-Plus coatings and the DS GTD-111 alloy substrate were investigated by measuring the width of the β+γ growth zones below the original interface of the coating. The layer growth calibration data measurements are plotted in Fig. 2 showing that the thickness of this zone increased in proportion to $t^{1/2}$. The parabolic growth constants at each temperature were estimated from linear regression of the data and fitted to a reciprocal temperature plot. To apply this model to a service-exposed component, it is assumed that the engine operates under constant (i.e. base load) firing conditions. The value of the growth constant is determined by using the number of operating hours.

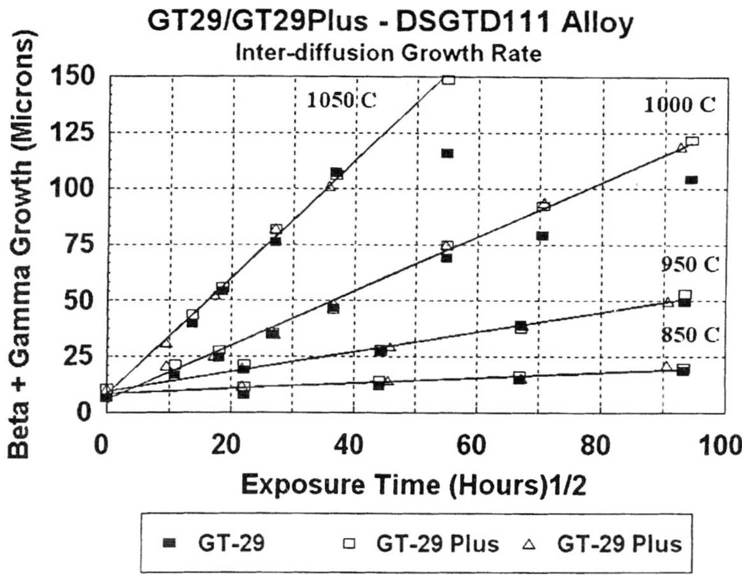

Fig. 2   Beta plus gamma growth as a function of temperature.

FINITE ELEMENT MODELING

A three-dimensional finite element model of the blade airfoil is created from measurements taken from the component. The mechanical and physical property data required for the analysis are obtained from published information and other sources. Material behaviour is assumed to be linear elastic. The effect of stress stiffening is included in the element loading. The finite element model is appropriately constrained, loaded with the operating rotational speed and estimated temperatures for material property definition and the stresses are calculated. Allowable stresses are also derived for the metal temperatures of interest and required design life (100 000 hours is typical). Suitable margins are applied to account for scatter in the properties as well as provide some allowance for the extrapolation of the measured data to times beyond the test conditions. In Fig. 3, the calculated centrifugal stresses in a cooled airfoil are shown and compared against the allowable stresses for the alloy. In the creep range, the allowable stresses are highly temperature dependent. Thus, the blade metal temperature profile has to be established as closely as possible.

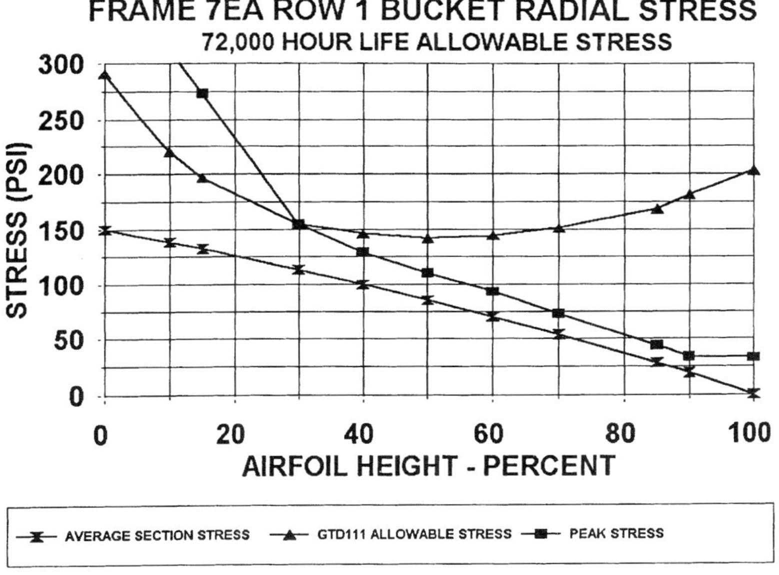

Fig 3   Comparison between calculated and allowable stresses for a particular blade.

Metal temperature prediction represents one of the most difficult problem areas in the analysis of gas turbine components. The gas path temperatures and surface heat transfer coefficients being subject to the most uncertainty. Local gas path temperature depends on the temperature and flow patterns leaving the combustion system and how these diffuse through the turbine. The combustor profile changes with load, condition of fuel nozzles and baskets, fuel type and emission control systems. Where available, however, estimates of metal temperature obtained from gamma prime and/or coating inter-diffusion growth considerations serve as a crucial check on the calculated data based on gas flow/heat transfer analysis. Based on these comparisons, boundary conditions in the thermal model can be adjusted until reasonable agreement with the metallurgically derived blade temperature data is obtained.

The temperature gradients in the airfoil section induce thermal stresses due to the differing levels of thermal expansion. The thermal stresses can be quite high and add to the stress due to the centrifugal load. However, at high temperatures over periods of time, the material tends to creep and relieve the thermal stresses. If the material has adequate ductility, the long-term effects of the thermal stresses will be small. On the other hand, if the turbine experiences frequent start-stop cycles, the high strain, low cycle fatigue behaviour becomes important.

STRESS RELAXATION TESTING (SRT)

The detailed test procedure has been described previously.[2,3] For the results described below the tests were loaded and held at either 0.4% or 0.8% total strain. The former represents a strain just beyond the elastic limit for most alloys so that the stress vs. creep rate provides a measure of the current creep strength with no additional hardening or softening. The latter involves approximately 0.5% nonelastic strain on loading. These data are used principally for calibration with times to 0.5% creep strain.

IN738

Several extensive studies have been conducted on this alloy; many of them on service exposed blades and vanes. A series of calculated stress vs. creep rate curves based on SRT tests at 0.4% strain on material taken from the root section of a serviced blade is shown in Fig. 4. A direct comparison of duplicate test specimens is included. The small differences for the two tests at 900°C may be real since there was a small but significant difference in measured modulus. The values used for E were 145 000 (5L), 140 000 (6L), 140 000 (7L)

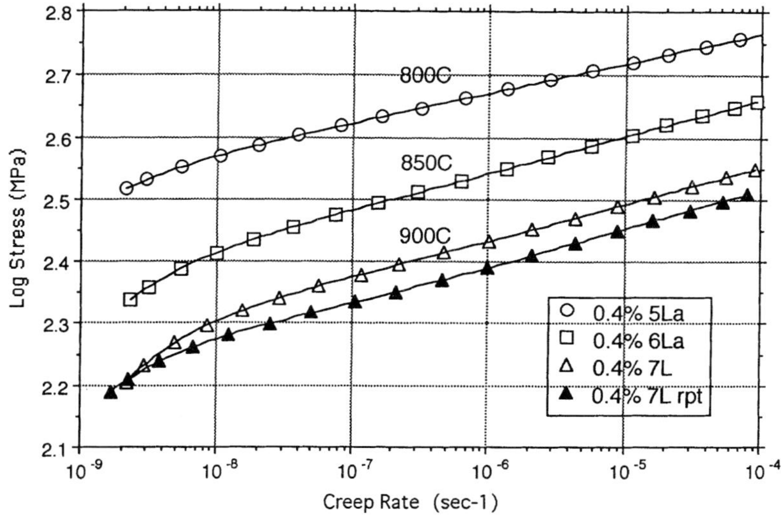

Fig. 4 Stress vs. creep rate curves for IN738 from 0.4% strain as a function of test temperature.

arc 135 000 MPa (7L repeat.). These are averages of the values measured for several loadings at 800, 850, 900°C and 900°C, respectively. The values were typically repeatable to within 10%, which correspondingly leads to a variation in creep rate and estimated time of 10%.

Using an intercept at log stress=2.5, and an average creep rate for the duplicate tests at 900C, it was possible to develop an Arrhenius type scaling law from Fig. 4. This allowed a master plot to be constructed covering 10 decades in creep rate and extrapolation of the lower temperature data. Thus, Fig. 5 shows possible design points at 800°C and 850°C based on creep rates of 3×10^{-11} s^{-1} (1% in 100 000hr.).

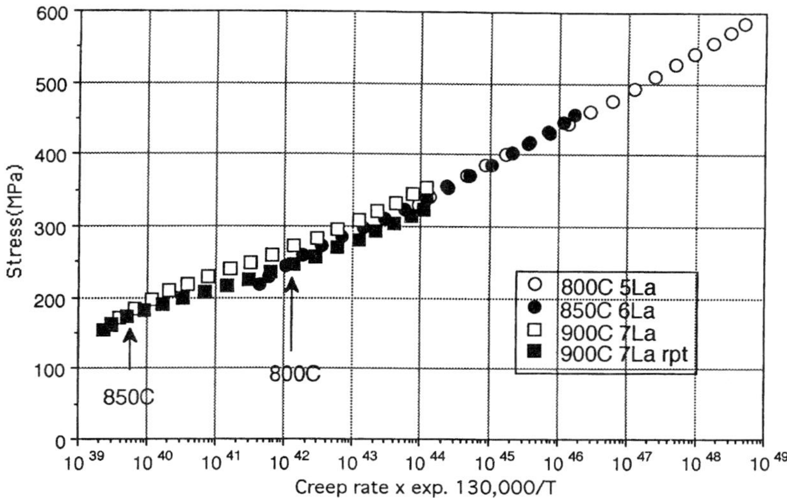

Fig. 5   Stress vs. parameter plot for IN738 showing possible design points at 800°C and 850°C.

Using the 0.8% SRT tests, a pseudo time was calculated for various creep rates and 0.5% creep (assuming constant creep rates to this strain), and plotted as projected times in Fig. 6 for the three temperatures. For example, the lowest creep rate of 10^{-9} s^{-1} gives a pseudo time of 1,390 hours. The resulting curves are shown as linear plots on a semi-logarithmic scale in Fig. 6. Also shown in the figure for comparison are times to 0.5% creep for IN738 taken from a different heat. The differences are well within the range of heat to heat or location to location variation.

The variation that may exist in a serviced blade and vane is shown in Fig. 7. This figure summarises the results on miniature specimens taken from a serviced blade and vane. A comparison of the blade and vane miniature specimens in terms of stress-creep rate from 0.4% strain at 850°C shows that 1v and 2v have significantly higher creep strength than all the other airfoil specimens. However, miniature specimen 4b and standard specimen 6L taken from the blade root are both a little stronger than 1v and 2v. The two bands of data in Fig. 7 are separated by roughly an order of magnitude in creep rates at a given stress at 850°C. This may be a response to a larger grain size and dendrite arm spacing in the root section of the blade compared with the leading edge area of the blade. However, it is

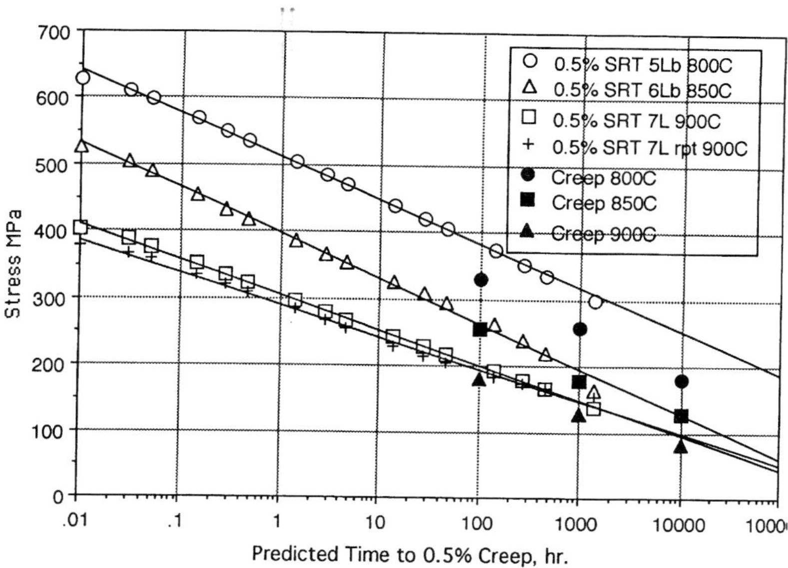

Fig. 6 Stress vs. predicted time for 0.5% creep for IN738 compared with actual long time creep data.

possible that it is primarily an orientation effect. The higher elastic modulus associated with the higher strength also suggests that local grain orientations make a major contribution to creep strength variation. The two minimum creep rate points fall on the lower bound of the data.

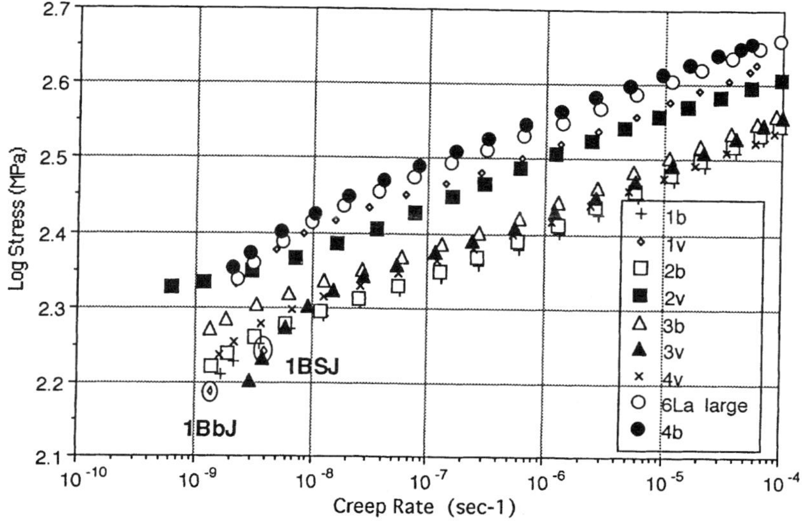

Fig. 7 Stress vs. creep rate for IN738 for specimens taken from different locations from a serviced blade and vane at 850°C.

These results on specimens 4b and 6L indicate that the creep strength of a particular microstructure (same grain size, dendrite spacing and modulus) should be independent of test specimen size to a good approximation. However, the creep strength of different sections within a component may vary considerably. Because of this, we cannot evaluate unambiguously the effect of engine operation on creep strength. Evidence was presented to indicate that major thermomechanical excursions (i.e. prior SRT testing) may have little effect on creep strength,[10] implying that modifications in precipitate distribution and dislocation structure are less important than the starting microstructural conditions identified above. It would clearly be of great value to do a comprehensive evaluation of a new blade.

GTD111

GTD111 is a cast nickel based alloy developed by General Electric for use in land based gas turbine hot section blades. It was derived from another GE alloy, René 80, which has seen extensive service in jet engine blades, and has closely similar creep rupture properties. Both conventionally cast and directionally solidified blades of GTD111 have been in service for several years, and the equiaxed version is believed to offer a creep strength advantage equivalent to about 20°C compared with IN738. The directional solidification process provides higher ductility and lower elastic modulus in the direction of alignment. However, ductility may be poor in other orientations and, in many alloys, grain boundary tougheners such as hafnium are added to increase transverse fracture resistance. Contrary to common belief, although the time to failure may be highest in the longitudinal direction because of the higher ductility, the creep strength, expressed in terms of time to a specific creep strain or minimum creep rate, may be higher in transverse or diagonal orientations, depending on the test conditions.[11]

Round miniature test specimens were taken from a new directionally solidified GTD111 small hollow gas turbine blade. To study the effect of orientation, specimens were taken from the blade root in the longitudinal, diagonal and transverse directions relative to the grain alignment. The specimens were 41mm long with a reduced section of 32mm by 1.5mm.

The transverse and diagonal properties were very similar and are expected to be comparable to the equiaxed condition. For the tests at 900°C, there were differences. Figure 8 compares the data for 0.8% total strain and shows that the longitudinal orientation has higher creep strength at low stresses but lower creep strength at higher stresses. The equivalent creep rate for 0.5% creep strain in 100 000 hours is $1.4{\times}10^{-11}s^{-1}$. Thus, a design stress could be taken from Fig. 8 in terms of creep rate.

The additional advantage of the creep rate format is that it may be used in detailed creep design. The particular representation shown in Fig. 8 is semi-logarithmic or exponential. It may be that a power function is more appropriate for some alloys. The fit is about as good for the data shown as the exponential fit and yields stress exponents of 17.62 and 11.88 for the longitudinal and transverse orientation specimens, respectively. These are appropriate exponents to be used, for example, in crack tip creep deformation analysis, rather than exponents calculated from minimum creep rate data.

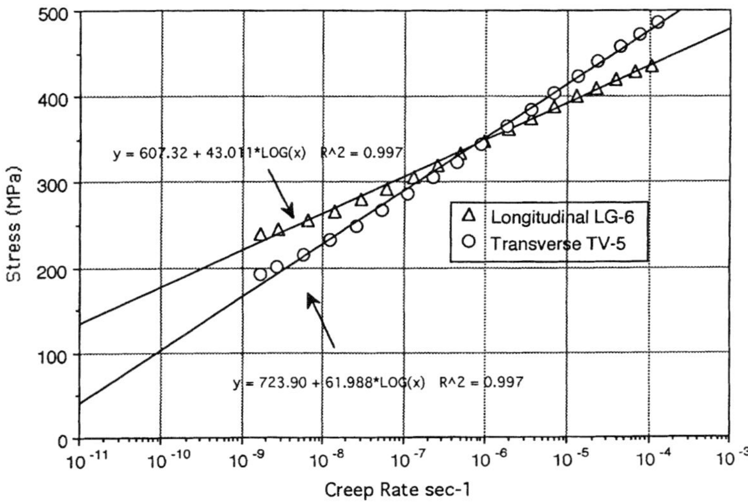

Fig. 8   Stress vs. creep rate for GTD111 from 0.8% strain for two orientations.

Since there are few data published on GTD111, estimates for creep rupture lives were made based on parametric comparison curves for a number of alloys including equiaxed GTD111.[12] From this parametric representation for GTD111, a stress rupture curve was estimated at 900°C to give a good fit on a semi-logarithmic plot and compared in Fig. 9 with the SRT prediction for the transverse orientation. Since the time to rupture is

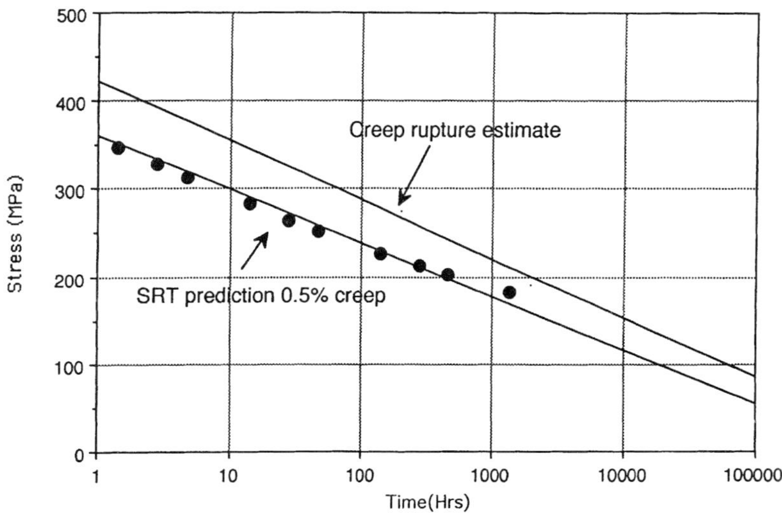

Fig. 9   Stress vs. projected time to 0.5% creep for transverse GTD111 compared with estimated time to rupture at 900°C.

typically two to three times the time to 0.5% creep in these cast alloys (except for directionally solidified alloys tested in the longitudinal orientation where the increased ductility may lead to higher ratios), the comparison is quite good This suggests that the SRT data could be used directly in standard design procedures either in terms of derived times or, preferably, in terms of stress vs. creep rate from the original analysis.

In a separate study of conventionally cast GTD111 taken from various locations in a blade individual specimens were first tested at 850°C then at 900°C. The specimens are identified:

#1 lower airfoil near fracture after 23 000 hours service
#2 lower airfoil of unbroken blade after 23 000 hours service
#3 root section of serviced blade and furnace exposed at 900°C for 5000 hours
#4 root section, presumed in original condition
#5 repeat of #4 from same location.

Differences were greatest at 850°C, and Fig. 10 shows the various responses from 0.4% strain at this temperature. The unexposed root section clearly had the highest creep strength. Furnace aging appreciably reduced the strength in this location to levels comparable to those in the service-exposed airfoil. The evidence for appreciable loss in creep strength due to the exposure is therefore quite strong in this case. A detailed analysis of these results with predictions of remaining life based on creep at the indicated rates for specific strains to failure has been published.[13]

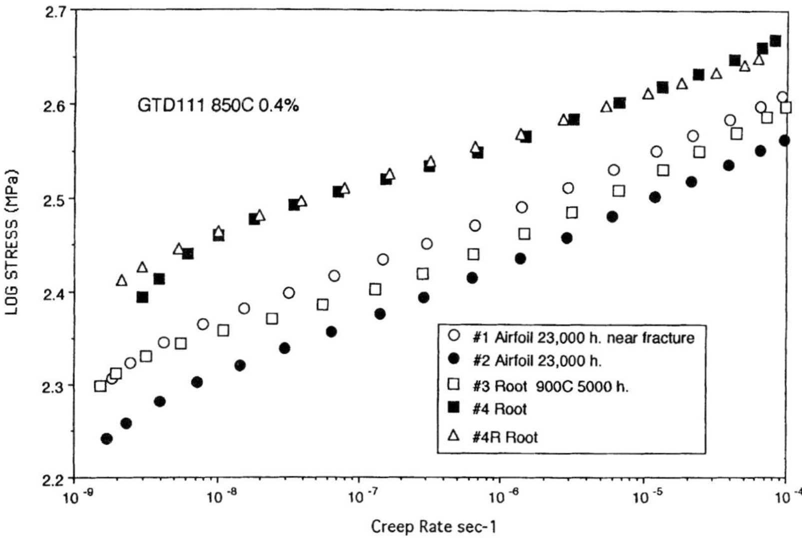

Fig. 10   SRT results from CCGTD111 taken from various locations for various conditions at 850°C and 0.4% strain.

CMSX-4

The first generation monocrystalline nickel based superalloys to attain turbine engine application, such as PWA 1480, René N4 and CMSX-2, have similar creep rupture strengths, although differing significantly in other critical properties. Operating experience with these alloys resulted in process improvements and chemistry modifications, which were introduced in the second-generation monocrystals. The principal chemistry change was the addition of 3–6% Re, with an associated lowering of Cr to maintain acceptable phase stability, and a partial replacement of Ti with Ta. The published chemistry of CMSX-4 as a second-generation alloy is: 6.5%Cr, 9%Co, 6%Mo, 6%W, 6.5%Ta, 3%Re, 5.6%Al, 1%Ti, 0.1%Hf, Bal. Ni.

In the tests conducted all specimens were within 8 degrees of [001] and were fully heat-treated.[2] Specimens were exposed in air for 20 hours at 1000°C, 1050C° and 1100°C. Figure 11 is a plot of the analyzed creep data from 1000°C for these exposed specimens compared with the unexposed condition. Three of the curves are closely similar but the test on the intermediate exposed sample (20 h. at 1050°C) shows significantly higher creep strength. The moduli were measured as 80GPa, 81GPa, 87.5GPa and 81GPa for the unexposed, 1000°C, 1050°C and 1100°C exposures, respectively. Since thermal exposure is unlikely to have a significant effect on elastic modulus, it is concluded that an orientation difference resulted in the higher modulus and strength level for this specimen. This interpretation also leads to the conclusion that exposures up to 1100°C have no significant effect on the creep strength at 1000°C.

Comparisons may be made between the SRT data and conventional creep results either directly in terms of creep rates or indirectly in terms of time. Duplicate SRT tests at temperatures between 850°C and 950°C showed excellent agreement and compared well

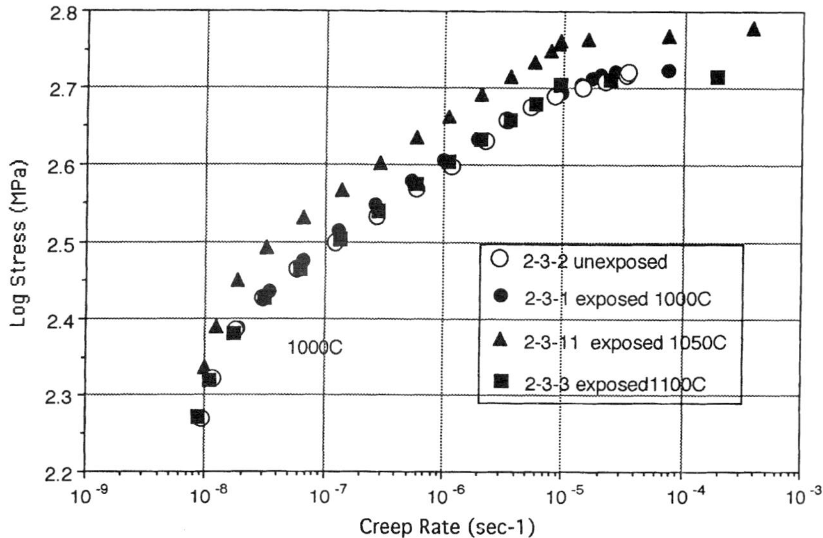

Fig. 11 Stress vs. creep rates from 0.8% strain at 1000°C for CMSX-4 for various exposures.

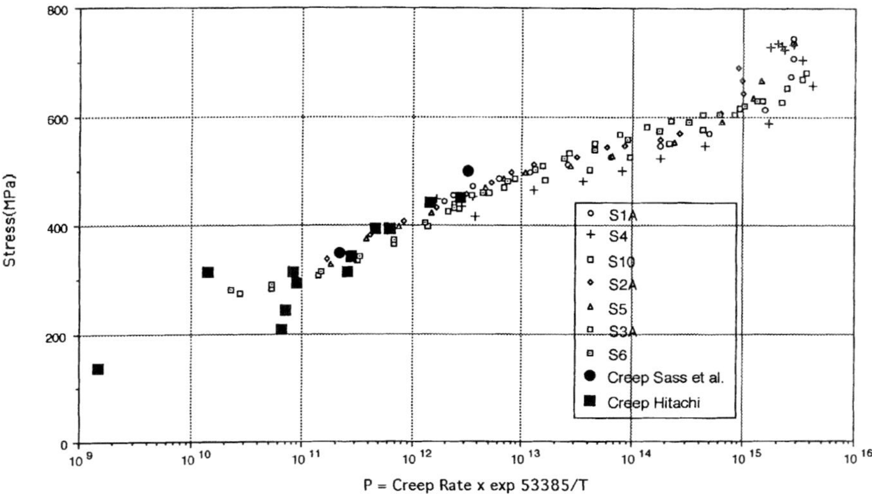

Fig. 12   Stress vs. parameter plot comparing SRT and long time creep data.

with minimum creep rate data as shown in Fig. 12. In fact, the creep data indicated much more scatter.

CONSTANT DISPLACEMENT RATE TESTING (CDR)

Rupture life is primarily a measure of creep strength; fracture resistance should better be identified with a separate measure that reflects the concern with embrittlement phenomena, which may lead to component failure. Most engineering alloys lose ductility during high temperature service. At a fixed strain rate for example, ductility first decreases with increasing temperature. This is believed due to grain boundaries playing an increasing role in the deformation process leading to the nucleation of intergranular cracks. At still higher temperatures, processes of recovery and relaxation at local stress concentrations lead to an improvement in ductility. Maximum embrittlement generally occurs in a critical range of temperature and strain rate.

With cast superalloys gas phase embrittlement (GPE) resulting from intergranular penetration of aggressive gaseous species is of greatest concern. Such penetration is accelerated by stress and is invariably a major factor in sustained load cracking (creep crack growth) and time-dependant fatigue crack growth.[6] The embrittlement that occurs due to this intergranular penetration of oxygen is illustrated in Figs 13 and 14. In these figures, the specimens were smooth miniature specimens suitable for removing from the thin sections of gas turbine blades. The embrittlement in IN738 due to air exposure at 1000°C is clearly shown in Fig. 13. Figure 14 shows the same effect for directionally solidified GTD111. For specimens cut transversely to the growth direction the grain boundaries embrittle. For specimens cut longitudinally there is little effect of exposure since very few boundaries intersect the surface.

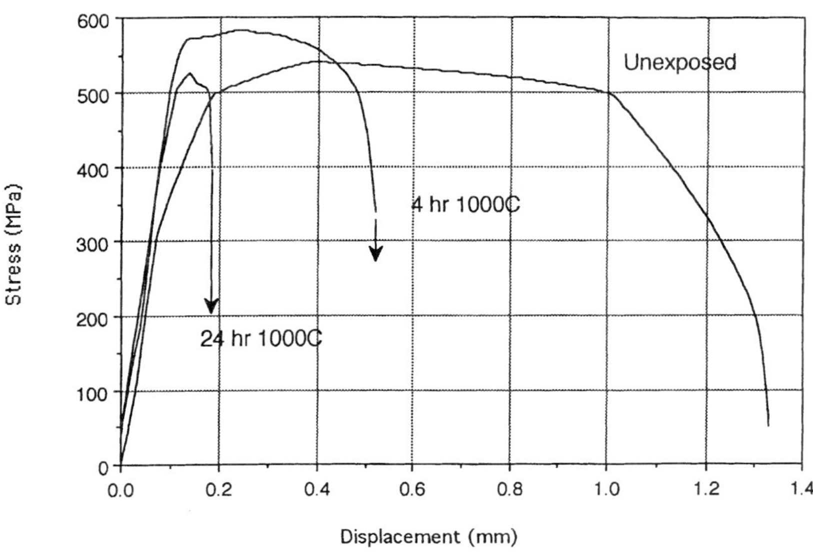

Fig. 13   Effect of oxygen embrittlement for various exposures in air at 1000°C in tests at 800°C on CDR data for IN738.

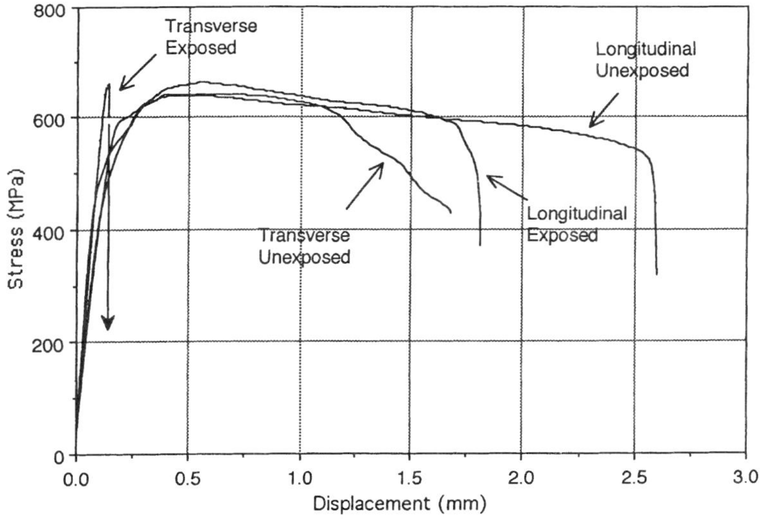

Fig. 14   Effect of exposure in air at 1000°C for 24 hours on fracture of longitudinal and transverse GTD111 at 800°C.

Finally, some notch CDR results on the monocrystalline alloy CMSX-4 are shown in Fig. 15. In this case, exposures in air for 20 hours at 1000°C, 1050°C and 1100°C have had no significant effect on notch stregth or displacement at failure. (Note that the abscissa is time, which is directly proportional to displacement). All three tests also show about the same amount of controlled unloading before final fracture. For this high strength

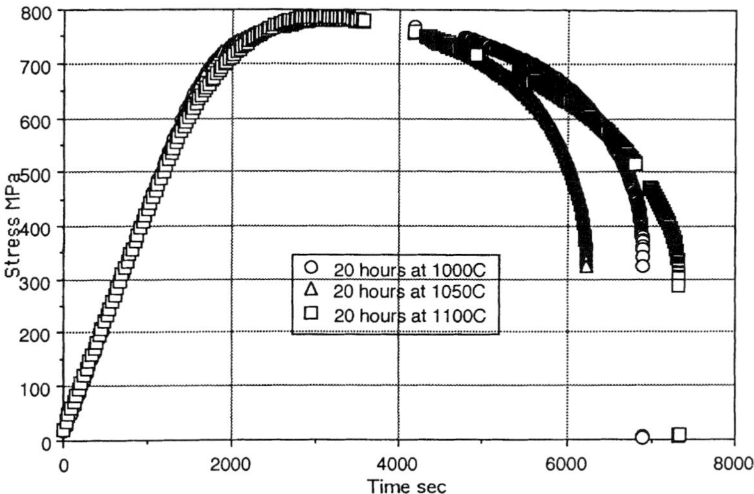

Fig. 15   CDR results for CMSX-4 at 1000°C exposed in air at the indicated temperatures.

monocrystal there is not only no measurable affect of high temperature thermal exposure on notch strength (see also Fig. 11 for confirmation of no effect on creep strength) but there is no sign of gas phase embrittlement in the absence of grain boundaries.

DISCUSSION

The most important general observations in the high temperature alloys are: repeatability is very good, the sensitivity to prior thermal exposures is often quite small for the very high strength alloys, there are no major effects of test section size for specimens taken from the same location, there may be appreciable effects of part section thickness for same size specimens taken from different locations.

There are several ways in which the SRT data may be used in optimising processing, as a basis for creep design, and for life management decisions. Since all the curves are derived from the basic stress vs. creep rate curves, it is recommended that these curves should form the basis of any design and analysis procedure. Moreover, creep rate may be used as a state variable, whereas time requires a zero reference point and cannot be usefully employed as a state variable. Also, the SRT data for 0.4% strain involves minimal change in state during the relaxation run. It thus provides a true measure of the creep strength for a given state. In contrast, in the traditional long time creep testing the state is changing continuously as a function of strain and time.

A possible foundation for a new design methodology based on the SRT test is presented in terms of the parametric representation of Fig. 5 for IN738. This master curve may be used, based on three one day tests, to compare processes, heat treatments, service exposures etc. It may also be used as a basis for creep design. As an example, possible design points at different temperatures are identified on the figure at creep rates of 3×10^{-11} s^{-1}. This is a rate corresponding to 1% in 100 000 hours. However, calibration with current design practice might lead to the choice of a different creep rate.

One aspect of the problem of remaining life assessment of turbine blades which has not been addressed is the difficulty in knowing the extent to which the measurements reflect differences which were present when the components were first put in service. It has become common practice to make measurements on the thick section blade root to serve as the undamaged condition for comparison. However, there is increasing evidence suggesting that very significant differences exist initially between the properties of these thick sections and the thin wall airfoil sections. In cast blades, the thicker section with lower solidification rates results in larger grain sizes and dendrite arm spacing. This in turn results in higher creep strength. It is worth noting that basic design data are normally measured on large cast to size specimens or on specimens taken from large cast blocks. There is no formal procedure to use properties unique to the local solidification conditions in analyzing blade creep. Clearly, resolution of these issues requires a systematic study of the properties of the material at different locations in new blades.

Although, as noted above, it may not be possible to determine whether significant loss in creep strength has occurred in service, it is still possible to determine whether the current strength is acceptable. This may be done directly in terms of creep rate or indirectly in terms of time to specific creep strain. It is also possible to assume a constant strain to rupture (e.g. 10% for equiaxed IN738) and provide a measure of rupture life.[13] This is done in reference to the estimated stress and temperature and in terms of current design practice. All of these approaches require the establishment of what is the minimum acceptable performance level for the application.

REFERENCES

1. R. Viswanathan and J.R. Foulds, 'Accelerated Stress Rupture Testing for Creep Life Prediction', Journ. *Pressure Vessel Technology, ASME,* 1998, **120**, 105–115.
2. D.A. Woodford, 'Accelerated Testing for High Temperature Materials Performance and Remaining Life Assessment', EPRI Report, WO 9005–19, 1999.
3. D.A. Woodford, 'Test Methods for Accelerated Development, Design and Life Assessment of High Temperature Materials', *Materials & Design,* 1993, **14** (4), 231–242.
4. N.F. Fiore, 'Mid-Range Ductility Minimum in Ni-Base Superalloys', *Reviews on High Temperature Materials,* 1975, **2** (4), p. 373–408.
5. J.J. Pepe and D.C. Gonyea, 'Constant Displacement Rate Testing at Elevated Temperature', *Int. Conf. Fossil Power Plant Rehabilitation,* ASM Int., 1989, 39.
6. D.A. Woodford and R.H. Bricknell, 'Environmental Embrittlement of High Temperature Alloys by Oxygen', *Treatise on Materials Science and Technology,* Academic Press, vol. 25, 1983, 157–199.
7. R.T. Holt and W. Wallace, 'Impurities and Trace Elements in Nickel-Base Superalloys', *Int. Metals Reviews,* March, 1976, 1–24
8. G. Chen et al., 'Grain Boundary Embrittlement by μ and σ Phases in Iron-Base Superalloys', *Superalloys 1980,* J. K. Tien *et al.* eds, ASM, 1980, p. 323–333.
9. K.A. Ellison, J.A. Daleo and D.H. Boone 'Metallurgical Temperature Estimates Based on Interdiffusion Between CoCrAlY Overlay Coatings and a Directionally Solidified Nickel-Base Superalloy Substrate', *Material for Advanced Power Engineering,* 1998.

10. D.A. Woodford, 'Stress Relaxation Testing of Service Exposed IN738 for Creep Strength Evaluation', ASME paper, IGTI Turbo Expo Conference, Indianapolis, 1999.
11. D.A. Woodford and J.J. Frawley, 'The Effect of Grain Boundary Orientation on Creep and Rupture of IN738 and Nichrome', *Met. Trans.*, 1974, 5, 2005–2013.
12. P.W. Schilke, A.D. Foster, J.J. Pepe and A.M. Beltran, 'Advanced materials Propel Progress in Land-based Gas Turbines', Advanced Materials and Processes, 1992, 4, 22–30.
13. J.A. Daleo, K.A. Ellison and D.A. Woodford, 'Application of Stress Relaxation Testing in Metallurgical Life Assessment Evaluations of GTD111 Alloy Turbine Blades', *Journal of Engineering for Gas Turbines and Power*, January 1999, 121.

Microstructural Evaluation of MCrAlY/Superalloy Interdiffusion Zones

K.A. ELLISON, J.A. DALEO and D.H. BOONE

BWD Turbines Ltd, 1-601 Tradewind Dr., Ancaster, Ontario L9G 5V5
Tel: 905-648-9262, Fax: 905-648-9264, email:ellison@bwdturbines.com

ABSTRACT

Overlay coatings of the (Ni,Co)CrAlY type are used extensively for protection of hot section components in gas turbine engines. Recently, the rates of interdiffusion between a CoCrAlY coating and a directionally solidified nickel-base alloy substrate were quantified and used to predict the service operating temperatures of a turbine rotor blade. The metallurgical temperature estimates are used to calibrate finite element heat transfer models that support component life and repair calculations. A qualitative understanding of the diffusional interactions and microstructural features is required before this type of analysis can be extended to other commercial coating/substrate systems. In this paper, the interdiffusion zones between (Ni,Co)CrAlY(Re,Si) coatings and nickel-base superalloys were evaluated after isothermal exposures at 1000°C. The multi-phase diffusion zones that formed in each case were characterized by optical and scanning electron microscopy. This information, coupled with available ternary phase diagrams and recent experimental work on similar coatings, was used to construct schematic 1000°C isothermal sections for the Ni–Co–Al–Cr quaternary system. The interdiffusion zone microstructures were interpreted by constructing schematic diffusion paths on existing Ni–Cr–Al and Ni–Co–Al ternary phase diagrams.

1 INTRODUCTION

Overlay coatings of the (Ni,Co)CrAlY type are used extensively for protection of hot section components in gas turbine engines.[1] Commercial (Ni,Co)CrAlY coatings are based on the γ-FCC and β-(Ni,Co)Al phases, but they may in addition contain phases such as γ', α-Cr and σ-(Cr$_x$Co$_y$).[2,3] The nickel-based superalloys used for rotating blades are primarily comprised of the γ and γ'-(Ni$_3$Al) phases. Interdiffusion between the coatings and substrates results in complex microstructural transformations which vary with the alloy compositions, temperature and time.[4]

Recently, the rates of interdiffusion between a CoCrAlY coating and a nickel-based superalloy substrate were quantified and used to predict the service operating temperatures of a combustion turbine blade airfoil, Fig. 1.[5] The metallurgical temperature estimates are used to calibrate finite element heat transfer models that support turbine hot section component life and repair calculations. It would be quite useful if this type of temperature estimation model could be extended more generally to other commercial coating/substrate systems. However, a qualitative description and understanding of the diffusional interactions and microstructural transformations that occur at turbine operating temperatures is required before this can be done. In this investigation, the interdiffusion zones between (Ni,Co)CrAlY(Re,Si) coatings containing 12–23 at.% Al and 17–30 at.% Cr and nickel-based superalloys were evaluated after laboratory exposures at 1000°C.

311

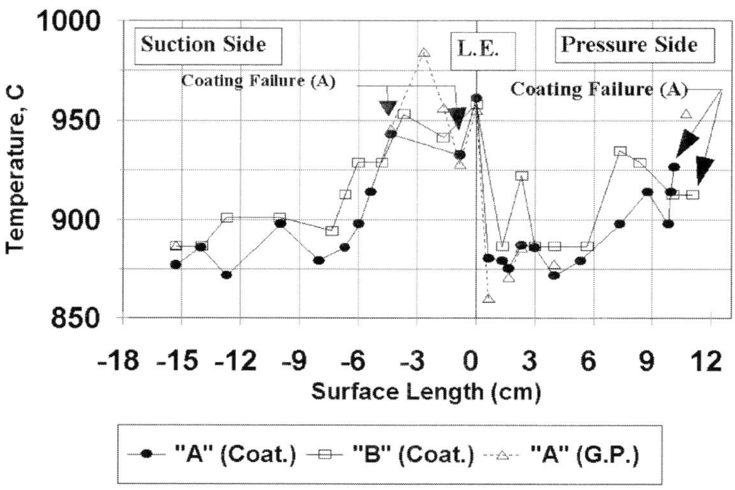

Fig. 1   Metallurgic temperature estimates from two first stage turbine blades ('A' and 'B') based on the coating/substrate interdiffusion (Coat.) and gamma prime (G.P.) coarsening models.

2   EXPERIMENTAL PROCEDURES

The nominal chemical compositions of the MCrAlY coatings and nickel-based substrate alloys used in this investigation are given in Table 1. The sample material was sectioned from commercially-processed gas turbine blades and blades. Four coating substrate combinations, applied by vacuum plasma spray (VPS) were examined in this work: (1) GT-29*/GTD-111, (2) GT-33/GTD-111, (3) Sicoat 2231†/IN-738, and (4) Sicoat

Table 1   Nominal Coating and Base Ally Compositions.

Alloy	Composition, wt %								
	Ni	Co	Cr	Al	Ti	Mo	W	Ta	Other
(Ni,Co)CrAlY Coatings									
GT-29		Bal	29	6.5					0.3 Y
GT-33	32	Bal	22	10					0.5 Y
Sicoat 2231	32	Bal	27	7.5					0.5 Si, 0.5 Y
Sicoat 2453	Bal	9	24	10					3 Re, 0.5 Y
Ni-Base Alloy Substrates									
GTD-111	Bal	9.5	14	3	4.9	1.5	3.8	2.8	0.1 C, 0.05 Zr, .015 B
IN-738	Bal	8.5	16	3.4	3.4	1.75	2.6	1.75	0.13 C, 0.12 Zr, 0.012 B 0.85 Nb (Cb)

---

* GTD-111, GT-29 and GT-33 are designations of General Electric Co., Schenectady, NY.
† Sicoat 2231 and Sicoat 2453 are designations of Siemens AG, KWU, Mülheim, Germany

2453/IN-738. Each of the coatings used for this investigation were approximately 200 μm thick. After deposition by VPS, the coatings were diffusion bonded to the alloy substrates by annealing in vacuum at 1120°C for 2–4 hours.

The coatings were isothermally exposed in still air for up to 10 000 hours at 1000°C. The samples were removed from the furnace and rapidly quenched in air to preserve the high temperature microstructures. The coating interdiffusion zones were examined by optical and scanning electron microscopy. The microstructural features were revealed by backscatter electron (BSE) imaging or optical and secondary electron (SE) imaging after etching in a 1% chromic acid solution at 5VDC. The various phases in the coatings were identified on the basis of energy dispersive spectroscopy (EDS) results obtained from unetched specimens using a LINK Analytical QX2000 analyser attached to a Phillips 515 scanning electron microscope operated at 20 keV.

2.1   Results

Samples of each coating type were examined after exposure times of approximately 1000, 3000, 5000 and 8000 h at 1000°C. Due to space limitations, the presentation and discussion of the results is limited to specific cases, as described below.

2.1.1   *GT-29/GTD-111 System*

The microstructural changes that occurred within the GT-29 CoCrAlY coating and the underlying GTD-111 substrate alloy after a thermal exposure of 1350 h at 1000°C are shown in Fig. 2. Interdiffusion between the coating and substrate resulted in the formation of a $\gamma + \beta$ growth zone in the GTD-111 alloy below the original coating/substrate interface (O.I.*) EDS analysis revealed that the large, elongated plates in the $\gamma + \beta$ growth zone were a nickel-rich (Ni,Co)Al type, Table 2. In contrast, the smaller β precipitates within the central coating zone were cobalt-rich. The volume fraction of the beta phase precipitates in the $\gamma + \beta$ growth zone below the O.I. was higher (~25%) than that within the 'bulk' coating zone (15%).

The interface between the a $\gamma + \beta$-(Ni,Co)Al growth zone and the $\gamma + \gamma'$ substrate was non-planar. The β-(Ni,Co)Al plates at the lower edge of this zone were often surrounded by pockets of $\gamma + \gamma'$.

Below the narrow $\gamma + \gamma' + \beta$ growth zone, there was a gradual reduction in the volume fraction of γ' precipitates moving in the direction of the bulk GTD-111 alloy. The growth of these layers below the O.I. continued with increasing exposure time until the point at which all of the β phase in the coating above the O.I. had been consumed. At 1000°C, this occurred somewhere between 3030 and 4970 hours of exposure.

In addition to the β-(Ni,Co)Al precipitates, a minor amount of a white (in BSE imaging mode) globular phase was present in the $\gamma + \beta$ growth zone. EDS analysis revealed that the white precipitates were rich in Ta and Ti, but no carbon, oxygen or nitrogen was detected

---

* The position of the O.I. was revealed in most of the coating samples by occasional alumina inclusions that resulted from the original grit blasting of the substrates prior to the application of the CoCrAlY overlay coatings. In the SEM, the O.I. was further highlighted, in backscattered electron imaging (BSE) mode, by a line of small white Ta_xTi_{1-x} precipitates.

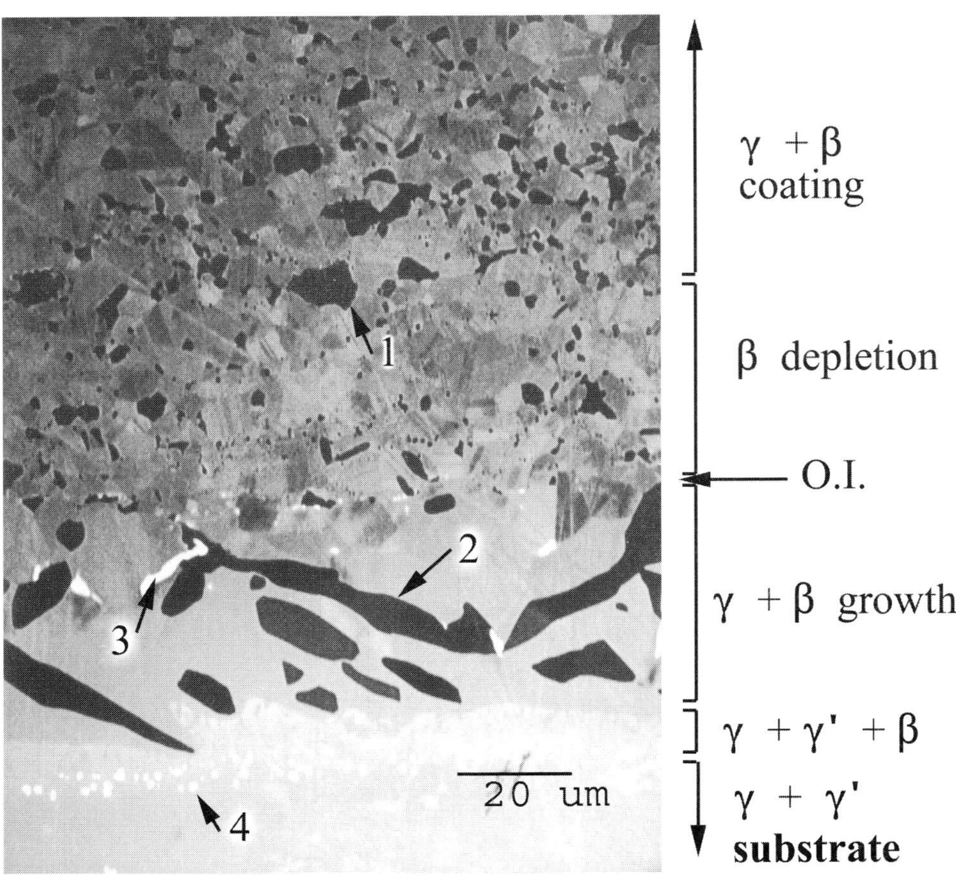

Fig. 2   Microstructural features within the GT-29/GTD-111 interdiffusion zone after exposure for 1340 h at 1000°C.

Table 2   EDS Results for Phases Identified in Fig. 2.

Position	Composition (at %)								Description
	Ni	Co	Cr	Al	Ti	W	Mo	Ta	
1	14	37	14	33	1				β
2	43	18	8	28	3				β
3	3	2	2		54	3	3	33	Ta_xTi_{1-x}
4	9	3	80	1	1	2	3		α

by light element analysis. A thin line of smaller (0.2–0.8 µm) Ta and Ti-rich precipitates also formed along the O.I. In ternary Ni-Ti-Al[6] and Ni-Ta-Al[7] alloys, Ta and Ti tend to partition to the γ' phase where they substitute for aluminum.[8,9] The solubilities of Ta and Ti in γ' in the ternary alloys are 7 at.% and 16 at.%, respectively, vs. 2.5 at.% and 12 at.%

ir the γ phase. However, the γ' phase is destabilized below the O.I. in the $\gamma + \beta$ growth zone due to the inward diffusion of Co from the GT-29 coating. The formation of the Ta,Ti-rich precipitates is most likely caused by the supersaturation of Ta and Ti in the $\gamma + \beta$ zone, causing these elements to precipitate out of solution. In the binary system, Ta and Ti form a complete series of BCC solid solutions above 882°C.[10] Thus, it is suspected that the white Ta and Ti-rich precipitates may be the BCC Ta_xTi_{1-x} phase. Above the O.I., the central CoCrAlY coating microstructure consisted of a γ (FCC) matrix with β-CoAl precipitates, Fig. 2, Table 2.

2.1.2 GT-33/DS GTD-111 and Sicoat 2231/IN-738 Systems

At 1000°C, the GT-33 central coating layers were comprised of β-(Ni,Co)Al precipitates (30 vol. %) in a γ matrix. Similar to the GT-29/GTD-111 system, the dominating feature of the GT-33/GTD-111 interdiffusion zone was the formation of a $\gamma + \beta$-(Ni,Co)Al growth layer below the O.I., Fig. 3, Table 3. However, the volume fraction of the β-(Ni,Co)Al phase in the $\gamma + \beta$ growth zone was relatively small (5 vol. %). As may seen by comparing Figs 2 and 3, the rate of $\gamma + \beta$ growth below the O.I. was significantly lower for

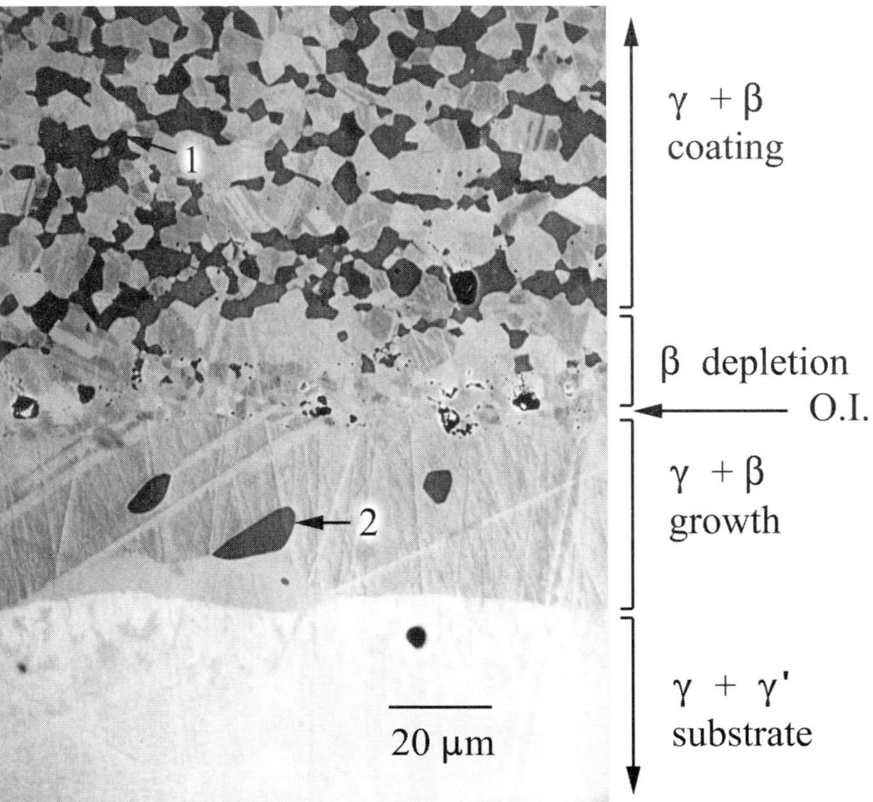

Fig. 3 Microstructural features of the GT-33/GTD-111 interdiffusion zone after 3030 h at 1000°C.

Fig. 4   Microstructure and phase regions of the Sicoat 2231/IN-738 interdiffusion zone after 3055 h at 1000°C (BSE image).

the GT-33/GTD-111 system compared with that of the GT-29/GTD-111 system. The thickness of the $\gamma + \beta$ zones was roughly the same in each case (40 μm) however the GT-33 coating was exposed 2.25X longer than the GT-29 sample. The microstructural features of the Sicoat 2231/IN-738 interdiffusion zone were qualitatively similar to those of GT-33/ GTD-111 system, Fig. 4.

Table 3   EDS Results for Phases Identified in Fig. 3.

Position	Composition (at %)					Description
	Ni	Co	Cr	Al	Ti	
1	45	11	8	35	1	β in coating
2	47	10	7	34	1	β in γ + β growth zone

2.1.3 Sicoat 2453/IN-738 System

At 1000°C, the bulk Sicoat 2453 NiCoCrAlYRe coating consisted of three principal phases: the γ matrix, β-(NiCo)Al and Cr/Re-rich α precipitates, Fig. 5, Table 4. Interdiffusion between this coating and the IN-738 alloy substrate resulted in depletion of the β precipitates above the O.I., leaving only the γ and α phases. This 30 μm thick zone was followed by a narrower (15 μm), single-phase γ zone adjacent to the γ + γ′ IN-738 alloy. Occasional α precipitates could also be observed along the γ/γ + γ′ interface. At the outer surface of the coating, another β-depletion zone was observed (not shown in Fig. 5), resulting in the formation of a two-phase γ + α zone between the air environment and the bulk coating.

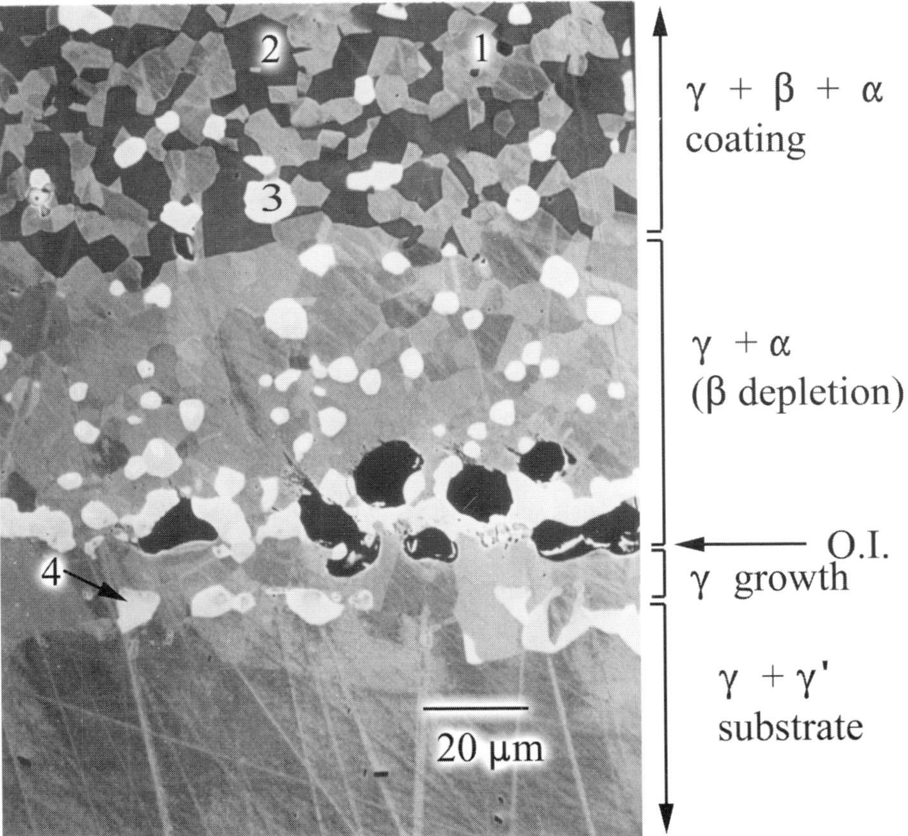

Fig. 5 Microstructure and phase regions of the Sicoat 2453/IN-738 inter-diffusion zone after 9110 h at 1000°C (BSE image).

3 DISCUSSION

3.1 INTERPRETATION OF COATING AND INTERDIFFUSION ZONE MICROSTRUCTURES

Interdiffusion in the (Ni,Co)CrAlY(Re,Si)/Ni-based superalloy systems resulted in the formation of multi-phase diffusion zones at the coating/substrate interface. These

Table 4   EDS Results for Phases Identified in Fig. 5.

Position	Composition (at %)						Description
	Ni	Co	Cr	Al	Re	Mo	
1	51	11	28	8			γ in coating
2	60	5	7	27			β in coating
3	5	2	90	1	2		α in coating
4	5	2	90	1	2	1	α in diffusion zone

microstructural transformations can be represented in a compact way and related to the governing phase equilibria by plotting diffusion paths on the appropriate phase diagrams.[11] A schematic isothermal phase diagram for the Ni–Co–Cr–Al quaternary system at 1000°C is shown in Fig. 6. This diagram is hypothetical but is based on available data and on microstructural observations made in the present study. The 1000°C isothermal sections for the Ni–Co–Al[12] and Co–Cr–Al[13,14] systems and the closest sections for the Ni–Cr–Al[15,16] (at 1025°C) and Ni–Co–Cr[17] (at 1200°C) phase diagrams were used to represent the four ternary faces.

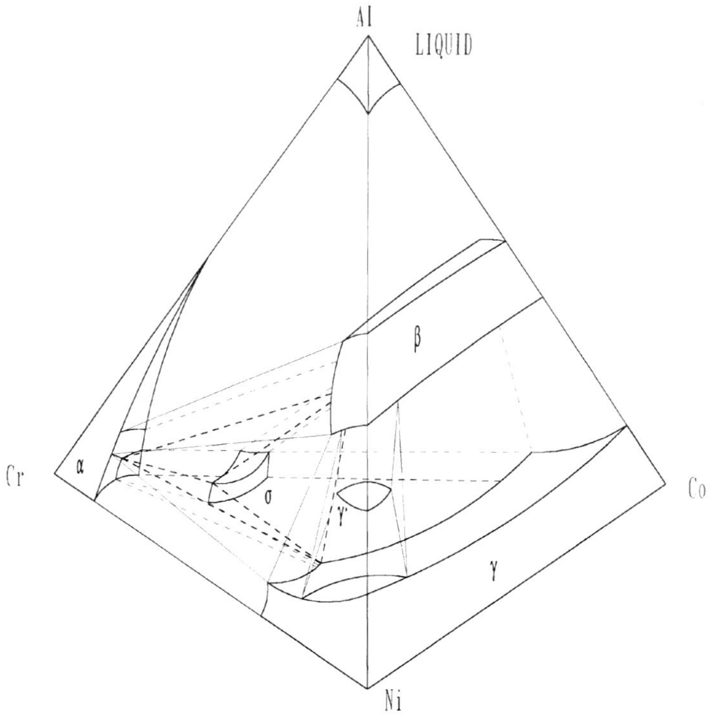

Fig. 6   Schematic 1000°C isothermal section for the Ni–Co–Cr–Al phase diagram (hypothetical).

Recently-published information on similar NiCoCrAlY(Re) coatings supports the hypothesis that a $\alpha + \beta + \gamma + \sigma$, four-phase field exists in the quaternary Ni–Co–Cr–Al system at 1000°C.[2,3] The individual phase constituents within three different coating alloys were analyzed by SEM/EDS and the results from the two earlier publications are summarized in Table 5. Under isobaric, isothermal conditions, four-phase equilibrium in a 'true' quaternary system is represented by a tie-tetrahedron, the four corners of which touch the four one-phase regions involved in the equilibrium.[18] To a first approximation, the multi-component NiCoCrAlY(Re) coating alloys can be approximated as quaternary alloys by treating the chromium concentration of the α and σ phases as an equivalent (Cr_{eq} = Cr + Re). This is a reasonable approximation since the rhenium tends to partition strongly with the chromium in the α and σ phases.[2,3] As can be seen by inspection of Table 5, the equivalent compositions of the four individual phases identified within the three different NiCoCrAlYRe coatings were essentially invariant. This supports the view that each of the three bulk alloy compositions resided within the same $\alpha + \beta + \gamma + \sigma$

Table 5 Composition of individual phases identified within NiCoCrAlYRe coatings annealed at 1000°C.

Composition (at %) of phases in NiCoCrAlYRe Coatings						
Ni	Co	Cr	Al	Re	Cr_{eq} (Cr + Re)	Y
Bulk Coating Composition: NiCoCrAl(16at.%)YRe(3)				Reference: Czech et al. (1995)		
α 3	4	83		10	93	
β 47	9	7	34			
γ 39	22	28	9	2	30	
σ 13	17	53		17	70	
Bulk Coating Composition: NiCoCrAl(24at.%)YRe(3)				Reference: Czech et al. (1995)		
α 4	4	84		8	92	
β 46	12	7	34			
γ 34	26	30	8	1	31	
σ 11	21	55		12	67	
Bulk Coating Composition: NiCoCrAl(12wt.%)YRe(3)				Reference: Beele et al. (1997)		
α 5	4	81		9	90	
β 50	11	7	32			
γ 39	24	28	8	1	29	
σ 11	19	50	2	8	58	9

four-phase tie-tetrahedron in the equivalent Ni–Co–Cr–Al system. The location of this tie-tetrahedron is indicated by heavy dashed lines on Fig. 6.

Alloy compositions and diffusion paths that lie within the interior of the pyramidal Ni–Co–Cr–Al diagram of Fig. 6 are difficult to represent graphically on a two-dimensional drawing. However, to a first approximation, the coating and substrate compositions of the present study were located close to, and on either side of the 15 at.% Al isoconcentration plane. To facilitate their representation, the coating and substrate alloys were mapped onto the schematic Ni–Co–Cr–Al$_{0.15}$ quasi-ternary plane as shown in Fig. 7. The intersection of the 15 at.% Al isoconcentration plane with the α + β + γ + σ tie tetrahedron would result in a triangular section, as shown by the solid lines in Fig. 7.[19] Too little data are available to accurately determine the location of the remaining phase boundaries, which are indicated as dashed lines on the 15% Al quasi-ternary section. These boundaries were therefore were drawn roughly, conforming to the rules of phase diagram construction[18] and in agreement with the constitution of the coating and substrate alloys of the present investigation.

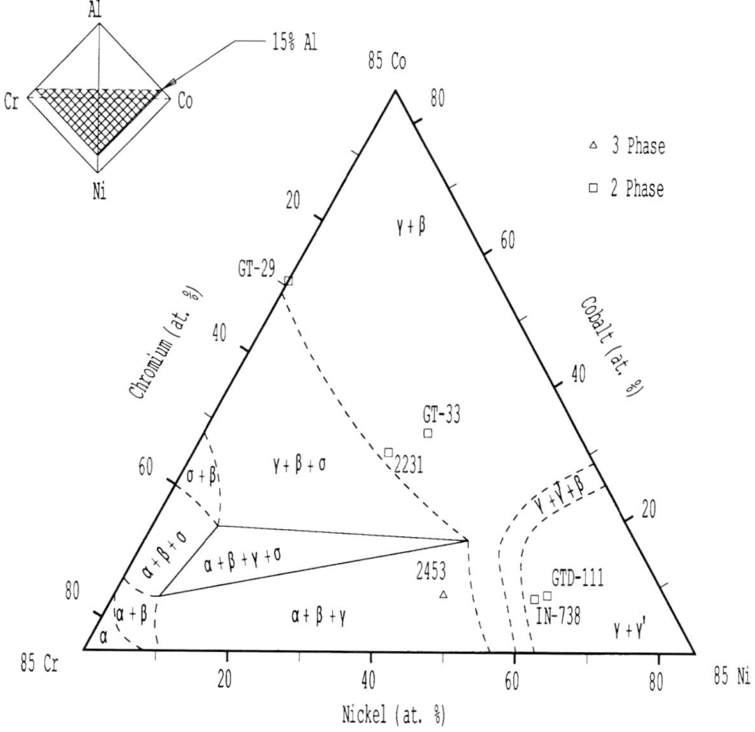

Fig. 7 Schematic 15 at.% Al section of the Ni–Co–Cr–Al quaternary phase diagram at 1000°C (hypothetical) showing coating and substrate compositions.

The alloy and coating compositions were projected onto the 15% Al section by maintaining their Ni:Co:Cr ratios. The two γ + γ' nickel-base substrate alloys were approximated as quaternary alloys by expressing the aluminum content as an aluminum equivalent, $Al_{eq} = Al + Ti + Ta + Nb$. This is also a reasonable approximation since these elements tend to partition to and stabilize the γ' phase in $\gamma + \gamma'$ alloys.[9] Note that, the presence of minor phase constituents such as Ta_xTi_{1-x}, MC, $M_{23}C_6$, Ni_5Y and Y_2O_3, which are present in the commercial alloys are not represented within the Ni–Co–Cr–Al system.

The $NiCoCrAl_{0.15}$ section of Fig. 7 shows that the constitution of NiCoCrAlY coatings at 1000°C may be quite sensitive to the Cr content and Ni:Co ratio. At Cr concentrations higher than about 20–25 at.%, $NiCoCrAl_{0.15}Y$ coatings may form σ and/or α, depending on the Ni:Co ratio. On the basis of Fig. 7, alloys with Ni:Co ratios above about 2.3:1 would form α in addition to $\gamma + \beta$, while lower ratios could result in the formation of σ. Supporting evidence for the latter conclusion can be found in the example of another rhenium-containing alloy, designated Sicoat 2412.[20,21] This alloy was reported have a bulk composition 30.9Ni–25.2Co–21.0Cr–16.4Al–3.1Re–3.1Si–0.2Y (at.%) and, after annealing at 1080°C, consisted of three phases which, on the basis of the EDS results, can be identified as γ, β and σ. The compositions of the individual phases in SICOAT 2412 were reported asfollows (at %): (γ) 32.8Ni–29.1Co–23.7Cr–7.8Al–2.3Re–4.4Si; (β) 44.0Ni–16.1Co–8.17Cr–27.7Al–0.7Re–3.4Si and (σ) 16.8Ni–19.5Co–33.7Cr–7.9Al–15.0Re–7.1Si. The Ni:Co ratio of Sicoat 2412 is 1.13:1. Finally, $(Ni,Co)Cr_{0.25}Al_{0.15}$ alloys with Ni:Co ratios close to 2.3:1 may form both α and σ in addition to γ and β. As noted earlier, four-phase alloys of this type have recently been described in the coating literature.[2,3]

Returning to the results of the present investigation, it is apparent that not all of the experimentally observed microstructural features can be represented by constructing diffusion paths on the $NiCoCrAl_{0.15}$ section of Fig. 7. In particular, variations in the volume fraction and/or disappearance of β phase within certain regions of the interdiffusion zones are most likely related to changes in aluminum concentration. To a first approximation, however, the observed microstructural features can be accounted for by drawing schematic diffusion paths on the Ni–Cr–Al and Co–Ni–Al faces of the Ni–Co–Cr–Al quaternary tetrahedron. These diagrams are shown in Figs 8 and 9.

The paths marked as 1 and 2 on Fig. 8 are for the case of the GT-33 and GT-29 coating samples, respectively. In the case of surface coatings, the diffusion paths are not stationary, but translate with time due to the finite nature of the coating composition.[22] In particular, Al and Cr are continually being lost from the coating system as a result of interdiffusion with the substrate and oxidation effects at the outer surface. Once the diffusion distances produced by a given time-temperature treatment begin to approach the physical dimensions of the coating, the terminus of the diffusion path representing this end of the diffusion couple begins to move away from its original position. In contrast, the composition over most of the underlying alloy substrate remains constant throughout the exposures, due to its much larger size. The terminus of the diffusion path representing the substrate alloy composition is therefore represented as a stationary point. In the case of the GT-29 sample exposed at 1000°C for 1340 h, the terminus of the path representing the composition at the outer surface is located within the γ single-phase field. This represents the narrow zone at the coatings outer surface which was depleted of β-CoAl due to the

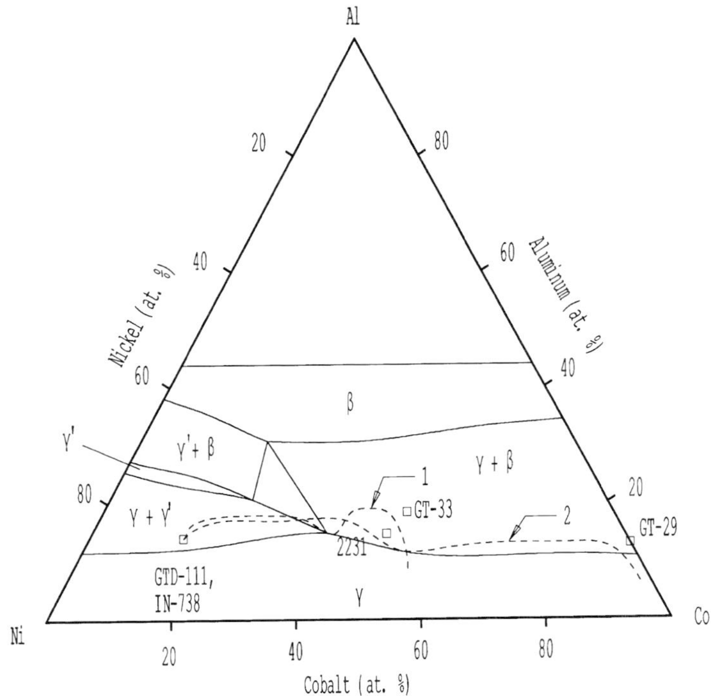

Fig. 8 Schematic diffusion paths for the (1) GT-33/GTD-111 and (2) GT-29/GTD-111 systems exposed at 1000°C for 1340 h and 3030 h, respectively.

selective oxidation of aluminum. The diffusion path then rises into and crosses the γ + β, two-phase field, cutting tie lines (not shown). As it crosses the γ + β phase field, the diffusion path turns down towards the γ single-phase field (i.e. to lower aluminum levels). The latter feature is representative of the narrow β-(Ni,Co)Al depletion zone just inside the coating above the O.I. The diffusion path then reverses course and rises back into γ + β field before reaching the O.I. This represents the formation of the γ + β phase field below the O.I. The increase in β-(NiCo)Al volume fraction within this zone, which implies that 'uphill' diffusion of Al is taking place, may be related to strong cross-term diffusion effects between Al and Cr. Steep Cr gradients are known to cause uphill diffusion of Al in γ + β/γ + γ' diffusion couples.[23]

Before reaching the GTD-111 bulk alloy composition in the γ + γ' phase field, the diffusion path crosses and cuts tie lines (not shown) within the three-phase γ + γ' + β region. Finally, as it cuts tie lines in the γ + γ' phase field, the path descends towards the γ' single phase field (i.e. to lower aluminum levels), representing the decrease in γ' volume fraction observed immediately below the γ + γ' and γ + γ' + β growth zones.

The diffusion path for the GT-33/GTD-111 system after 3030 h is similar to the one for GT-29, except that, after dropping down towards the γ single-phase field, the path stays close to the γ/γ + β boundary (within the γ + β field) until it crosses into the γ + γ' + β, three-phase field. This difference reflects the much lower volume fraction of β phase observed within the γ + β growth zone on the substrate side of the O.I.

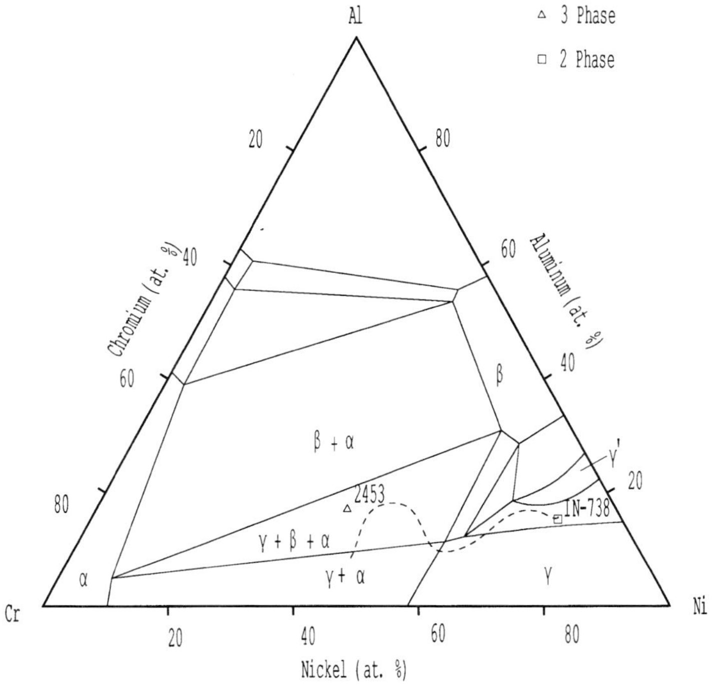

Fig. 9 Schematic diffusion path for a NiCoCrAlYRe (Sicoat 2453) coating and IN-738 alloy after 9110 h at 1000°C.

The schematic diffusion path of Fig. 9 is for the case of the Sicoat 2453/IN-738 system after 9110 h at 1000°C. Oxidation at the outer coating surface resulted in depletion of the β phase, so the initial part of this path is represented as residing within the γ + α, two-phase field. The path then moves up into the γ + β + α, three-phase field representing the central coating microstructure with its higher aluminum content. As it approaches the coating/substrate interface, the path turns and re-enters the γ + α phase field created by the depletion of aluminum from the coating caused by diffusion into the substrate. The path then enters the γ single-phase field before rising upwards into the γ + γ' field and terminating at the alloy substrate composition.

3.2   Implications for the Development of Temperature Estimation Models Based on Coating-Substrate Interdiffusion

As described in Ref 5, the temperature estimation model for service-exposed CoCrAlY (GT-29) coatings on Ni-based (GTD-111) substrates was based on the rate of growth of the well-defined γ + β zone below the O.I. This general feature of the interdiffusion zone was (1) present over the temperature range of interest to the analysis of service-exposed turbine blades and (2) growing at a rate sufficient to allow temperature estimates to be made with reasonable accuracy. A review of the Ni-Co-Al ternary system[12] confirms that the γ + β phase field is, in fact, stable over a very wide temperature range (150°C to 1300°C), although the extent of the two-phase region varies considerably with temperature. (The latter effect was accounted for in the temperature estimation model by incorporating a constitutional term within the total activation energy for interdiffusion based on $\gamma + \beta$ growth).

In this investigation $\gamma + \beta$ growth was also observed at 1000°C in the case of CoNiCrAlY(Si) (GT-33, Sicoat 2231)/Ni-base systems, although the growth rates were reduced, as compared to the CoCrAlY/Ni-base system. This implies that the same general temperature estimation model may be applicable to certain CoNiCrAlY/Ni-base systems, provided that the composition of the coating and interdiffusion zone remains within the $\gamma + \beta$ two-phase field over a useful range of operating temperatures. The alloy-specific material constants in the interdiffusion model would also need to be calibrated, as described in Ref. 5. Referring again to Fig. 7, it is apparent that CoNiCrAlY coatings with 'higher' chromium levels (e.g. Sicoat 2231) are closer to the $\gamma + \beta$ phase boundary at 1000°C than coatings with 'lower' chromium levels (e.g. GT-33). The interdiffusion zone microstructures for CoNiCrAlY coatings with 'high' chromium levels therefore have greater potential to undergo constitutional changes at lower temperatures than those with 'lower' chromium content.

A similar difficulty appears to arise in the case of the NiCoCrAlYRe coating (Sicoat 2453). A review of the Ni-Cr-Al[16] system shows that there is a four-phase, quasi-peritectoid invariant reaction at approximately 990°C ($\gamma + \beta \leftrightarrow \gamma' + \alpha$). The ternary approximation of the Sicoat 2453 composition (Fig. 9) places it in a region of the Ni-Cr-Al phase diagram that is affected by this transformation. Specifically, the bulk coating composition is predicted to undergo a transformation $\gamma + \beta + \alpha \rightarrow \gamma' + \beta + \alpha$ on cooling to temperatures below that of the invariant transformation. This again, could lead to constitutional changes within the coating and interdiffusion zone at temperatures below 1000°C and make temperature estimation more difficult using the existing model. Experience with CoNiCrAlY(Si) and NiCoCrAlY(Re) coatings on service-exposed IN-738 turbine blades has demonstrated that the coating/substrate interdiffusion zones can, in fact, become quite complex in these systems, Fig. 10. However, in the case of the Sicoat 2453/IN-738 system, it may be possible to use the $\gamma + \beta + \alpha \rightarrow \gamma + \alpha$ transformation (β depletion) above the O.I. as an alternative to $\gamma + \beta$ growth in the temperature estimation model.

4 CONCLUSIONS

This investigation has shown that the microstructural transformations caused by interdiffusion between (Ni,Co)CrAlY(Re,Si) coatings and Ni-based superalloys can be quite variable. In the case of the CoCrAlY (GT-29) and CoNiCrAlY (GT-33, Sicoat 2231) coatings at 1000°C, a $\gamma + \gamma' \rightarrow \gamma + \beta$ transformation occurred on the alloy substrate side of the original coating interface. The growth kinetics of this zone was significantly faster for the CoCrAlY coating than for the CoNiCrAlY coatings. In the case of the NiCoCrAlYRe (Sicoat 2453) coating, the growth of a single phase γ zone below the original interface was even more sluggish, although a more pronounced $\gamma + \beta + \alpha \rightarrow \gamma + \alpha$ transformation was

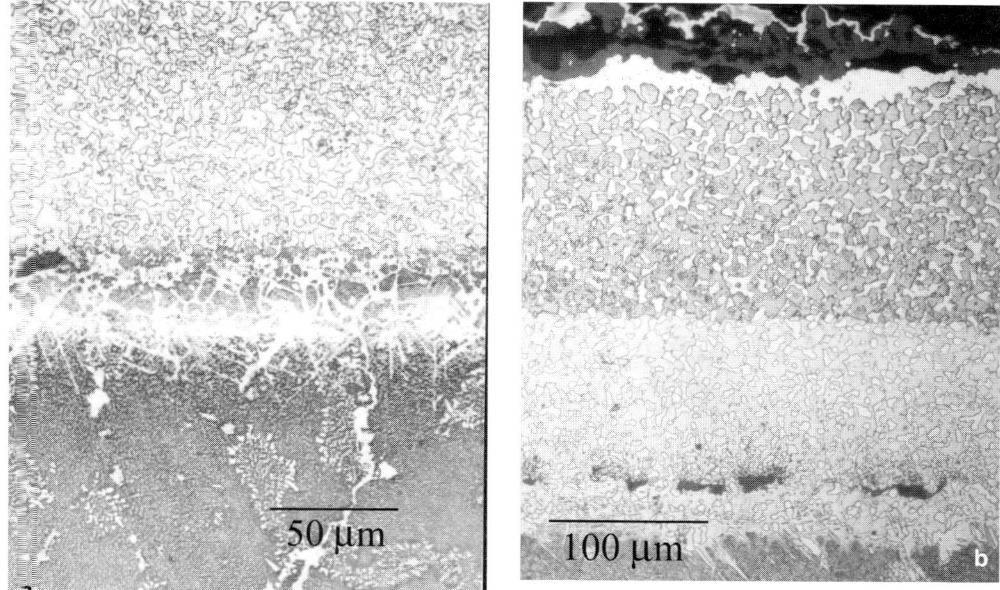

Fig. 10 Optical photomicrographs illustrating the microstructure of (a) Sicoat 2231 and (b) Sicoat 2453 coatings observed on service-exposed IN-738 alloy combustion turbine blades.

observed within the coating itself. It has been shown that the microstructural transformations that occur between the (Ni,Co)CrAlY(Re) coatings and Ni-based superalloy substrates at 1000°C can be qualitatively interpreted through the application of diffusion paths on the available Ni–Cr–Al and Co–Ni–Al ternary phase diagrams.

5 ACKNOWLEDGEMENTS

Financial support for this work was provided, in part, by the National Research Council of Canada, Industrial Research Assistance Program.

REFERENCES

1. Proceedings of the Conference on 'Protective Coating Systems for High-Temperature Gas Turbine Components', *Materials Science and Technology*, 1986, **2**, 193–326.
2. W. Beele, N. Czech, W.J. Quadakkers and W. Stamm, 'Long-term oxidation tests on a Re-containing MCrAlY coating', *Surface and Coatings Technology*, 1997, **94–95**, 41–45.
3. N. Czech, F. Schmitz and W. Stamm, 'Microstructural Analysis of the Role of Rhenium in Advanced MCrAlY Coatings', *Surface and Coatings Technology*, 1995, **76–77**, 28–33.
4. P. Mazars, D. Manesse, and C. Lopvet, 'Degradation of MCrAlY Coatings by Interdiffusion', ASME Paper No. 87-GT-58, presented at the Gas Turbine Conference and Exhibition, Anaheim, California, May 31–June 4, 1987.

5. K.A. Ellison, J.A. Daleo and D.H. Boone, 'Metallurgical Temperature Estimates Based on Interdiffusion Between CoCrAlY Overlay Coatings and a Directionally Solidified Nickel-Base Superalloy Substrate', *Materials for Advanced Power Engineering 1998*, J. Lecomte-Beckers et al., eds, Forschungszentrum Jülich GmbH, 1998, 1523–1534.

6. P. Nash and W.W. Liang, 'Phase Equilibria in the Ni–Al–Ti System at 1173 K', *Metall. Trans. A*, 1985, **16**, 319–322.

7. P. Nash and D.R.F. West, 'Phase Equilibrium in the Ni–Ta–Al System', *Met. Sci.*, 1979, **13**, 673.

8. E.W. Ross and C.T. Sims, 'Nickel-Base Alloys', *Superalloys II*, C.T. Sims, N.S. Stoloff, and W.C. Hagel, eds, John Wiley & Sons, 1987, 97–133.

9. C.C. Jia, K. Ishida and T. Nishizawa, 'Partition of Alloying Elements Between (A1), '(L1$_2$) and (B2) Phases in Ni–Al Base Systems', *Metallurgical and Materials Transactions A*, 1994, **25A**, 473–485.

10. M.Hansen and P. Anderko, *Binary Alloy Systems*, McGraw-Hill, 1958.

11. J.S. Kirkaldy and L.C. Brown, 'Diffusion Behaviour in Ternary, Multiphase Systems', *Canadian Metallurgical Quarterly*, 1963, **2** (1), 89–115.

12. J. Schramm, 'Das Dreistoffsystem Nickel-Kobalt-Aluminium', *Zeitschrift für Metallkunde*, 1941, **33**, 403–412.

13. B.G. Livshits and N.N. Myuller, 'Investigations of the Phase Equilibria in the Co–Cr–Al System' (in Russian), *Nauch. Doklady Vyssh. Shkoly*, 1958, **3**, 201–206.

14. B.G. Livshits and N.N. Myuller, 'Phase Equilibria in the Co–Cr–Al System' (in Russian), *Sbornik Moscov. Inst. Stali*, 1960, **39**, 267–283.

15. A. Taylor and R.W. Floyd, 'The Constitution of Nickel-Rich Alloys of the Nickel–Chromium–Aluminium System', *J. Inst. Met.*, 1952–1953, **81**, 451–464.

16. S.M. Merchant and M.R. Notis, 'A Review: Constitution of the Al–Cr–Ni System', *Materials Science and Engineering*, 1984, **66**, 47–60.

17. W.D. Manly and P.A. Beck, NACA Technical Note 2602, February 1952.

18. F.N. Rhines, *Phase Diagrams in Metallurgy*, McGraw-Hill Book Company, 1956, 220–272.

19. V. Raghavan, 'The Two-Dimensional Topology of Quaternary Invariant Reactions', *Experimental Methods of Phase Diagram Determination*, J.E. Morral et al., eds, TMS, 1994, 3–18.

20. N. Czech, F. Schmitz and W. Stamm, 'Improvement of MCrAlY Coatings by Addition of Rhenium', *Surface and Coatings Technology*, 1994, **68/69**, 17–21.

21. L. Peichl and D.F. Bettridge, 'Overlay and Diffusion Coatings for Aero Gas Turbines', *Materials for Advanced Power Engineering 1994*, D. Coutsouradis, et al., ed., Kluwer Academic Publishers, 1994, 717–740.

22. K.L. Luthra and M.R. Jackson, 'Coating/Substrate Interactions at High Temperature', *High Temperature Coatings*, M. Khobaib and R.C. Krutenat, eds, TMS, 1987, 85–100.

23. L.A. Carol, 'A Study of Interdiffusion in β + γ/γ + γ′ Ni–Cr–Al Alloys at 1200°C', Ph.D. Thesis, Michigan Technological University, February 1985. Available as NASA CR-174852.

The Development of a Process Model for the Weldability of Nickel-Base Superalloys for Gas Turbine Applications

S.M. ROBERTS, O. HUNZIKER, D. DYE and R.C. REED

Rolls–Royce University Technology Centre, Department of Materials Science and Metallurgy, University of Cambridge, Pembroke St., Cambridge CB2 3QZ

ABSTRACT

Improvements in engine efficiency and power output require the use of alloys with strength and creep resistance at increased operating temperatures. For combuster liner and casing applications these components have traditionally been joined and repaired using the TIG (Tungsten Inert Gas) welding process. Unfortunately the modern successor alloys, which demonstrate the high temperature properties required, have proved difficult to weld, with a susceptibility to cracking and therefore possessing only limited 'weldability'.

In this paper we investigate the coupling of a thermo-mechanical model for the development of stress during welding with a microstructural model for the diffusion of the alloying elements. This approach enables the weld engineer to predict the welding conditions under which a given alloy might be susceptible to either solidification or liquation cracking.

IN718 sheet material has been TIG welded, in order to provide model validation. *In situ* thermocouples have provided transient temperature profile verification and post weld distortion has been characterised by needle probe analysis. Residual stress measurements have been performed using both hole drilling and neutron diffraction techniques.

INTRODUCTION

The IN718 and C263 alloys currently used for combustor liner and casing applications do not have adequate mechanical properties at the higher temperatures required in new generation aeroengines. A major requirement for candidate replacement alloys is that they must be weldable. Unfortunately, alloys such as MAR-M-002 which have suitable higher temperature properties are known to be very difficult to weld. There is therefore a need for a better understanding of the effects of alloy composition and of the conditions required for a weld to be successful. Significant experimentation and practical experience is required to determine the appropriate paramaters to successfully weld a given component. This is both expensive and time consuming, and becomes increasingly so as the weldability of the alloy decreases. Modern computational modelling methods offer the possibility of a reduced cost route to the correct parameters. Approaches such as the Finite Element Method (FEM) predict both distortion and residual stress[1,2,3] and may be coupled with rules defining the onset of cracking.[4]

Tungsten Inert Gas welding (TIG) is commonly employed in both the manufacture and repair of modern jet engine combustors. Hot cracking during nickel-base superalloy welding occurs by two principal mechanisms:

327

- Solidification cracking. Such cracking occus when the liquid–solid region (mushy zone) experiences tensile strain during solidification. In regions with a high fraction of solid, such as at the trailing edge of the melt pool, the liquid film remaining may be insufficient to feed the opening interdendritic space.[5,6] The result is the development of an interdendritic crack.
- Liquation cracking. The generation of liquid at the interface between a secondary phase, such as a carbide, and the matrix at the eutectic point in the heat affected zone (HAZ)[7] is known as constitutional liquation. This liquid may migrate along grain boundaries away from the precipitate to form a film on HAZ grain boundaries. Microcracks are then formed as a result of tension on those liquid films during the post weld cooling period.

The approach chosen for prediction of these cracking phenomena is to link a thermal model to both a mechanical analysis via the coefficient of thermal expansion and models for microstructure evolution. A cracking criterion can then be identified when tensile stress occurs in a region where liquid is present. The chart in Fig. 1 summarises this approach.

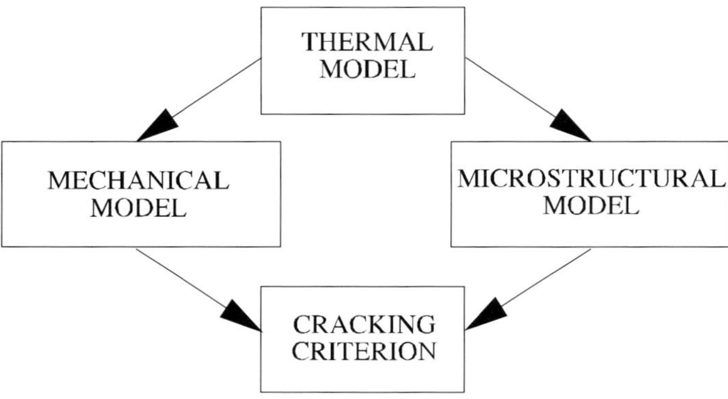

Fig. 1   Schematic of interdependence of cracking model.

In this paper we present thermal, mechanical and microstructural models. The confidence that can be attributed to such predictions is critically dependent on the aquisition of accurate and appropriate validatory data. These models have therefore been extensively validated via *in situ* thermocouple readings, post weld distortion and residual stress measurement and electron probe microanalysis (EPMA) for microsegregation.

In this study a bead-on-plate weld was performed along the central portion of a 200 × 100 × 2 mm thick rectangular plate of solution heat treated, rolled IN718 (Fig. 2). The welding parameters used in this study were a traverse speed of 1.59 mm s^{-1}, a torch-plate potential of 9V and a square-wave D.C. power source of 576W.

THERMAL MODEL

The accurate modelling of the temperature field during welding is vital as it provides the driving force for both the mechanical and microstructural analyses. Fluid flow within the

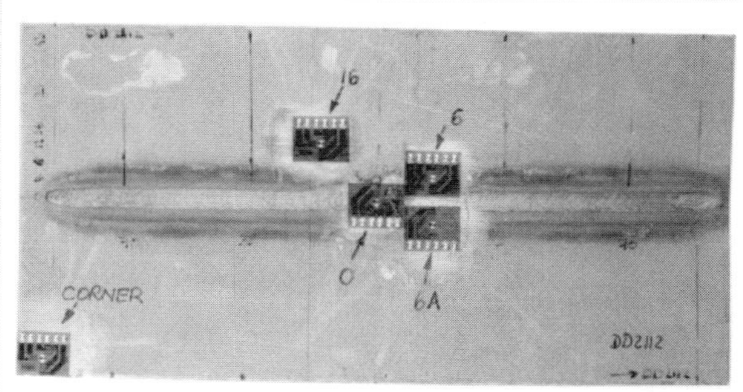

Fig 2 IN718 plate used for bead on plate weld experiment, illustrating the positions of the strain gauges.

molten zone defines the shape of the melt pool; however for the present purposes the modelling of the complicated physics occuring in this region is considered unnecessary. Through careful use of an appropriate heat source expression this shape can be replicated computationally.

For TIG welding a surface disc source has been found to represent the source adequately. The surface flux Q is given by

$$Q = \eta Q_0 / \pi a^2$$

where Q_0 is the power from the power source (576 W), η is the weld efficiency and a the disc radius. Typically a is of the order of the weld pool half width whilst η is around 50 to 60% and may be evaluated from the weld pool profile or peak temperatures mesured by the thermocouples. In practise, a series of steady state (Eulerian) analyses are performed in which these two parameters were systematically varied until a good match to the measured fusion boundary profile was achieved. Figure 3 illustrates such a match together with graphs indicating the variation of profile with process parameter. Whilst a full transient description is required for an accurate mechanical analysis, a steady state thermal solution is sufficient for the microstructural model, and as such a semi-analytical model based on Rosenthal[8] has been written to provide the temperatures which feed into this model. In a similar way the steady state FE solution could also be used to feed the microstructural model.

THERMAL VALIDATION

Measurements of the thermal cycle in the heat affected zone during welding have been performed using thermocouples welded on the surface of the IN718 plate shown in Fig. 2. These results are used to determine the heat source properties, notably the weld efficiency and convective cooling coefficient, and to validate both the FEM and analytical thermal models. Satisfactory agreement has been obtained between the measured and the predicted profiles, as can be seen in Fig. 4.

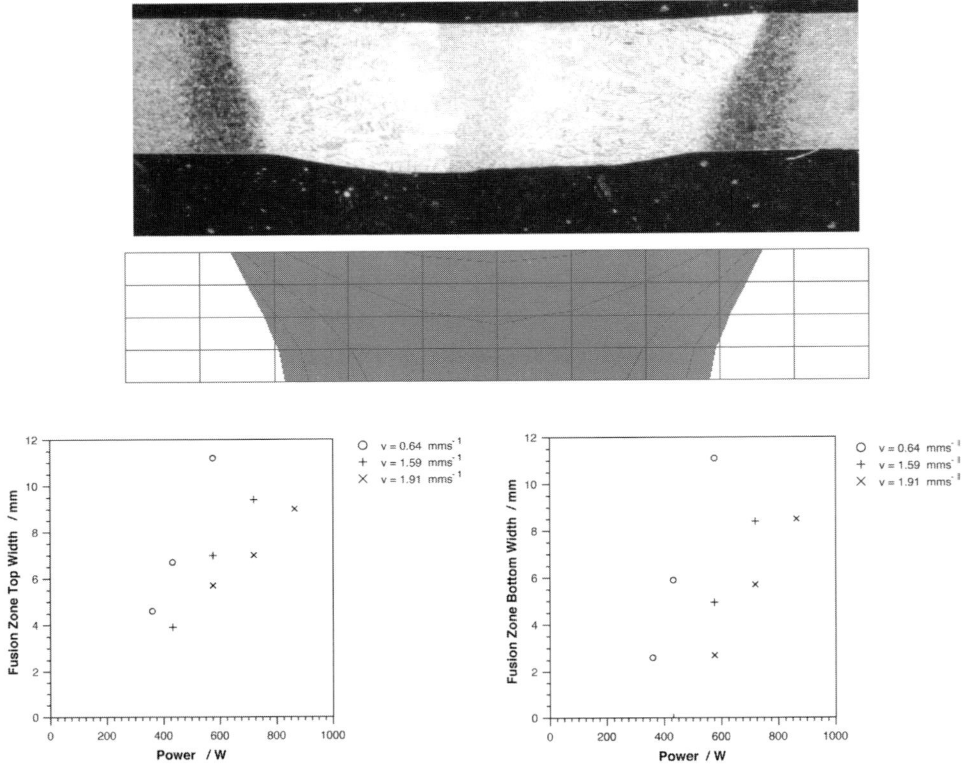

Fig. 3 Comparison of fusion boundary profile and steady state melt isotherm and variation of weld bead profile with weld parameter.

MECHANICAL MODEL

Two different methods are under development for the structural analysis. Elasto-plastic 3D finite element transient analysis has been performed and validated against both distortion and residual stress measurements, following the methodology of Roberts *et al.*[1,9] A major advantage of this approach is that it enables real combuster geometries and jigging arrangements to be considered. The advantages of such a technique are well known, however, such precision may not be required to give the weld engineer an indication of the likelihood of cracking under given welding conditions in a specific alloy. As a result a second method which involves coupling a 2D linear plane stress mechanical solution to the analytical thermal model is being generated. This model will allow the rapid testing of the influence of alloy composition and weld parameter on the propensity for cracking.

MECHANICAL VALIDATION

Post weld distortion measurements were performed using a Leitz needle probe coordinate instrument with a 3 mm ball tip. The distortion of the plate is characterised by a

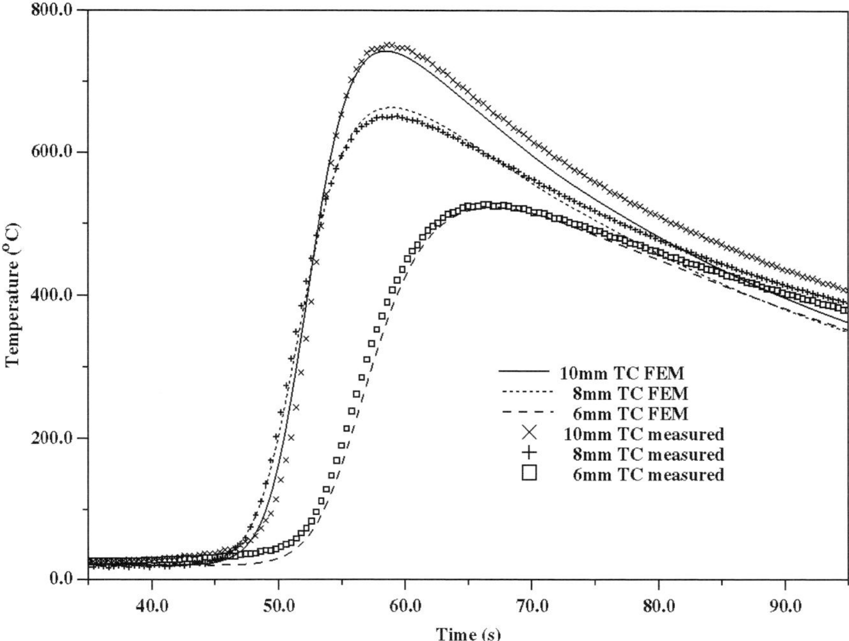

Fig. 4   Comparison of measured thermocouple temperatures and predicted values.

butterfly fold about the weld axis and a camber along the length of the plate in the weld direction, yielding a characteristic saddle shape. Figure 5 compares the Leitz results for distortion out of the plane of the plate with the 3D elastoplastic FEM predictions, excellent agreement has been found.

Two forms of validation were employed for residual stress characterisation, incremental hole drilling and neutron diffraction. The hole drilling technique relies on measuring the relaxation around a hole drilled in stressed material. A strain gauge rosette is mounted on the surface of the specimen, as shown in Fig. 6, and a hole drilled in the centre. The relaxation of material during drilling is measured by the rosette and the stress state deduced from techniques such as those employed by Schajer.[10] Both the destructive nature of the technique and its inherently large sampling areas must be acknowledged, however Stone et al.[11] have shown, at least for electron beam welds (EBW) in Ni-base superalloys, that across a monotonically varying stress gradient the average stress measured is not unreasonable.

The second technique employed was neutron diffraction. These measurements were made using the ENGIN instrument on the PEARL beamline at the ISIS facility, Rutherford Appleton Laboratory, Didcot, UK. The purpose of the measurements was to examine the variation in residual strain along the length of the plate and to map the distribution of longitudinal stress in the transverse direction away from the weld in the heat affected zone. Neutron diffraction allows the non-destructive evaluation of strains and therefore stresses

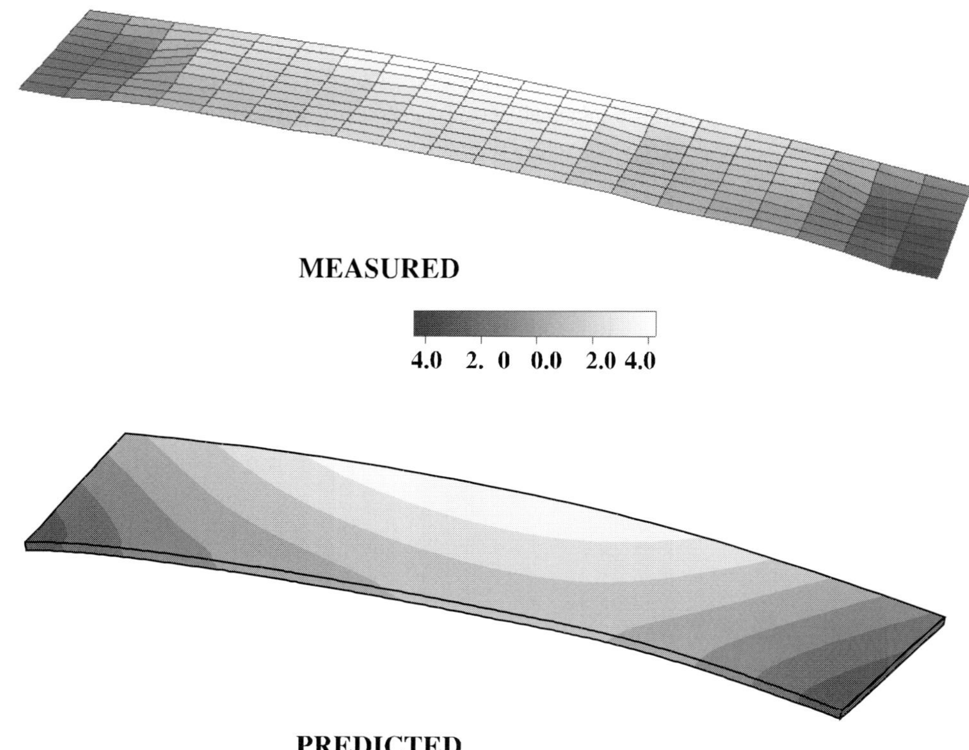

Fig. 5 Comparison of displacement out of the plane of the plate (NB only half of plate is shown. leading edge = weldline).

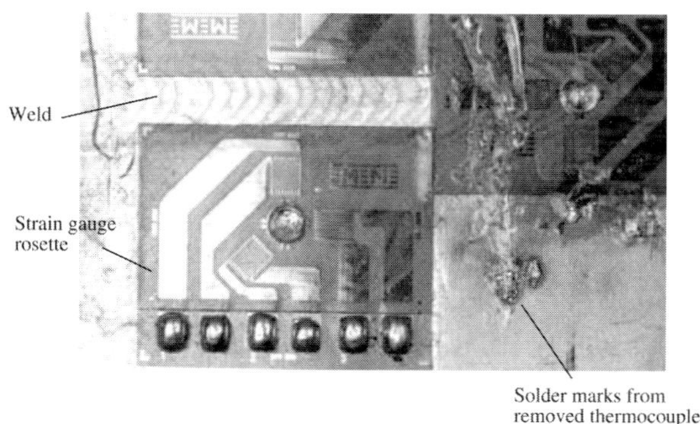

Fig. 6 Strain gauge arrangement for incremental hole drilling residual stress evaluation.

at depth within a sample because of the deep penetration of thermal neutrons in most metals. Furthermore, such penetration offers the possibility of small gauge volumes and hence a more accurate definition of the stress state. The determination of residual strain via diffraction techniques in a crystalline material relies on the use of the interplanar lattice spacing as an internal strain gauge. In order to evaluate the bulk engineering strain from the lattice strain of an individual crystallite, a simple Reitveld fit to a single lattice parameter was used. The details of the validity of this approach and the experimental detail including the conversion of strain to stress are explored by Dye et al. elsewhere.[12]

The results show that within the central 140 mm of the weld, which is 180 mm long, the variation in stress is small. The width of the tensile region is approximately 11 mm, the maximum compressive longitudinal stress is ~ −170 MPa, while the maximum tensile longitudinal stress is ~300 MPa, which occurs in the heat-affected zone 8 mm from the weld centreline. Stress equilibrium across the plate is satisfied, so the results are self-consistent. The transverse and through-thickness stresses are much smaller than the longitudinal stress.

Figure 7 illustrates the agreement between the measured values from both methods and the predicted residual stresses from the 3D elastoplastic finite element analysis. The values shown are the longitudinal stresses, which are the biggest, at the mid plane of the plate, at the mid height. The model accurately predicts both the peak level of stress and the width of the tensile zone. At the same time the measurement techniques are in close agreement.

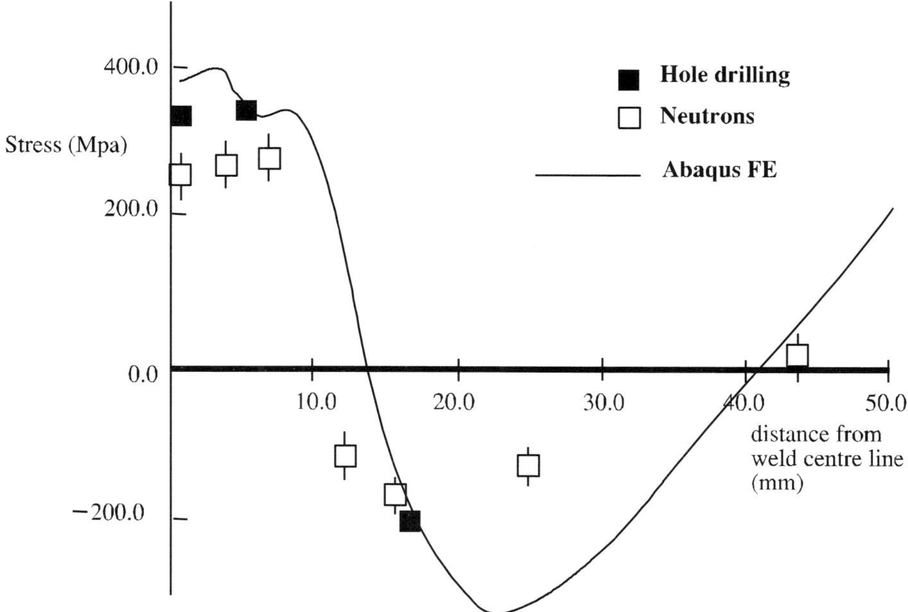

Fig. 7   Longitudinal residual stress measured vs. finite element predicted.

MICROSTRUCTURE MODEL

To date the development of the microstructural model has concentrated on a treatment of solidification cracking. The techniques described below are rather generic, however, and as such a similar approach can be used to describe liquation.

The two simplest approaches to solute redistribution during solidification are the classic 'lever' and 'Scheil'[13,14] techniques. In the former, diffusion in both the liquid and solid is considered to be so rapid as to maintain thermodynamic equilibrium throughout the solidification process. The Scheil approach differs in that no diffusion is allowed within the solid. This lack of diffusion in the solid leads to an over-prediction of the amount of eutectic and the degree of segregation. Lighter elements are likely to 'back diffuse' into the solid significantly, as a result Scheil tends to underpredict the amount of solid above approximately 50% volume fraction of solid. Such inaccuracies are particularly true in Ni-base superalloys for elements with large partitioning coefficients such as Ta and Nb in the liquid or Re and Fe in the solid. However, as a first approximation, the Scheil approach produces a conservative answer, since it over predicts the amount of liquid remaining.

The results of coupling the analytical thermal solution and the microsegregation analysis for the weld under consideration are shown in Fig. 8. From the fraction of solid prediction it is clear that some interdendritic liquid still persists a significant distance (at least double the melt pool radius) behind the weld. The solidification cracking hypothesis is that if this trailing mushy zone experiences an external tensile stress then cracking is likely. For the case shown the stresses within the mushy zone are solely compressive and, as such, no solidification cracking would be anticipated. This result is reassuring as no such cracking is observed in practise with this alloy under these conditions.

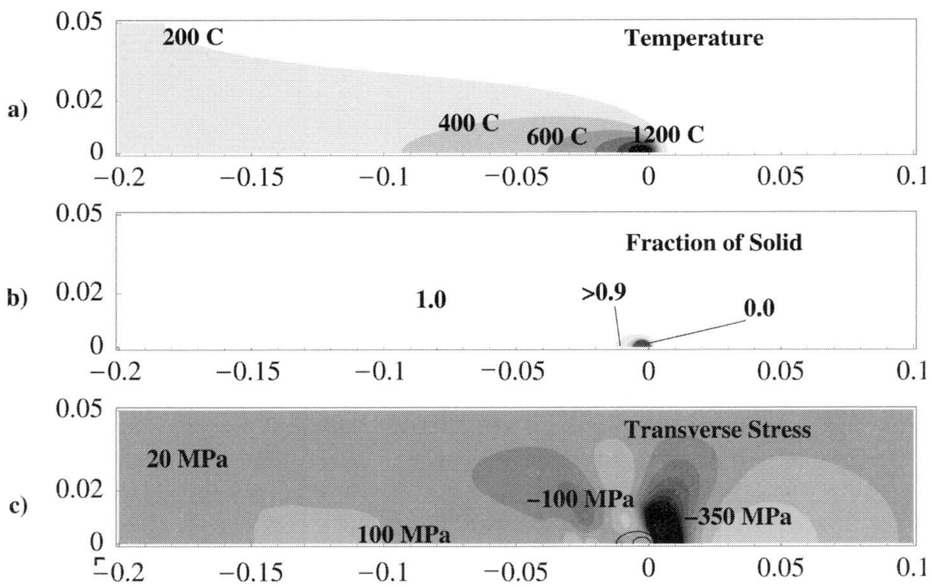

Fig. 8 Steady state temperature field, fraction of solid predicted from Scheil and transverse stress (with fs = 0 and fs = 1 boundaries overlaid).

MICROSTRUCTURE VALIDATION

In order to validate the microsegregation calculation a map of the microsegregation around a dendrite has been aquired. Such maps for Fe and Nb are shown in Fig. 9. The maps were produced by EPMA. The composition was mapped over an area 60×60 μm with a spacing of 2 μm. The dendrite arms are clearly visible with the Nb migrating to the interdendritic region whilst the Fe remains in the dendrite core. In order to sort the elements into a one dimensional plot the concentrations Fe and Nb were subtracted for each position. The highest values were used to define the first solidifying point (fs = 0.0) whilst the lowest defined the eutectic. Figure 10 compares the contents measured with those predicted by the Scheil approximation. Clearly the trends are correct. It is hoped that a more accurate prediction of the microsegregation kinetics might allow even more realistic prediction of the likelihood of cracking.

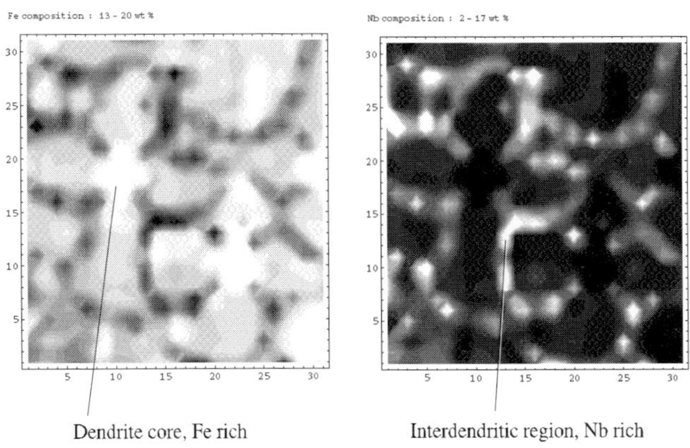

Fig. 9   Composition profiles of Fe and Nb around a dendrite arm.

CONCLUSIONS

A first attempt at a coupled mechanical and microstructural approach to define the 'weldability' of Ni-base superalloys has been presented. Both transient and steady state thermal solutions have been shown to adequately reproduce the *in situ* thermocouple measurements of temperature in the HAZ. The transient FEM thermal analysis has been used to drive a 3D transient mechanical analysis of the TIG welding process. Post weld distortion measurements have shown the validity of the elasto-plastic, transient, 3D approach adopted with both the form and the degree of deformation accurately predicted. Residual stress measurements, made via both neutron diffraction and hole drilling, show promising agreement with those predicted also. Having established the accuracy of both the thermal and mechanical aspects a microstructural model capable of predicting the microsegregation during solidification has been applied. Whilst the Scheil approach adopted has some limitations, this model shows promising agreement with measured microsegregation. The present authors believe that the coupling of these three models should provide a methodology for prediction of cracking phenomena in Ni-base superalloys.

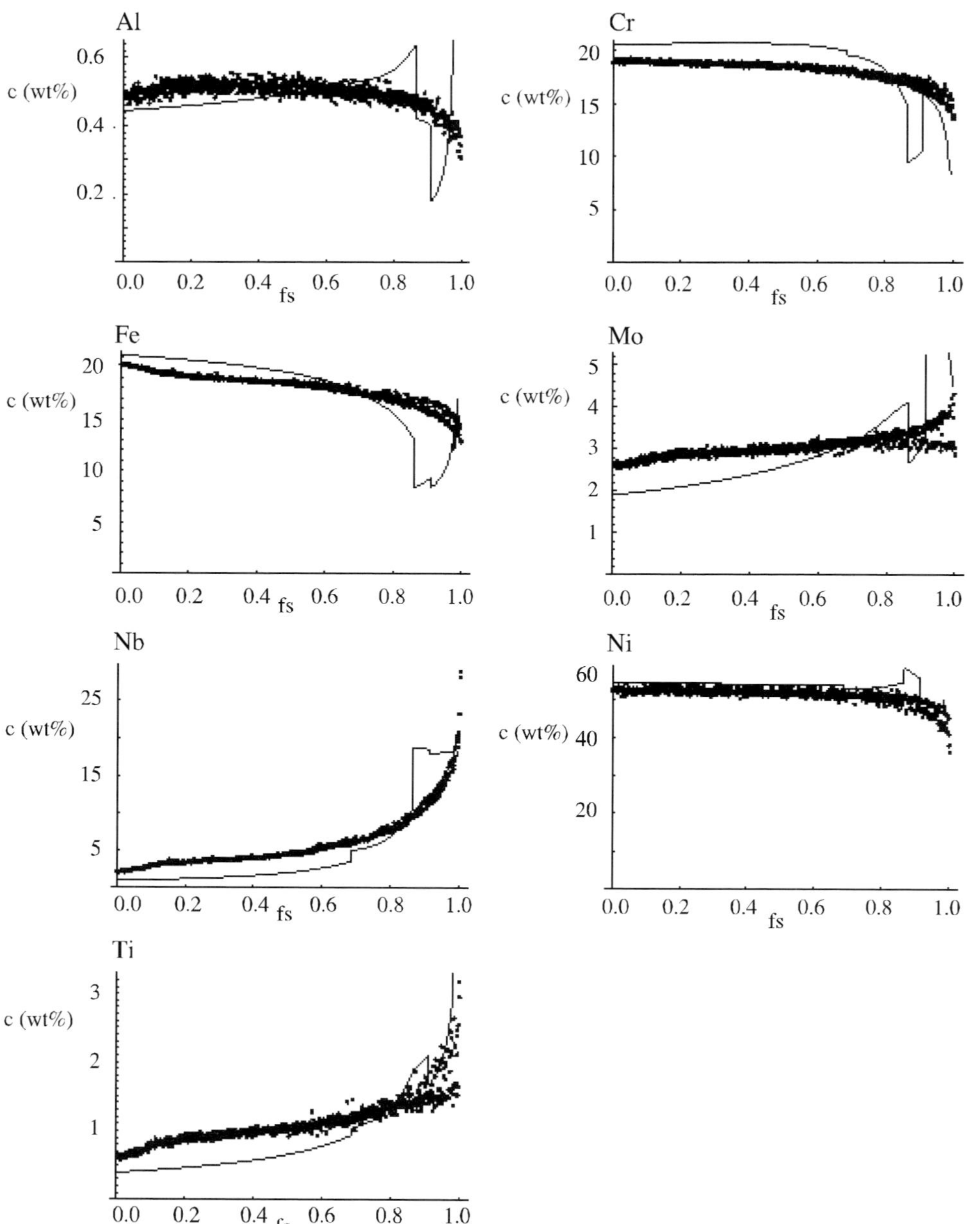

Fig. 10 Comparison of measured profiles and Schiel (line) predicted for all elements tested.

ACKNOWLEDGEMENTS

The authors acknowledge sponsorship from the Engineering & Physical Science Research Council (EPSRC), the Defence and Evaluation Research Agency (DERA), Farnborough and Rolls–Royce plc.

Mark Roberts and Roger Reed are employed in the UTC, whilst Dave Dye is sponsored by EPSRC and Rolls–Royce and Olivier Hunziker is employed on an EPSRC/DERA joint awards scheme grant – GR/L 66380.

Informative talks with Paul Andrews, Bryan Benn, Douglas Crooke and others at Rolls–Royce, Howard Stone at the UTC, Mike Henderson at DERA, Gary Schajer at UCB, John Coldak at Carelton and Alan Oddy and Moyra McDill of OMNIS are also acknowledged.

For the neutron diffraction residual stress measurements the use of the ISIS facility (RAL) and the help of Mark Daymond is also appreciated, as was the hole drilling performed by Phil Whitehead at Stresscraft.

REFERENCES

1 S.M. Roberts, H.J. Stone, J.M. Robinson and R.C. Reed, 'Characterisation and Modelling of the Electron Beam Welding of Waspaloy', *Mathematical Modelling of Weld Phenomena 4*, H. Cerjak and H.K.D.H. Bhadeshia, eds, IOM Communications, London, 1998, 631–648.

2 J.A. Goldak, 'Coupling Heat Transfer, Microstructure Evolution and Thermal Stress Analysis in Weld Mechanics', *Mechanical Effects of Welding*, IUTAM Symposium, L. Karlsson *et al.*, Springer-Verlag, 1992, 1–30.

3 L. Karlsson and L.E. Lindgren, 'Combined Heat and Stress–Strain Calculations', *Modeling of Casting, Welding and Advanced Solidification Processes* V, M. Rappaz *et al.*, The Minerals, Metals and Materials Society, (1991).

4 S.A. David, S.S. Babu, J.M. Vitek, 'Weldability and Microstructure development in Nickel-base Superalloys' *Mathematical Modelling of Weld Phenomena 4*, H. Cerjak and H.K.D.H. Bhadeshia, eds, IOM Communications, London, 1998, 269–289.

5 H.W. Bergmann and R.M. Hilbinger, 'Numerical simulation of Centre Line Hot Cracks in Laser Welding of Aluminium Close to the Sheet Edge', *Mathematical Modelling of Weld Phenomena 4*, H. Cerjak and H.K.D.H. Bhadeshia, eds, IOM Communications, London, 1998, 658–668.

6 M. Rappaz, J.-M. Drezet, M. Gremaud, 'A New Hot Tearing Criterion', *Met. Trans. A*, 1999, **30A**, 449–455.

7 B. Radhakrishnan and R.G. Thompson, 'A Model for the Formation and Solidification of Grain Boundary Liquid in the Heat-Affected Zone (HAZ) of Welds', *Met. Trans. A.*, 1992, **23A**, 1783–1799.

8 D. Rosenthal, 'The Theory of Moving Sources of Heat and Its Application to Metal Treatments', *Trans. ASME*, 1946, 849–866.

9 H.J. Stone, S.M. Roberts, T. Holden, P.J. Withers and R.C. Reed, 'The Development and Validation of a Model for the Electron Beam Welding of Aero-engine Components', *Trends in Welding Research*, AWS, USA, 1998.

10 G.S. Schajer, 'Measurement of Non-Uniform Residual stresses using the Hole-Drilling Method. Part 1 – Stress Calculation Procedures', *Trans. ASME*, 1988, **110**, 338–349.

11 H.J. Stone, P.J. Withers, T. Holden, S.M. Roberts and R.C. Reed, 'Characterisation of the Residual Stresses and Strains in an Electron Beam Welded Waspaloy Plate Using Neutron Diffraction, X-Ray Diffraction and Hole-Drilling, *Metall. Mater. Trans.*, 1999.

12 D. Dye, S.M. Roberts, P.J. Withers and R.C. Reed, 'The Determination of the Residual Strains and Stresses in a Tig-welded Sheet of IN718 Superalloy Using Neutron Diffraction', *J. Strain Analysis*, 1999

13 G.M. Gulliver, *Metallic Alloys*, Griffen, London 1922.

14 E. Scheil, *Z. Metallkd.*, 1942, **34**, 70.

Author Index

Subject Index